Recent Advances in Robotics and Intelligent Robots Applications

Recent Advances in Robotics and Intelligent Robots Applications

Editors

Qi Song
Qinglei Zhao

MDPI

Basel • Beijing • Wuhan • Barcelona • Belgrade • Novi Sad • Cluj • Manchester

Editors

Qi Song
Electrical Engineering and Artificial Intelligence
Chinese Academy of Sciences
Suzhou
China

Qinglei Zhao
Space Optics Department
Chinese Academy of Sciences (CIOMP)
Changchun
China

Editorial Office
MDPI
St. Alban-Anlage 66
4052 Basel, Switzerland

This is a reprint of articles from the Special Issue published online in the open access journal *Applied Sciences* (ISSN 2076-3417) (available at: www.mdpi.com/journal/applsci/special_issues/6270XG52D8).

For citation purposes, cite each article independently as indicated on the article page online and as indicated below:

Lastname, A.A.; Lastname, B.B. Article Title. *Journal Name* **Year**, *Volume Number*, Page Range.

ISBN 978-3-7258-1294-3 (Hbk)
ISBN 978-3-7258-1293-6 (PDF)
doi.org/10.3390/books978-3-7258-1293-6

Contents

About the Editors

Qi Song

Qi Song received his Ph.D. in Electrical Engineering and Computer Science from the University of California, Irvine, in 2013. After 9 years of industry experience with Broadcom Inc. and Maxlinear Inc., he joined the Suzhou Institute of Biomedical Engineering and Technology at the Chinese Academy of Sciences as a professor in 2019. He also serves as CEO of Soleilware Photonics LLC, USA. His research interests include theoretical and experimental investigations of metamaterials and metasurfaces, photonic-integrated circuits (PICs), optoelectronic chips, sold-state LiDAR, deep-learning models, and computer vision techniques, with applications to autonomous vehicles and robotics. He has over sixty journal and conference publications, two books, and more than ten patents.

Qinglei Zhao

Qinglei Zhao received his Ph.D. degree in Mechanical and Electronic Engineering from the University of Chinese Academy of Sciences in 2016. Since 2010, he has served as an associate professor in the Space Department at the Changchun Institute of Optics, Fine Mechanics, and Physics of the Chinese Academy of Sciences. He is engaged in the research of embedded imaging system design and intelligent control of space robots, image processing, and artificial intelligence applications for remote sensing and tunable metasurface system development. He has published more than 10 papers and 10 patents. Moreover, he serves as an editor for two scientific journals.

Preface

Robotics has witnessed a transformative evolution over the past decade, driven by unprecedented advancements in artificial intelligence, machine learning, and material science. This Special Issue, "Recent Advances in Robotics and Intelligent Robot Applications", aims to provide a comprehensive overview of the latest research and developments that are shaping future robotics research in corresponding areas. In recent years, the confluence of deep learning algorithms with sophisticated sensor technologies has enabled robots to achieve levels of autonomy and efficiency. From autonomous vehicles navigating in complex urban environments to humanoid robots capable of intricate human–robot interactions, the scope of what robots can achieve has expanded rapidly. This Special Issue delves into these advancements, exploring how they are revolutionizing sectors such as bionics and soft robots, motion and path planning, manipulator control, image classification, and processing algorithms. Moreover, the advent of soft robotics, inspired by the flexibility and adaptability of biological organisms, represents a significant leap forward. Through detailed case studies and expert analyses, this book sheds light on the innovative designs and applications of soft robots. One of the most exciting aspects of space robotics research is the integration of robots into space exploration environments. The development of space robots highlights a paradigm shift where robots and astronauts work side by side, supporting scientific missions and safety in space. This reprint examines some technological breakthroughs about machine vision in space and robot motion planning. "Recent Advances in Robotics and Intelligent Robot Applications" is a collaborative effort, bringing together contributions from active leading researchers in the field, providing in-depth discussions, solid scientific proof, and forward-thinking perspectives. We hope that this compilation is informative for seasoned robotics professional and academic researcher and that it also inspires the next generation of innovators to push the boundaries of robotics research. Finally, I would like to express my deepest gratitude to my wife Chao Li and my children Norah Song and Roman Song for their unwavering support and encouragement throughout this endeavour.

Qi Song and Qinglei Zhao
Editors

applied sciences

MDPI

Editorial

Recent Advances in Robotics and Intelligent Robots Applications

Qi Song [1,*] and Qinglei Zhao [2]

[1] Suzhou Institute of Biomedical Engineering and Technology, Chinese Academy of Sciences, Suzhou 215163, China
[2] Changchun Institute of Optics, Fine Mechanics and Physics, Chinese Academy of Sciences (CIOMP), 3888 Dongnanhu Rd., Jingkai District, Changchun 130033, China; coldsun@ciomp.ac.cn
* Correspondence: songq@sibet.ac.cn

Robotics research has a unique allure for both academia and the industry due to its potential for groundbreaking innovation and real-world applications [1,2]. Robotics research and applications encompass a broad range of topics, challenges, and opportunities. The topics in this Special Issue represent just a small fraction of the diverse and interdisciplinary field of robotics, which intersects with areas such as materials science and mechatronics, computer science, hardware engineering, robot kinematics, and bionics [3,4]. Fast-paced development in sensor hardware, robot perception, smart decision-making strategy, and gripper designs allows robots to react in "real-time" to the environment and thus work intelligently alongside human beings [5]. More crucially, artificial intelligence (AI) integrated into robotic applications has the potential to revolutionize various aspects of human life, offering a wide range of benefits and support.

This Special Issue of *Applied Sciences*, entitled "Recent Advances in Robotics and Intelligent Robots Applications", has 14 research papers, covering topics from bionics (contribution 1) and soft-material robot designs (contribution 14), infrared image algorithms (contribution 2), target tracking algorithms (contribution 3 and 7), hyperspectral image classification (contribution 4), manipulator control (contribution 5 and 8), and space image denoise and analysis methods (contribution 6 and 10) to motion and path planning (contribution 9, 11, 12, and 13). The demonstrated robot designs and algorithm development include novel low-cost robot designs with leaping abilities (contribution 1) and resilient meta-materials (contribution 14); an accurate model that identifies the thermal target (contribution 2); sophisticated neutral network models that track solar positions with time scales (contribution 3); pixel-level hyperspectral image classification with neural networks (contribution 4); space manipulator control models in orbit (contribution 5); and image denoise methods for sky-based backgrounds (contribution 6); visual object tracking with convolution and correlation (contribution 7); pick-and-place models with small dataset training (contribution 8); mathematic models for continuum robots (contribution 9); analyses of the MTF of remote sensing cameras (contribution 10); time-varying method for global path planning (contribution 11); low-cost localization models with a sparse modular point matrix (SMPM) (contribution 12); and deep reinforcement learning combined with RRT for path planning (contribution 13). The spectrum of the contributed research spans a wide range of topics, from traditional image processing methods and robot designs to robot kinematic models and path and motion planning with deep learning approaches.

Soft robotics focuses on the research and development of robots with compliant and flexible structures that are inspired by natural organisms. These robots can better adapt to dynamic environments and are often used in applications such as medical robotics, search and rescue tasks, and outdoor exploration [6].

Similarly, bio-inspired robotics draws inspiration from biology to design robots that mimic the structure and behavior of living organisms. This includes biomimetic locomotion,

check for updates

Citation: Song, Q.; Zhao, Q. Recent Advances in Robotics and Intelligent Robots Applications. *Appl. Sci.* **2024**, *14*, 4279. https://doi.org/10.3390/app14104279

Received: 24 April 2024
Revised: 14 May 2024
Accepted: 16 May 2024
Published: 18 May 2024

sensory systems inspired by animal perception, and soft actuators inspired by muscles and tendons [7].

Manipulation focuses on the development of robotic arms and hands that are capable of grasping, manipulating, and interacting with objects in a dexterous and precise manner. Primary applications include industrial automation, warehouse logistics, and assistive robotics for space applications [8].

Motion planning involves the algorithms and techniques used to plan the path of a robot from the current state to a desired state while avoiding obstacles and adhering to constraints without any human intervention. Robot motion planning involves designing feedback systems to regulate the robot's motion, and this ensures that it executes its planned actions intelligently and safely [9].

Perception in robotics refers to the ability of robots to sense and understand their environment using various sensors, such as cameras, LiDAR, millimeter-wave radars, and ultrasonic sensors. This includes tasks such as object detection, recognition, localization, and mapping (well known as SLAM—simultaneous localization and mapping) [10].

Autonomous navigation involves enabling robots to move and navigate complex and dynamic environments without human intervention. This includes developing algorithms for localization, path planning, obstacle avoidance, and decision making under uncertainty [11].

Currently, machine learning and AI techniques are growing very fast, and incredibly, they are integrated into robotics, which enables them to learn using data and allows them to adapt to dynamic environments. This includes reinforcement learning for robotic control, deep learning for perception tasks, and probabilistic modeling for decision making. Recently, the transformer-based generative model (GPT) has been recognized as a technical revolution and is expected to fill the long-existing gap between robots and artificial general intelligence (AGI) [12].

On the other hand, space robotics technology involves the development and deployment of robotic systems for exploration, maintenance, construction, and other tasks in space environments [13].

In space, robotic systems are utilized for the maintenance and servicing of spacecraft and satellites in orbit. These robots can perform tasks such as refueling, repairing, and upgrading satellites; extending their operational lifetimes; and reducing the need for costly replacements. Robotic arms equipped with various tools and instruments are essential for performing precise tasks in space, such as capturing payloads, deploying instruments, and conducting repairs. These arms are often mounted on spacecraft, landers, or rovers. Space robots also require advanced navigation and localization systems to accurately determine their position and orientation relative to celestial bodies, obstacles, and other spacecraft [14]. This involves integrating sensors, such as cameras, LiDAR, GPS, and inertial measurement units (IMUs).

Space robotics technology plays a crucial role in advancing our understanding of the universe, enabling scientific exploration, supporting space missions, and laying the groundwork for future human exploration and settlement of space [15].

In summary, more novel research efforts have been emerging with more innovative topics, including humanoid robotics, bio-inspired design, collaborative robotics, and ethical considerations in robot deployment. Robotics will continue to drive advancements in automation, efficiency, safety, and human well-being, with ongoing efforts to address societal challenges and shape the future of technology.

Even though it is impossible to cover all research areas of robotics and related applications, this Special Issue provides a humble collection of selected topics with cutting-edge research, and it hopes to show recent primary achievements from intuitive aspects. We would like to take this opportunity to thank all the contributing authors and reviewers of this Special Issue for their excellent research work and valuable time spent on providing comments for the collected papers. Finally, the unconditional and generous support from the editorial staff of *Applied Sciences* is also a key to this high-quality Special Issue.

Conflicts of Interest: The authors declare no conflict of interest.

List of Contributions:

1. Bai, X.W.; Kong, D.Y.; Wang, Q.; Yu, X.H; Xie, X.X. Bionic Design of a Miniature Jumping Robot. *Appl. Sci.* **2023**, *13*, 4534. https://doi.org/10.3390/app13074534.
2. Yang, M.Y.; Xu, L.; Tan, X.; Shen, H.H. A Method Based on Blackbody to Estimate Actual Radiation of Measured Cooperative Target Using an Infrared Thermal Imager. *Appl. Sci.* **2023**, *13*, 4832. https://doi.org/10.3390/app13084832.
3. Dai, S.; Song, K.F.; Wang, Y.L.; Zhang, P.J. Two-Dimensional Space Turntable Pitch Axis Trajectory Prediction Method Based on Sun Vector and CNN-LSTM Model. *Appl. Sci.* **2023**, *13*, 4939. https://doi.org/10.3390/app13084939.
4. Gao, L.Y.; Xiao, S.L.; Hu, C.H.; Yan, Y. Hyperspectral Image Classification Based on Fusion of Convolutional Neural Network and Graph Network. *Appl. Sci.* **2023**, *13*, 7143. https://doi.org/10.3390/app13127143.
5. Yang, T.; Xu, F.; Zeng, S.; Zhao, S.J.; Liu, Y.W.; Wang, Y.B. A Novel Constant Damping and High Stiffness Control Method for Flexible Space Manipulators Using Luenberger State Observer. *Appl. Sci.* **2023**, *13*, 7954. https://doi.org/10.3390/app13137954.
6. Ma, Z.L.; Zhao, Q.L.; Che, X.; Qi, X.D.; Li, W.X.; Wang, S.X. An Image Denoising Method for a Visible Light Camera in a Complex Sky-Based Background. *Appl. Sci.* **2023**, *13*, 8484. https://doi.org/10.3390/app13148484.
7. Liu, L.D.; Long, Y.J.; Li, G.N.; Nie, T.; Zhang, C.C.; He, B. Fast and Accurate Visual Tracking with Group Convolution and Pixel-Level Correlation. *Appl. Sci.* **2023**, *13*, 9746. https://doi.org/10.3390/app13179746.
8. Kee, E.; Chong, J.J.; Choong, Z.J.; Lau, M. Development of Smart and Lean Pick-and-Place System Using EfficientDet-Lite for Custom Dataset. *Appl. Sci.* **2023**, *13*, 11131. https://doi.org/10.3390/app132011131.
9. Li, Y.F.; Wang, Q.H.; Liu, Q. Developing a Static Kinematic Model for Continuum Robots Using Dual Quaternions for Efficient Attitude and Trajectory Planning. *Appl. Sci.* **2023**, *13*, 11289. https://doi.org/10.3390/app132011289.
10. Yu, J.; Zhang, Y.; Qi, B.; Bai, X.T.; Wu, W.; Liu, H.X. Analysis of the Slanted-Edge Measurement Method for the Modulation Transfer Function of Remote Sensing Cameras. *Appl. Sci.* **2023**, *13*, 13191. https://doi.org/10.3390/app132413191.
11. Jia, L.F.; Zeng, S.; Feng, L.; Lv, B.H.; Yu, Z.Y.; Huang, Y.P. Global Time-Varying Path Planning Method Based on Tunable Bezier Curves. *Appl. Sci.* **2023**, *13*, 13334. https://doi.org/10.3390/app132413334.
12. Lin, H.Y.; Quan, P.K.; Liang, Z.; Wei, D.B.; Di, S.C. Low-Cost Data-Driven Robot Collision Localization Using a Sparse Modular Point Matrix. *Appl. Sci.* **2024**, *14*, 2131. https://doi.org/10.3390/app14052131.
13. Cai, R.G.; Li, X. Path Planning Method for Manipulators Based on Improved Twin Delayed Deep Deterministic Policy Gradient and RRT*. *Appl. Sci.* **2024**, *14*, 2765. https://doi.org/10.3390/app14072765.
14. Muñoz-Barron, B.; Sandoval-Castro, X.Y.; Castillo-Castaneda, E.; Laribi, M.A. Characterization of a Rectangular-Cut Kirigami Pattern for Soft Material Tuning. *Appl. Sci.* **2024**, *14*, 3223. https://doi.org/10.3390/app14083223.

References

1. Garcia, E.; Jimenez, M.A.; Santos, P.G.D.; Armada, M. The evolution of robotics research. *IEEE Robot. Autom. Mag.* **2007**, *14*, 90–103. [CrossRef]
2. Murphy, R.R. *Introduction to AI Robotics*; MIT Press: Cambridge, UK; London, UK, 2019.
3. Cadena, C.; Carlone, L.; Carrillo, H.; Latif, Y.; Scaramuzza, D.; Neira, J.; Reid, I.; Leonard, J.J. Past, present, and future of simultaneous localization and mapping: Toward the robust-perception age. *IEEE Trans. Robot.* **2016**, *32*, 1309–1332. [CrossRef]
4. Nof, S.Y. *Handbook of Industrial Robotics*; Wiley Press: New York, NY, USA, 1999.
5. Kleeberger, K.; Bormann, R.; Kraus, W.; Huber, M.F. A survey on learning-based robotic grasping. *Curr. Robot. Rep.* **2020**, *1*, 239–249. [CrossRef]
6. Iida, F.; Laschi, C. Soft robotics: Challenges and perspectives. *Procedia Computer Science* **2011**, *7*, 99–102. [CrossRef]
7. Pfeifer, R.; Lungarella, M.; Iida, F. Self-organization, embodiment, and biologically inspired robotics. *Science* **2007**, *318*, 1088–1093. [CrossRef]

8. Hsia, T.C.S.; Lasky, T.A.; Guo, Z.Y. Robust independent joint controller design for industrial robot manipulators. *IEEE Trans. Ind. Electron.* **1991**, *38*, 21–25. [CrossRef]
9. Song, Q.; Zhao, Q.L.; Wang, S.X.; Liu, Q.; Chen, X.H. Dynamic path planning for unmanned vehicles based on fuzzy logic and improved ant colony optimization. *IEEE Access* **2020**, *8*, 62107–62115. [CrossRef]
10. Wang, S.; Chen, X.H.; Ding, G.Y.; Li, Y.Y.; Xu, W.C.; Zhao, Q.L.; Gong, Y.; Song, Q. A Lightweight Localization Strategy for LiDAR-Guided Autonomous Robots with Artificial Landmarks. *Sensors* **2021**, *21*, 4479. [CrossRef] [PubMed]
11. Gasparetto, A.; Boscariol, P.; Lanzutti, A.; Vidoni, R. Path planning and trajectory planning algorithms: A general overview. In *Motion and Operation Planning of Robotic Systems: Background and Practical Approaches*; Springer: Berlin/Heidelberg, Germany, 2015.
12. Vaswani, A.; Shazeer, N.; Parmar, N.; Uszkoreit, J.; Jones, L.; Gomez, A.N.; Kaiser, Ł.; Polosukhin, I. Attention is all you need. *Adv. Neural Inf. Process. Syst.* **2017**, *30*.
13. Flores-Abad, A.; Ma, O.; Pham, K.; Ulrich, S. A review of space robotics technologies for on-orbit servicing. *Prog. Aerosp. Sci.* **2014**, *68*, 1–26. [CrossRef]
14. Ma, O.; Dang, H.; Pham, K. On-orbit identification of inertia properties of spacecraft using a robotic arm. *J. Guid. Control. Dyn.* **2008**, *31*, 1761–1771. [CrossRef]
15. Wei, L.; Zhang, L.; Gong, X.; Ma, D.M. Design and optimization for main support structure of a large-area off-axis three-mirror space camera. *Appl. Opt.* **2017**, *56*, 1094–1100. [CrossRef] [PubMed]

applied sciences

MDPI

Article

Bionic Design of a Miniature Jumping Robot

Xianwei Bai [1,2], Deyi Kong [1,*], Qiong Wang [1,2], Xianhai Yu [3] and Xiaoxuan Xie [1,2]

1 Hefei Institutes of Physical Science, Chinese Academy of Sciences, Hefei 230031, China
2 Science Island Branch, Graduate School of University of Science and Technology of China, Hefei 230026, China
3 School of Microelectronics, Hefei University of Technology, Hefei 230601, China
* Correspondence: kongdy@iim.ac.cn

Abstract: In response to the problem of low energy storage density in the structure of existing miniature jumping robots, this study designed a parallel single-degree-of-freedom double six-link jumping robot by imitating the physiological structure and jumping mechanism of wax cicadas. The designed six-link mechanism was first mathematically modeled, and to accommodate the jumping structure of this robot, a six-link mechanism with a smaller cam pushrod stroke was obtained by optimizing the linkage size and position parameters in the model. The dynamics of the robot's jumping process were then analyzed utilizing the second type of Lagrange equation to determine the joint angles of the robot's jumping phase. The results were compared with an ADAMS-based jumping simulation to verify the validity of the analysis of the dynamics. The feasibility of the structural design was then validated using ADAMS simulations. Finally, a physical prototype of the jumping robot was produced and tested; the findings revealed that the robot had good jumping performance, was stable in the air, fully discharged 600.2 mJ of energy, and was able to overcome obstacles measuring 220 mm in height and 330 mm in distance. The design of the jumping robot provides a novel approach to improving energy storage density and serves as a foundation for future research on footed jumping robots.

Keywords: imitation wax cicada; parallel single-degree-of-freedom double six-linkage mechanism; optimization; Lagrange equation; ADAMS

Citation: Bai, X.; Kong, D.; Wang, Q.; Yu, X.; Xie, X. Bionic Design of a Miniature Jumping Robot. *Appl. Sci.* **2023**, *13*, 4534. https://doi.org/10.3390/app13074534

Academic Editors: Qi Song and Qinglei Zhao

Received: 4 March 2023
Revised: 31 March 2023
Accepted: 1 April 2023
Published: 3 April 2023

1. Introduction

The rapid advancement of science and technology has necessitated the development of efficient, intelligent, and multifunctional micro-robots. Micro-robots are currently plagued by two main problems: the complex motions and long execution times of footed robots [1–3] traversing small target obstacles, and the restricted movement of wheeled tracked robots [4–6] with strong maneuverability in complex terrain. It becomes more difficult for micro-robots to navigate efficiently in natural unstructured environments as their size decreases [7]. Jumping robots are an excellent solution to these problems. With a small contact area and high energy efficiency, they can overcome obstacles several times their size [8,9] in a short period of time and adapt to unstructured terrain, which holds great potential for future applications in rescue and military operations and exploration.

In recent years, several research teams both domestically and abroad have developed some miniature jumping robots by drawing inspiration from insects including grasshoppers, fleas, and foam cicadas. Using two torsion springs as energy storage elements, the Swiss Federal Institute of Technology in Lausanne has designed a locust-like jumping robot with excellent jumping capabilities [10], which weighs only 7 g and is capable of jumping to a height of 1.38 m, which is about 27 times its own dimensional height. The Sant' Anna Institute of Higher Studies in Italy designed a cicada-like jumping robot with two linear springs for energy storage by optimizing the leg length ratio [11], which enables stable landings and continuous jumps with a jump height of approximately 100 mm. Based on the principle of frog-leg jumping, Changlong Ye et al. from Shenyang University of Aeronautics and Astronautics designed a jumping robot with a deformed wheel structure that integrates wheeling and jumping [12], which can achieve jumping direction controllability, and tests have shown that a 300 g robot can achieve a maximum jump height of 285 mm. Riccy Kurniawan et al. from the University of Washington, U.S. proposed a Yak jumping robot [13]. The robot has a symmetrical mechanical structure with two shape memory alloy (SMA) spring actuators in the upper and lower parts, which are driven by radio transmission to store energy by converting the linear displacement of the SMA springs into a bending moment. According to the test results, the 216 mg robot has a vertical jump height of 138 mm, making it one of the top insect-grade wirelessly powered robots for jumping performance. The University of California proposed a crawling jumping robot driven by shape memory alloy actuators and aluminum foil [14] with magnet actuation to drive the whole system. The initial state of SMA has some pre-deformation, and jumping motion is promoted by induction heating of SMA and aluminum foil in a high-frequency alternating magnetic field. During the heating of the SMA berth, this design converts SMA from martensite to austenite. More elastic potential energy is stored, increasing the jump height, which is not found in conventional springs. A locust-like jumping robot developed by the Northwestern Polytechnic University in China mimicking the movements of the femur, tibia, and tarsus of the locust's hind legs [15] uses a linear spring to store energy and utilizes the negative rotation of the robot's body to counteract its tendency to flip in the air phase.

The aforementioned studies on miniature jumping robots have focused on the implementation and miniaturization of jumping functions but have neglected to design energy storage mechanisms with a greater energy storage density, the size of which directly affects the potential that a jumping robot can achieve. In this paper, the wax cicada, which has an excellent jumping ability, is used as a bionic prototype to design a jumping robot with a parallel single-degree-of-freedom six-link energy storage mechanism as the core, to solve the problem of insufficient energy storage in the existing energy storage mechanism of micro-robots and to provide a reference for the improvement of the energy storage density of future footed robots.

2. Design of the Wax-like Cicada Jumping Mechanism

2.1. Study of the Structure and Jumping Mechanism of the Hind Legs of a Bionic Prototype Wax Cicada

The wax cicada is an insect belonging to the family Cicadidae of the order Hemiptera, as shown in Figure 1a, its accustomed jumping movement pattern allows it to swiftly escape from threats posed by the outside world. Studies on wax cicadas have shown the presence of paired pleural arch structures on the hind thorax [16] that function as energy stores and contain a large number of tendon and arthroplasty elastic elements. It is this unique pleural arch that contributes the vast majority of the jumping energy source, as shown in Figure 1b [16].

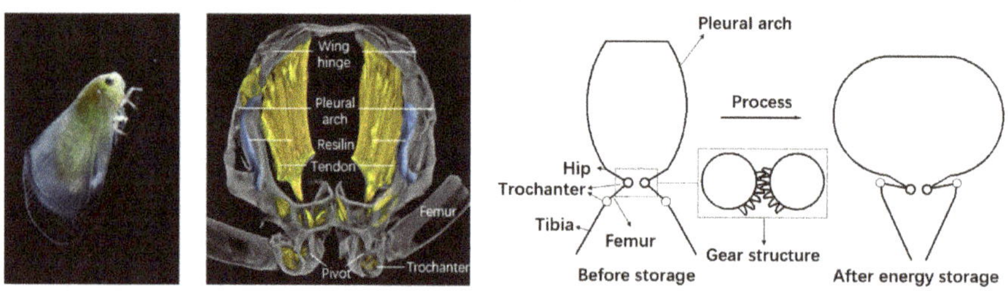

(**a**) (**b**) (**c**)

Figure 1. (**a**) Wax cicada in life; (**b**) Biological structure of the pleural arch [16]; (**c**) Simplified model of the jumping process.

Figure 1c shows a significant change in the posture of the hind legs of the wax cicada during the jump. Before energy storage, the angle between the femoral and tibial joints is in a large state. During energy accumulation, the elastic element contracts heavily to flex the pleural arch [17] and store large amounts of energy. The wax cicada has a cogwheel structure at the hip rotor and the released energy is transferred from the hip rotor to the end of the hindfoot. At the same time, the femoral and tibial joint stance gradually becomes smaller and reduces to a post-energy accumulation stance in which the wax cicada enters a pre-jump state, after which both feet are rapidly and synchronously extended due to the cogwheel structure at the hip rotor [18], producing a fast and powerful jump. The energy storage capacity of the pleural arch, which contains many elastic elements, is undoubtedly much higher than that of a single elastic element, and studies have shown that the jumping speed of the wax cicada is a staggering 5.5 m/s [19] compared to that of a flea with a jumping speed of 1 m/s [20].

2.2. Modeling of the Wax Cicada Jumping Mechanism

Based on the physiological structure and jumping mechanism of the wax cicada, the femur was used to simulate the wax cicada's femoral joint, the tibia to simulate the cicada's tibial joint, and two four-bar mechanisms consisting of two semi-pleural arch structures to simulate the cicada's two posterior pleural arch structures, respectively, and two linear springs were used to replace tendons and other elements to design the bionic energy storage structure model, as shown in Figure 2. In Figure 2, the connecting rods are articulated and the degree of freedom of the jumping leg mechanism is calculated as shown in Equation (1):

$$F = 3n - 2p_l - p_h = 3 \times 6 - 2 \times 8 - 1 = 1 \tag{1}$$

where n represents the number of moving parts and has a value of 6; p_l represents the number of low pairs in the kinematic pair and has the value of 8; and p_h represents the number of higher pairs in the kinematic pair and has the value of 1. The structure stores energy with the cam rotating counterclockwise as the prime mover, making contact with the rod CE and generating a collision force. As the mechanism only has one degree of freedom, the two rotating joints E and F of the four-rod mechanism will move in the curved slots, gradually changing the attitude of the four-rod mechanism, the four linear springs fixed in the two four-bar mechanisms will be stretched, and the two parallel six-link mechanisms will have a defined motion.

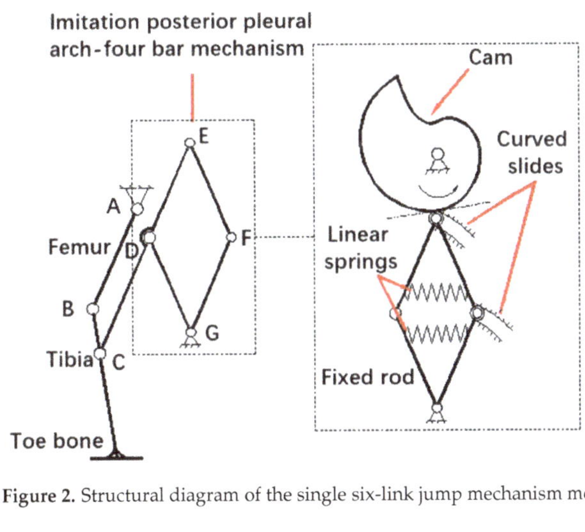

Figure 2. Structural diagram of the single six-link jump mechanism model.

2.3. Modeling of the Wax Cicada Jumping Mechanism

Based on the structural diagram of the jumping mechanism established above, a virtual prototype of the bouncing robot was created, as shown in Figure 3a. The robot is composed of four main parts: the front legs, the parallel double six-link mechanism, the body, and the energy control mechanism, which is shown in Figure 3b. The front legs of the robot provide support and attitude adjustment, the double six-link mechanism with four linear springs provides the power source for bouncing, the energy control mechanism stores and releases energy for the whole system, and the body part of the bouncing robot connects and balances the entire robot. In addition, the body is equipped with a servo control board to control the energy storage and release state of the entire system, as well as a battery that powers the entire robot.

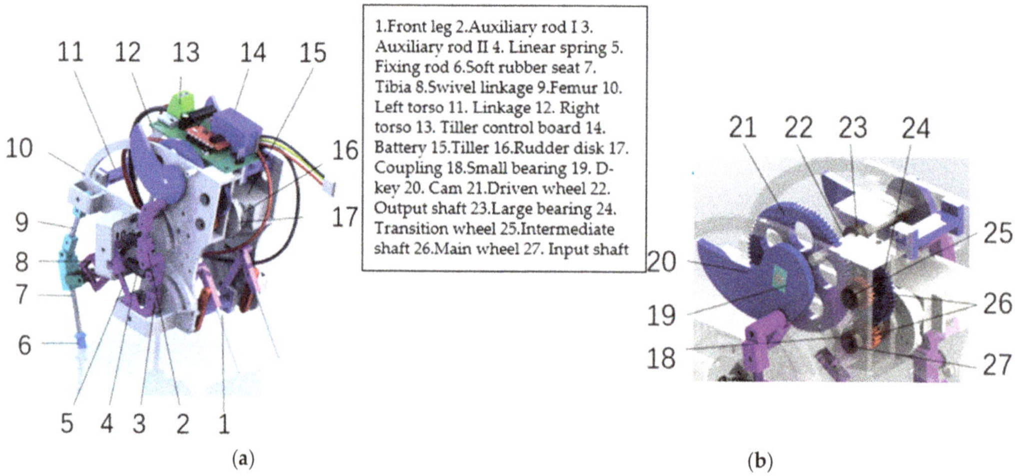

Figure 3. (**a**) Overall structure of the bouncing robot; (**b**) Energy control structures for robots.

The robot bouncing principle is as follows: During the energy storage phase of the robot, the driving rudder rotates clockwise to transmit power through the rudder disk and coupling to the input shaft in a predetermined ratio, and then to the two cams of the output shaft via the energy control mechanism. The two cams rotate counterclockwise until they collide with the rotating linkage of the sub-mechanism (a four-bar mechanism), resulting in a collision force that forces two of the rotating joints to make a circular motion. The femur and tibia undergo a specific regular attitude change with the movement of the four-bar mechanism. At the same time, the four linear springs fixed to the two four-bar mechanisms are forced to stretch, and the whole system begins to store energy. The cam continues to rotate until it reaches the maximal energy storage for the whole system, at which point there are no more collision forces between the cam and the rotating joint, and the femur and tibia are in the ready-to-jump position. The cams continue to rotate after the energy is released, and the energy stored throughout the robot is instantaneously and completely released. The entire robot will return to its original stance in a very short time and the energy stored in the four linear springs will be instantly transferred through the entire mechanism to the toe bones causing the robot to jump by the reaction force of the ground on the toe bones.

2.4. Power Conversion Principle

A secondary reduction mechanism was employed in this study to reduce the speed and load inertia and multiply the torque on the output shaft to achieve the minimum torque required to stretch the four linear springs to their maximum. Figure 4 illustrates the principle of how the servo amplifies the torque of the input shaft and reduces the speed of the input shaft to the output shaft. The process is as follows: The servo is coupled to the input shaft by a coupling and outputs with a torque of value τ_1, which is transmitted to the output shaft under the action of the reducer with a value τ_2. ω_1, ω_2 and ω_3 in the diagram represent the rotational speeds of the input, intermediate and output shafts, respectively. The theoretical variation of torque is related as:

$$\frac{\tau_2}{\tau_1} = \frac{Z_3}{Z_1} \cdot \frac{Z_2}{Z_1} = \frac{Z_3 \cdot Z_2}{Z_1^2} \tag{2}$$

In the above equation, Z_1, Z_2 and Z_3 correspond to the number of teeth on the driving gear, intermediate gear, and driven gear, respectively. Conversely, the theoretical relationship between the rotational speeds is expressed as follows:

$$\frac{\omega_3}{\omega_1} = \frac{Z_1^2}{Z_3 \cdot Z_2} \tag{3}$$

This multiplies the maximum torque of the output shaft and reduces the torque requirement of the rudder, while at the same time reducing the output shaft speed proportionally and increasing the overall energy storage time.

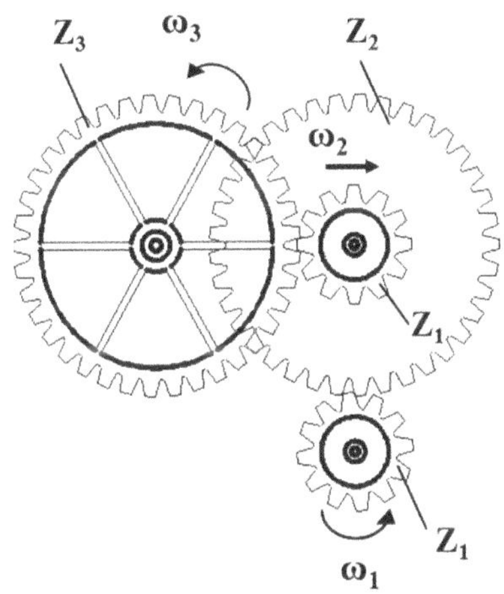

Figure 4. Power conversion diagram.

2.5. Bouncing Motion Process

The robot completing the entire bouncing motion is shown in Figure 5. To begin, the rudder control board drives the rudder to turn counterclockwise. The power from the rudder is transmitted via the energy control mechanism to the two cams on the output shaft, which are in contact with the two rotating links on the left and right, respectively, and the two six-link mechanisms in parallel begin to change stance in a synchronized manner. The four linear springs fixed to the four-bar mechanism are stretched, causing the robot's center of gravity to shift downwards gradually. At this point, the two cams continue to rotate, the attitude of the six-link mechanism changes, and the deformation of the linear spring gradually increases until the cams reach a critical point, at which point the robot's energy storage phase ends. During the release phase, the cams continue to rotate until they cross the critical point, at which moment the entire system instantly releases all its energy and the robot leaps up.

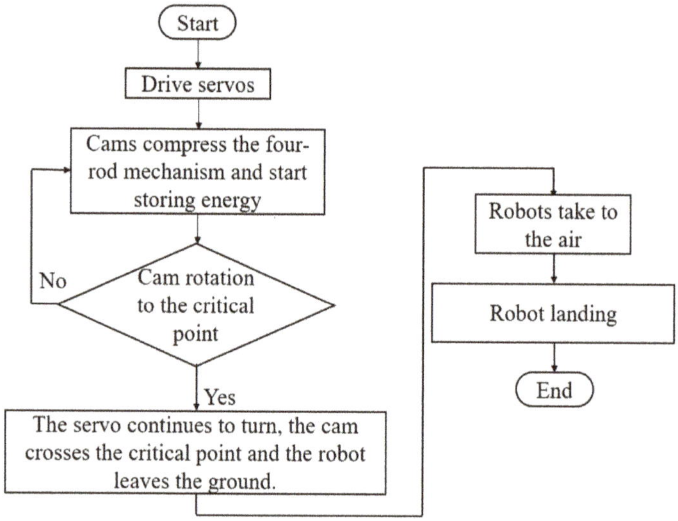

Figure 5. Bouncing movement flow chart.

3. Jumping Mechanism Analysis and Optimization

3.1. Design and Parameter Optimization of the Jumping Leg Mechanism

This section explores the effect of the angle of the variable C_1DG and the length of DG on the magnitude of the spring deformation, provided that the fixed rod DG and the length of DE are constants, to allow the double six-link mechanism to rotate to the limited position of maximum compression of the linear spring, as shown in Figure 6. To reduce the minimum torque required by the cam during compression, a cam with a small actuator stroke was first designed, followed by the calculation of the deformation of the spring when the rotating linkage is rotated clockwise to a critical state based on the profile of the cam, and the analysis of the variation law between the angle of the variable C_1DG and the length of DG and the deformation of the spring.

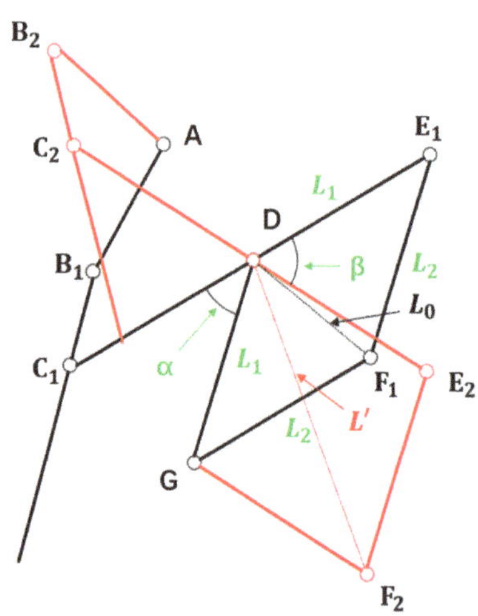

Figure 6. Mathematical model of a six-link mechanism.

In Figure 6, the six-rod mechanism $AB_1C_1DE_1F_1G$ represents the initial state of the jumping leg mechanism without energy storage, $AB_2C_2DE_2F_2G$ is the limit state of the jumping leg mechanism at maximum energy storage, the rotating linkage C_1E_1 rotates the fixed angle β to reach the critical state, and the DG rod is a fixed rod. According to the relationship between two linear spring positions with symmetry, $DE_1 = DG = L_1$ and is a constant, and $E_1F_1 = GF_1 = L_2$, L_2 is a variable. According to the given cam profile, $\beta = \pi/3$, let the initial length of the spring be L_0, the limit length at the maximum of the deformation variable is L', the angle between the rotating connecting rod C_1E_1 and the fixed rod DG is the variable α, and the value of α is in the range of (0, π). In the triangle DE_1F_1, according to the cosine theorem, we have:

$$cos\frac{\pi-\alpha}{2} = \frac{L_1^2 + L_0^2 - L_2^2}{2L_0L_1} \tag{4}$$

In the triangle DE_2F_2:

$$cos\frac{\pi-\alpha-\frac{\pi}{3}}{2} = \frac{L_1^2 + L'^2 - L_2^2}{2L'L_1} \tag{5}$$

Simplifying the above two equations yields:

$$\begin{cases} L_0^2 - 2L_1 sin\frac{\alpha}{2} L_0 + L_1^2 - L_2^2 = 0 \\ L'^2 - 2L_1 cos\left(\frac{\pi}{3} - \frac{\alpha}{2}\right) L' + L_1^2 - L_2^2 = 0 \end{cases} \tag{6}$$

From Equation (6), it follows that:

$$L' - L_0 = L_1\left[\cos\left(\frac{\pi}{3} - \frac{\alpha}{2}\right) - sin\frac{\alpha}{2}\right] + \sqrt{L_1^2 cos^2\left(\frac{\alpha}{2} - \frac{\pi}{3}\right) - \left(L_1^2 - L_2^2\right)} - \sqrt{L_1^2 sin^2\frac{\alpha}{2} - \left(L_1^2 - L_2^2\right)} \tag{7}$$

In Equation (7), α and L_2 are two independent variables that have no influence on one another. Using the control variables method, the mathematical analysis of this function leads to the conclusion that the smaller the angle of α, the larger the value of $L' - L_0$, and the larger the deformation of the spring, while L_2 remains constant. If α remains constant, the value of $L' - L_0$ tends to increase as L_2 decreases. We thus propose a design principle that is compatible with this six-rod mechanism: the linkage length L_1 remains constant, the initial spring length L_0 and the limit length L' satisfy the conditions for the use of the spring, and the size of α or is L_2 is reduced in the six-rod mechanism so that the size of the linkage is reduced and the deformation of the spring is quickly maximized. There is, however, a theoretical minimum value for L_2, which is the minimum value when E_1F_1G coincides with the position of the dead point of the mechanism, when $E_1F_1G = 2L_2$.

3.2. Dynamical Modeling and Analysis

A common approach used for establishing the dynamic equation of robots is the Lagrange method. During the take-off phase of the robot, its joint angles can be obtained by solving the analytical equations. Assuming that the robot's tibia is in a state where it is not slipping on the ground and that the tibia and the ground can be regarded as a rotational connection, the generalized angle θ_1 between the tibia and the ground is used as the coordinate of the system; the dynamics of the robot in the take-off phase is modeled based on the Lagrange approach, as shown in Figure 7.

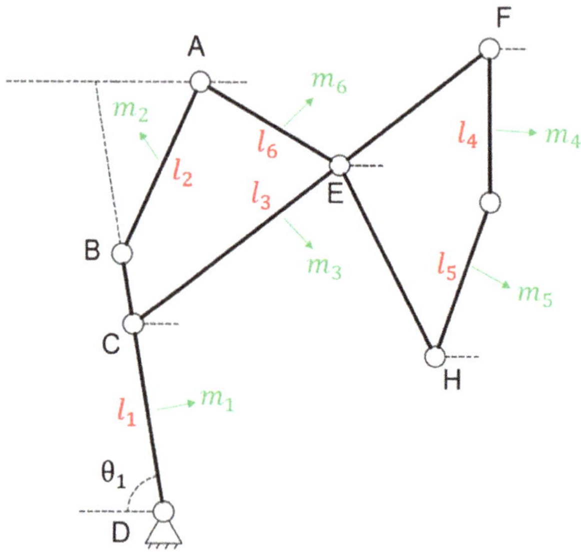

Figure 7. Overall dynamic model of the take-off phase.

In Figure 7, the point E is the center of mass of the body, m_i, l_i represent the mass and length of each bar, and the dynamic equation of the robot during the jumping phase is obtained by:

$$\begin{cases} \frac{d}{dt}\frac{\partial L}{\partial \dot{\Theta}_1} - \frac{\partial L}{\partial \Theta_1} = 0 \\ L = k - u \end{cases} \tag{8}$$

In the preceding equation, the dynamic energy k generated by each rod includes the linear velocity dynamic energy and angular velocity dynamic energy generated by the center of mass of that rod. Using the contact point between the tibia and the ground as the zero potential energy point, the potential energy u of the system can be divided into two parts, one part is the elastic potential energy of the spring and the other part is the total gravitational potential energy of the system composed of each rod, resulting in the dynamic energy and potential energy of the system when the robot jumps, as shown in Equation (9):

$$\begin{cases} k = \sum\limits_{1}^{6} k_i = \frac{1}{4} m_i v_i^2 + I_i \omega_i^2 \\ u = 4 \bullet \frac{1}{2} k_c \Delta x^2 + 2 \sum\limits_{1}^{6} m_i g h_i \end{cases} \tag{9}$$

In Equation (9), k_c is the spring's coefficient of elasticity, Δx is the spring deformation, g is the acceleration of gravity, v_i is the velocity of the center of mass of each linkage, h_i is the vertical coordinate of the center of mass of each linkage relative to the point of zero potential energy, and I_i is the rotational inertia of each linkage to the center of mass. Table 1 displays the parameters of the robot's connecting rods, such as mass, length, and inertia.

The dynamic equations were solved numerically using the Runge–Kutta method with an initial value of θ_1 of 88° and ω_1 of 0 rad/s. The final calculated output curves were compared with the simulation results in the ADAMS environment, as shown in Figure 8.

In Figure 8, the energy storage phase of the robot lasts from 0 to 5.02 s, during which the femur-tibia and tibia-ground angles gradually decrease, whereas the body-femur angle gradually increases. The curves of the body-femur, femur-tibia, and tibia-ground angles are approximately the same, indicating that there are minor errors between the simulation results and the calculation results, but the curves are identical, demonstrating the validity of the dynamic analysis.

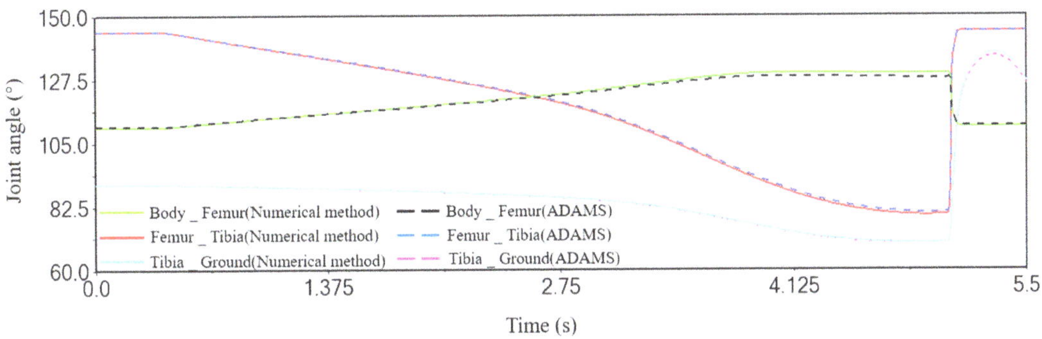

Figure 8. Joint angle during the take-off phase of the robot.

Table 1. Relevant parameters for robot components.

Components	Parts	Materials	Weight (g)	Length (mm)	Rotational Inertia (kg · mm^2)
Femur	-	Aluminum	1.4	36.8	0.63
Tibia	-	Aluminum	4.5	50.9	3.89
Rotary linkage	Connecting rod Connectors	Aluminum Photosensitive resin	5.6 0.4	75.5	2.85
Auxiliary rod I	Connectors Connecting rod	Photosensitive resin Aluminum	0.1 0.6	30.8	0.22
Auxiliary rod II	Connecting rod Connectors	Aluminum Nylon glass fiber	0.6 0.1	30.8	0.22
Body	Camshafts	Stainless steel	6	62	55.15
	Cams	Resin	4	-	
	Gear sets	Nylon glass fiber	6	-	
	Bearings	Steel	1.8		
	Servo		12		
	Steering wheel	Plastic	0.6		
	Couplings	Photosensitive resin	1.2		
	Forelegs	Nylon	5		
	Battery	-	16		
	Control panels	-	12		
	Other	-	75.4		
Total mass		-	166.6	-	-

4. ADAMS-Based Simulation Analysis of Bouncing Robot Motion

According to the structural design scheme of the imitation wax cicada jumping robot proposed in Section 2, the model was imported from SolidWorks into ADAMS for kinematic simulation analysis to predict the kinematic performance of the bouncing robot and to verify the feasibility of the overall mechanism design.

4.1. Component Type Parameters of the Bionic Bouncing Robot

Based on the 3D structure of the robot, a prototype model of the robot was assembled, of which the model parameters of the relevant components are shown in Table 2.

Table 2. Relevant component type parameters.

Components and Parameters	Numerical Values
Spring wire diameter ^ outside diameter ^ length (mm)	0.7 ^ 6 ^ 25
Modulus of active wheel ^ number of teeth (mm)	0.5 ^ 18
Modulus of intermediate wheel ^ number of teeth (mm)	0.5 ^ 42
Modulus of driven wheel ^ number of teeth (mm)	0.5 ^ 80
Small bearing bore ^ outer diameter ^ thickness (mm)	4 ^ 7 ^ 2.5
Large bearing bore ^ outer diameter ^ thickness (mm)	4 ^ 8 ^ 3
Servo mass (g) ^ max. torque (kg · cm)	12 ^ 5

4.2. Motion Simulation and Experiments

The process used for the simulation is as follows: according to the given material properties, add the material properties of each part in ADAMS, set the connection relationship between each part, add the collision force of the parts in contact, and finally add the driving force for simulation.

The trajectory of the robot's center-of-mass motion is determined from the simulation of the virtual prototype and is shown in Figure 9. It can be seen from the figure that the robot is relatively stable during its motion in the air, with no substantial overturning. At the end of the simulation, the bouncing performance data of the robot was obtained using the post-processing module of ADAMS, as shown in Figure 10a,b.

During the simulation, the robot saves energy from 0 to 5.02 s, the robot starts to store energy under the force of the cam, and the center of mass gradually drops to its lowest point. When the cam is ready to pass the contact boundary of the four-bar mechanism and the energy is about to be fully released, the system's energy storage achieves a maximum at t = 5.02 s. At t = 5.02~5.31 s, the robot releases energy, starts jumping, and gradually accelerates to the highest point with a jump angle of about 62°. The elastic potential energy accumulated in the robot during this process will be completely transformed into kinetic energy. The vertical height of the highest point is about 393 mm, and horizontal displacement will continue to increase until the robot hits the ground. When the robot begins to descend at t = 5.31~5.55 s, its gravitational potential energy at the highest point is completely converted into kinetic energy, and the robot lands at t = 5.55 s with a horizontal displacement of 595 mm, reaching the maximum. According to the center of mass displacement graph, the robot has a maximum bounce height of 306 mm and a maximum jump distance of 595 mm. The acceleration of the robot during the whole jumping process and the change of the torque of the rotating vice between the body and the femur and the femur and the tibia of the single six-link mechanism during the jumping process are shown in Figure 11a,b. During the gradual stretching of the linear spring and the energy storage process of the whole system, the center of gravity of the robot shifts downwards and has a smaller tendency of jumping up and down until 5.02 s and gains an instantaneous acceleration of a value of 275 m/s^2 during the jumping, after which the momentary acceleration changes to a gravitational acceleration of approximately 9.8 m/s^2. During the energy storage phase, the moment between the body and femur and femur and tibia in the six-link mechanism builds and then falls to zero at t = 5.02 s due to the immediate release of energy.

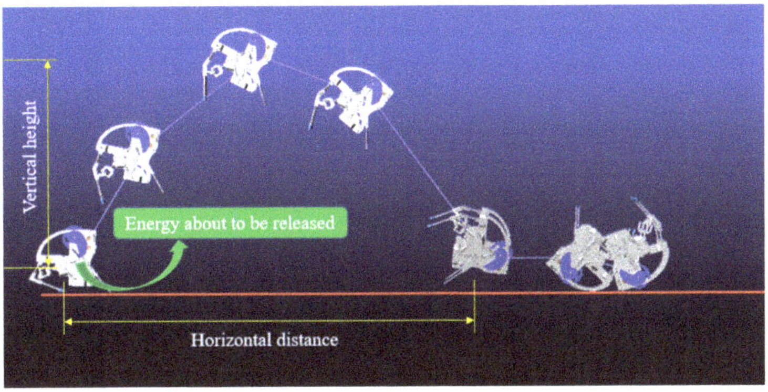

Figure 9. Movement diagram of a bouncing robot in a simulation environment.

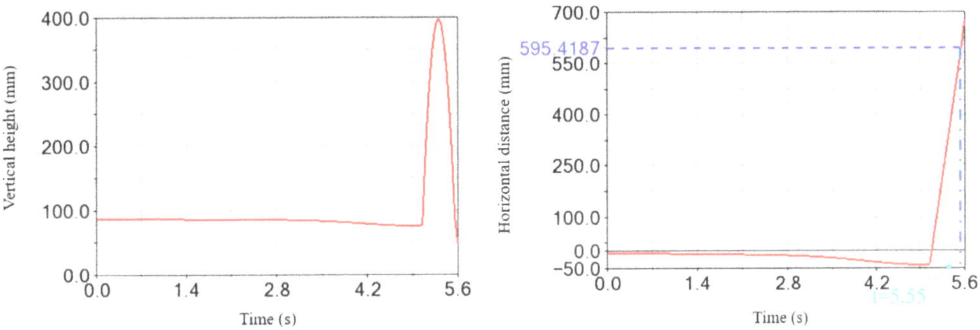

(**a**) (**b**)

Figure 10. (**a**) Vertical height displacement of the center of mass; (**b**) Horizontal distance displacement of the center of mass.

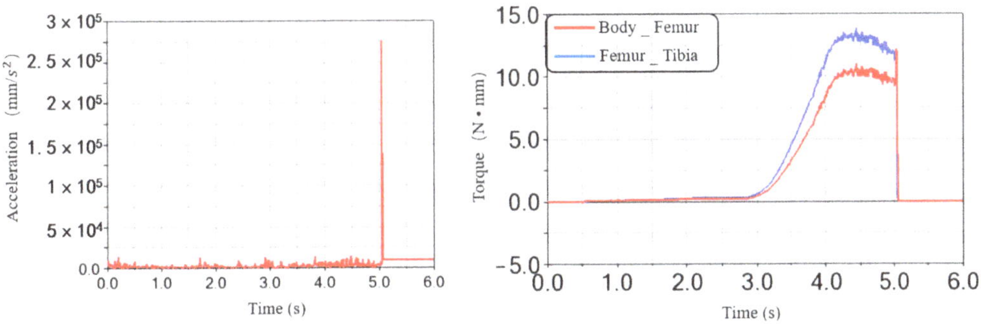

(**a**) (**b**)

Figure 11. (**a**) Acceleration of the robot's center of mass; (**b**) Graph of torque change during the jumping phase.

5. Prototyping and Experimental Testing

To measure the actual jumping performance of the bouncing robot, experimental tests were carried out on the prototype robot. The control board was a one-way control board with a DIY wireless servo. Aluminum alloy was utilized for the six-link mechanism, the joint parts and the gear set of the energy control mechanism were made of robust nylon fiberglass, and the rest of the components were made of resin. A high-speed camera was used to record the jumping process and to obtain the robot's attitude and position at the highest point and the landing point, as shown in Figure 12.

Figure 12. Robot prototype jump test.

With 600.2 mJ of energy fully released, the test results showed that the vertical height of the jumping robot was 220 mm, the horizontal distance was 330 mm, and its jumping speed was 2.21 m/s, of which the vertical speed was 2.11 m/s and the horizontal speed was 0.66 m/s. The overall energy consumption efficiency was 67%, with the majority of energy lost due to the friction of each rotating connection and air resistance during the jumping process. As the simulation cannot fully simulate the contact between the end of the robot's hind foot and the end of the forefoot and the ground, as well as the influence of realistic factors such as physical prototype production defect, there is a certain degree of discrepancy between the simulation results and the experimental results.

Many bouncing robots have a high jumping capacity, but researchers have rarely considered the impact of the bouncing robot's mass on the weight of the robot. For example, the MSU Jumper has a jump height of 872 mm and a distance of 898 mm at zero loads, but a 4 g increase in weight decreases jump performance by 5.8% and an 8 g increase in weight decreases jump performance by 20.3% (695 mm jump height and 678 mm jump distance at 8 g), which shows that the weight of the robot greatly affects its jump performance. This shows that the weight of the robot itself greatly affects its jumping performance. To better highlight the performance of the bouncing robot, this paper proposes a reference factor $\Gamma = m\left(v_x^2 + v_y^2\right)$ from the perspective of the bouncing performance of the robot per unit weight. The variable m represents the weight of the robot, the variable v_x and the variable v_y represent the horizontal and vertical velocity of the robot when it starts to jump, and it responds to the relationship between the bouncing performance of the robot and its weight. Next, the above equation can be simplified by the physics equation for oblique throwing motion as $\Gamma = m(4h^2 + x^2)/2h$, where: variable h represents the height of the robot's bounce and variable x represents the distance of the robot's bounce. Furthermore, some representative micro-robots and their characteristics are compared, as shown in Table 3 below.

The weight of the robot varies negatively with jumping performance, with larger values of Γ indicating better jumping performance. Although the introduced reference factors may not reflect the relationship between robot weight and jumping ability completely and accurately, they can reflect the robot's jumping performance per unit weight to some extent. After comparative analysis, it was found that the reference factors proposed in this paper have some reference value for measuring the bouncing ability of a robot per unit weight. Specifically, the energy storage density of the bouncing robot proposed in this paper is 600.2 mJ, and its reference factor Γ value is 11.454. Compared with existing miniature bouncing robots, the bouncing robot in this paper has a high energy storage density and a certain advantage in bouncing performance per unit weight.

Table 3. Comparison of the characteristics of miniature bouncing robots.

Bouncing Robot	Weight (g)	Jump Height (mm)	Jumping Distance (s)	Energy Value (mJ)	Parameters Γ
GRILLO III [11]	22	100	200	31.8	0.88
Miniature jumping robot [10]	7	1380	790	154	2.09
MSU Jumper [21]	23.5	872	898	-	5.185
Mini-Whegs [22]	191	18	-	-	0.688
JelloCube [23]	250	70	180	-	9.285
This article robot	166.6	220	330	600.2	11.454

6. Conclusions

This study has presented the overall structural design of a bouncing robot with high energy density based on the jumping mechanism of the wax cicada. An optimized parametric analysis model of the six-bar mechanism was established, and the influence of two variables of the six-bar mechanism—the angle of C_1DG and the length of DG—on the spring deformation variables was ultimately proposed.

The kinetic model of the jumping phase was established, and the joint angle of the robot was obtained using kinetic equations and compared to simulation results in the ADAMS environment to verify the validity of the kinetic equations.

Simulation of the jumping performance of the robot was carried out in the ADAMS virtual prototype environment. The feasibility of the entire implementation of the jumping robot was validated by analyzing the jumping motion of the robot in one cycle, and the jumping performance in the simulation environment was obtained. The simulation results show that the bouncing robot has good jumping ability and air posture stability.

A physical prototype was assembled using 3D printing and other techniques, and experimental tests were completed. The results show that with an energy storage capacity of 600.2 mJ, the robot can overcome obstacles with a height of about 220 mm and a distance of about 330 mm and does not flip significantly in the air. Compared to existing miniature bouncing robots, the bouncing robot developed in this paper has a higher energy storage density and better bouncing performance per unit weight.

In this paper, we have designed an energy storage mechanism with a high energy density of up to 600.2 mJ based on bionic principles. This design may provide a valuable reference for future research on energy storage in footed bouncing robots. However, we are aware of some shortcomings in the design of this paper, which need further research and improvement. Future research could explore higher energy density energy storage mechanisms and add other movement modes to expand the range of applications of the robot in different environments.

Author Contributions: Conceptualization: X.B.; Data collection: X.B.; Design methodology: X.B.; Theoretical analysis: X.B.; Software: X.B.; Experiments and testing: X.B.; Writing—original draft: X.B.; Writing—review and editing: X.B., D.K., Q.W., X.Y. and X.X. All authors have read and agreed to the published version of the manuscript.

Funding: This research was supported by Plan for Anhui Major Provincial Science & Technology Project, grant number 202203a06020002.

Institutional Review Board Statement: Not applicable.

Informed Consent Statement: Not applicable.

Data Availability Statement: The data used to support the findings of this study are available from the corresponding author upon request.

Conflicts of Interest: The authors declare no conflict of interest.

References

1. Zhao, J.; Chen, Y.; Wang, J. Fractional order-based control of leg impedance in electric wheeled and footed robots. *Trans. Beijing Inst. Technol.* **2019**, *39*, 187–192.
2. Sun, J.; Sun, Z.; Xin, P. A Review on the Development of Foot Robots for Deep Space Landing Exploration. *China Mech. Eng.* **2021**, *32*, 1765–1775.
3. Wang, S.; Shi, M.; Yue, B. Research on vibration isolation control of wheel-footed robots based on adaptive impedance control. *Trans. Beijing Inst. Technol.* **2020**, *40*, 888–893.
4. Zhu, L.; Fan, J.; Zhao, J. A multi-joint tracked robot for autonomous obstacle-crossing motion planning. *J. Harbin Inst. Technol.* **2012**, *44*, 88–93.
5. Song, Y.; Zhang, Y.; Yao, Q. A path-following control method for crawler robots based on heuristic dynamic planning. *Trans. Chin. Soc. Agric. Mach.* **2019**, *50*, 24–33.
6. Yao, Y.; Wang, S.; Cheng, J. Multi-mode adaptive differential tracked robot. *J. Nanjing Univ. Aeronaut. Astronaut.* **2017**, *49*, 757–765.
7. Kaspari, M.; Weiser, M.D. The size-grain hypothesis and interspecific scaling in ants. *Funct. Ecol.* **1999**, *13*, 530–538. [CrossRef]
8. Birch, M.C.; Quinn, R.D.; Hahm, G. Design of a cricket microrobot. In Proceedings of the IEEE International Conference on Robotics and Automation, San Francisco, CA, USA, 24–28 April 2000.
9. Laksanacharoen, S.; Pollack, A.J.; Nelson, G.M. Biomechanics and simulation of cricket for microrobot design. In Proceedings of the IEEE International Conference on Robotics and Automation, San Francisco, CA, USA, 24–28 April 2000.
10. Kovac, M.; Fuchs, M.; Guignard, A. A miniature 7g jumping robot. In Proceedings of the 2008 IEEE International Conference on Robotics and Automation, Pasadena, CA, USA, 19–23 May 2008.
11. Li, F.; Liu, W.; Fu, X. Jumping like an insect: Design and dynamic optimization of a jumping mini robot based on bio-mimetic inspiration. *Mechatronics* **2012**, *22*, 167–176. [CrossRef]
12. Ye, C.; Wang, B.; Wei, B. Modeling and Analysis of a Jumping Robot with Deforming Wheeled Mechanism. In Proceedings of the 2018 IEEE International Conference on Mechatronics and Automation (ICMA), Changchun, China, 5–8 August 2018.
13. Wang, D.; Sui, F.; Qiu, W. An Untethered Crawling and Jumping Micro-Robot. In Proceedings of the 2021 21st International Conference on Solid-State Sensors, Actuators and Microsystems (Transducers), Orlando, FL, USA, 20–24 June 2021.
14. Kurniawan, R.; Fukudome, T.; Qiu, H. An Untethered 216-mg Insect-Sized Jumping Robot with Wireless Power Transmission. In Proceedings of the 2020 IEEE/RSJ International Conference on Intelligent Robots and Systems (IROS), Las Vegas, NV, USA, 24 October 2020–24 January 2021.
15. Shen, Y.; Ge, W.; Mo, X. Design of a locust-inspired miniature jumping robot. In Proceedings of the 2018 IEEE International Conference on Robotics and Biomimetics (ROBIO), Kuala Lumpur, Malaysia, 12–15 December 2018.
16. Burrows, M.; Shaw, S.R.; Sutton, G.P. Resilin and chitinous cuticle form a composite structure for energy storage in jumping by froghopper insects. *BMC Biol.* **2008**, *6*, 41. [CrossRef] [PubMed]
17. Siwanowicz, I.; Burrows, M. Three dimensional reconstruction of energy stores for jumping in planthoppers and froghoppers from confocal laser scanning microscopy. *eLife* **2017**, *6*, e23824. [CrossRef] [PubMed]
18. Bonsignori, G.; Stefanini, C.; Scarfogliero, U. The green leafhopper, Cicadella viridis (Hemiptera, Auchenorrhyncha, Cicadellidae), jumps with near-constant acceleration. *J. Exp. Biol.* **2013**, *216*, 1270–1279. [CrossRef] [PubMed]
19. Burrows, M. Energy storage and synchronisation of hind leg movements during jumping in planthopper insects (Hemiptera, Issidae). *J. Exp. Biol.* **2010**, *213*, 469–478. [CrossRef] [PubMed]
20. Sutton, G.P.; Burrows, M. Biomechanics of jumping in the flea. *J. Exp. Biol.* **2011**, *214*, 836–847. [CrossRef] [PubMed]
21. Zhao, J.; Xu, J.; Gao, B.; Xi, N.; Cintron, F.J.; Mutka, M.W.; Xiao, L. MSU jumper: A single-motor-actuated miniature steerable jumping robot. *IEEE Trans. Robot.* **2013**, *29*, 602–614. [CrossRef]
22. Lambrecht, B.G.A.; Horchler, A.D.; Quinn, R.D. A small, insect inspired robot that runs and jumps. In Proceedings of the 2005 IEEE International Conference on Robotics and Automation, Barcelona, Spain, 18–22 April 2005.
23. Li, S.; Rus, D. JelloCube: A Continuously Jumping Robot with Soft Body. *IEEE ASME Trans. Mechatron.* **2019**, *24*, 447–458. [CrossRef]

applied sciences

MDPI

Communication

A Method Based on Blackbody to Estimate Actual Radiation of Measured Cooperative Target Using an Infrared Thermal Imager

Mingyu Yang [1,2], Liang Xu [1,2,*], Xin Tan [1] and Honghai Shen [1,3]

1 Changchun Institute of Optics, Fine Mechanics and Physics, Chinese Academy of Sciences, Changchun 130033, China; ymy1983@163.com (M.Y.)
2 Agricultural Hyperspectral Application & Information Database, Changchun 130033, China
3 Key Laboratory of Airborne Optical Imaging and Measurement, Changchun Institute of Optics, Fine Mechanics and Physics, Chinese Academy of Sciences, Changchun 130033, China
* Correspondence: xuliang_9981@163.com

Abstract: Infrared signature of targets is one important approach for target detection and recognition. When measuring the infrared signature of a target in the atmosphere, it is necessary to take the atmospheric transmittance and atmospheric radiation between the measured target and the observer into account. In this study, a blackbody-based approach for estimating atmospheric transmittance and atmospheric radiation is proposed to improve accuracy. Radiometric calibration is first carried out in the laboratory for the infrared thermal imager to determine the slope and offset used in the linear regression. With a set of different temperatures, radiance of the blackbody and digital number value of images are calculated. Finally, according to the analytical expressions derived, the atmospheric transmittance and atmospheric radiation are determined, and actual radiance for the cooperative target is calculated. Results demonstrate that the uncertainty of the actual radiance of measured cooperative target calculated via the proposed method is lower than that by MODTRAN, from MODTRAN at 5.7% and 16.7%, from proposed method at 2.56% and 10.2% in two experiments.

Keywords: infrared signature; atmospheric transmittance; infrared thermal imager; radiometric calibration

Citation: Yang, M.; Xu, L.; Tan, X.; Shen, H. A Method Based on Blackbody to Estimate Actual Radiation of Measured Cooperative Target Using an Infrared Thermal Imager. *Appl. Sci.* **2023**, *13*, 4832. https://doi.org/10.3390/app13084832

Academic Editor: Vittoria Guglielmi

Received: 5 February 2023
Revised: 17 March 2023
Accepted: 28 March 2023
Published: 12 April 2023

1. Introduction

With the further development of infrared focal plane arrays (IRFPA), infrared remote sensing techniques play an important role in many fields, especially in multiple military cases for aerospace vehicles [1–4]. Infrared signature has been demonstrated much during the last few years in feature acquisition and recognition of targets, which can provide signature information of targets such as radiance and intensity [5–7], for different vehicles usually represent different infrared signature.

At present, the research on the infrared radiation characteristics of targets mainly includes two ways, namely the simulation and the experiment. In the simulation, the mathematical radiation model is first established based on the state and the environment of targets. Then, according to the radiation formation mechanism of targets, each component that influences the radiation characteristic of the target is analyzed and calculated theoretically [8–13]. Although the simulation is widely used in the study of the infrared characteristics, due to the fact that they are not subject to the site and cost, the accuracy and effectiveness cannot be easily verified. In contrast, the experiment that measures and inverses the actual radiation of the target with a radiometrically calibrated infrared measuring system is the direct and the only means to obtain and verify the radiation characteristics of the measured target [14,15]. In experiments, infrared signature measurement of targets usually consists of three steps: (1) radiometric calibration for the infrared thermal imager, which aims at quantifying the relationship between the radiation received and the output digital number (DN); (2) estimation of the emission and reflection from the ground and the atmospheric effects, such as atmospheric transmittance and path radiation, which can be predicted by software like MODTRAN; (3) inversion of the radiation characteristics using a model of radiation measurement, according to the target material surface emissivity, transmittance and so on. In both approaches mentioned above, the influence caused by the atmosphere must be inevitably corrected, especially for long range targets. However, the estimation of atmospheric parameters and the atmospheric radiation, nowadays, mainly depends on atmospheric transmission radiative transfer software such as LOWTRAN, MODTRAN, and FASCODE, of which the uncertainty is merely approximately 20–30% which is far from enough in engineering application.

In this work, a method of infrared radiation measurement based on blackbody or an object served as a blackbody for cooperative targets is proposed, based on which the atmospheric transmittance and the atmospheric radiation are estimated and the actual radiation of the measured cooperative target is calculated with lower uncertainty. The proposed method can be considered as an alternative for practical scenes when measuring cooperative targets in engineering applications.

2. Materials and Methods

2.1. MODTRAN

Moderate Resolution Atmospheric Transmission radiative transfer code and algorithm (MODTRAN) [16], developed by the U.S. Air Force Research Lab, is widely used to calculate various atmospheric radiative transfer parameters under complex atmospheric conditions from 0 to 50,000 cm^{-1}, such as transmittance, et al. Models of atmosphere in MODTRAN can be defined by users according to the theoretical calculation or measured data, which makes the simulation and use of MODTRAN particularly flexible. In addition, MODTRAN includes representative models of atmosphere, aerosol, cloud, and rain, and various complex geometric conditions such as horizontal, vertical, inclined upward, and downward transmission, which makes MODTRAN a powerful tool in the calculation of atmospheric radiative transfer parameters.

Although MODTRAN software provides default parameters for most settings, it would be best if measured data is input in order to approximate the real situation. For atmospheric parameters collection, cloud and aerosol micro lidar to determine the visibility and aerosol extinction profile, automatic sun tracking photometer to measure total atmospheric transmittance and total water vapor, and meterological station to collect the temperature, humidity, wind, pressure, et al., are often employed in the process of atmospheric parameters measurement. Figure 1 shows the workflow for the correction of atmospheric transmission via MODTRAN using measured atmospheric parameters.

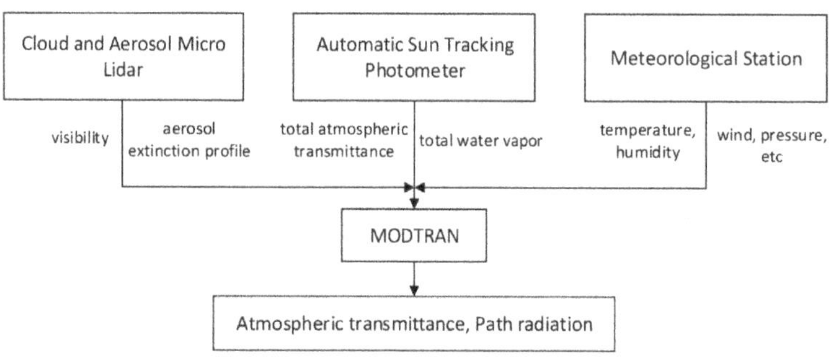

Figure 1. Workflow for MODTRAN software using measured atmospheric parameters.

2.2. Theory of Infrared Radiation

The radiation obtained by an infrared thermal imager mainly includes three parts: the self-radiation of the measured target, reflected environmental radiation, and atmospheric radiation [17]. The equivalent radiance from the surface of an opaque target by an infrared thermal imager can be expressed as:

$$L_\lambda(T_r) = \tau_{a\lambda}\varepsilon_\lambda L_{b\lambda}(T_0) + \tau_{a\lambda}(1-\alpha_\lambda)L_{b\lambda}(T_u) + \varepsilon_{a\lambda}L_{b\lambda}(T_a) \quad (1)$$

where T_r is the radiation temperature measured with the infrared thermal imager, T_0 is the surface temperature of the object, T_u is the ambient temperature, T_a is the atmospheric temperature, ε_λ is the surface emissivity of the measured object, $\varepsilon_{a\lambda}$ is the atmospheric emissivity, $\tau_{a\lambda}$ is the atmospheric transmittance, α_λ is the surface absorptance of the object, and $L_{b\lambda}$ denotes the radiance emitted by the blackbody at wavelength λ.

The corresponding irradiance of entrance pupil for the infrared thermal imager is:

$$E_\lambda = A_0 d^{-2} L_\lambda = A_0 d^{-2}[\tau_{a\lambda}\varepsilon_\lambda L_{b\lambda}(T_0) + \tau_{a\lambda}(1-\alpha_\lambda)L_{b\lambda}(T_u) + \varepsilon_{a\lambda}L_{b\lambda}(T_a)] \quad (2)$$

where A_0 is the visible area of the target corresponding to the minimum space angle of the thermal imager, d is the distance between the target and the thermal imager, and usually $A_0 d^{-2}$ is considered as a constant.

Radiation power of the infrared thermal imager received for a certain wavelength is:

$$P_\lambda = E_\lambda A_r \quad (3)$$

where A_r is the area of lens of the thermal imager.

Without considering stray radiation, which can be minimized via high projection material and cold aperture, the total radiation of the thermal imager received P_T is:

$$P_T = \tau_{op} P_\lambda \quad (4)$$

where τ_{op} is the transmittance of optical systems.

The response voltage signal V_s is referred according to [18] as:

$$V_s = g(R_\lambda P_T + V_0) \quad (5)$$

where R_λ is the spectral responsivity of the infrared thermal imager, which is a constant for a certain infrared thermal imager; V_0 is the DC bias voltage; and g is the response gain used to amplify the signal.

$$\begin{aligned} V_s = \; & A_r A_0 d^{-2} \tau_{op} g \{\tau_{a\lambda}[\varepsilon_\lambda \textstyle\int_{\lambda_1}^{\lambda_2} R_\lambda L_{b\lambda}(T_0) d\lambda + (1-\alpha_\lambda)\int_{\lambda_1}^{\lambda_2} R_\lambda L_{b\lambda}(T_u) d\lambda] \\ & + \varepsilon_{a\lambda} \textstyle\int_{\lambda_1}^{\lambda_2} R_\lambda L_{b\lambda}(T_a) d\lambda\} + g V_0 \end{aligned} \quad (6)$$

Let

$$K = A_r A_0 d^{-2} \tau_{op} g \quad (7)$$

$$f(T) = \int_{\lambda_1}^{\lambda_2} R_\lambda L_{b\lambda}(T) d\lambda \quad (8)$$

Equation (6) is then simplified for:

$$V_s = K\{\tau_{a\lambda}[\varepsilon_\lambda f(T_0) + (1-\alpha_\lambda)f(T_u)] + \varepsilon_{a\lambda} f(T_a)\} + g V_0 \quad (9)$$

When the surface of the measured target can be treated as a gray body, according to Kirchhoff's law ($\varepsilon_\lambda = \alpha_\lambda$ and $\varepsilon_{a\lambda} = 1 - \tau_{a\lambda}$) and especially when the measured target is a blackbody, Equation (9) is simplified for:

$$V_s = K\{\tau_{a\lambda} f(T_0) + \varepsilon_{a\lambda} f(T_a)\} + g V_0 \quad (10)$$

As the output of the infrared thermal imager, also called the digital number (DN), is usually proportion to V_s [19], Equation (10) is transformed into:

$$G_s = k\{\tau_{a\lambda}[\varepsilon_\lambda f(T_0) + (1-\alpha_\lambda)f(T_u)] + \varepsilon_{a\lambda} f(T_a)\} + G_0 \tag{11}$$

When the measured target and the infrared thermal imager are determined, the coefficients become fixed values; the actual radiance of the measured target is:

$$f(T_0) = \frac{\frac{\frac{G_s - G_0}{k} - \varepsilon_{a\lambda} f(T_a)}{\tau_{a\lambda}} - (1-\alpha_\lambda)f(T_u)}{\varepsilon_\lambda} \tag{12}$$

2.3. Model of Radiometric Calibration and Infrared Thermal Imager

One of the most commonly used approaches of radiometric calibration for infrared thermal imager is that setting a blackbody source of large area to completely cover the field-of-view (FOV) of the infrared thermal imager [20–22], as shown in Figure 2.

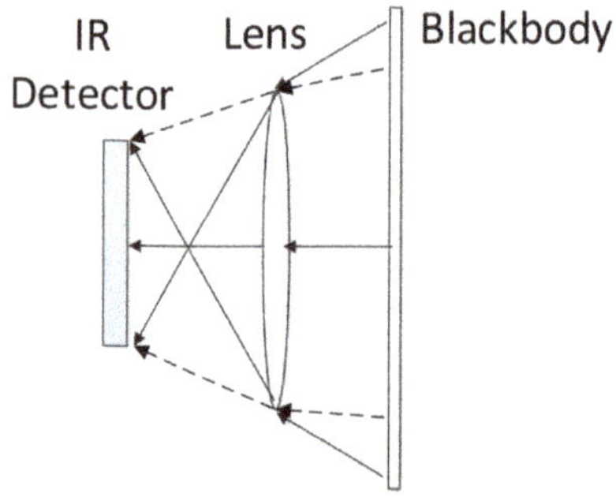

Figure 2. Schematic diagram of the radiometric calibration.

Since the calibration is usually carried out in the laboratory and the thermal imager is very close to the blackbody, the influence of the atmosphere can be ignored, $\varepsilon_\lambda = \alpha_\lambda = 1$, $\tau_{a\lambda} = 1$ and $\varepsilon_{a\lambda} = 0$. As a result, Equation (11) is simplified as:

$$G_s = kf(T_0) + G_0 \tag{13}$$

where $f(T_0)$ can be calculated according to Planck's law of radiation:

$$f(T_0) = \frac{1}{\pi}\int_{\lambda_1}^{\lambda_2} \frac{C_1\lambda^{-5}}{e^{C_2/\lambda T_0} - 1} d\lambda \tag{14}$$

In Equation (14), C_1 and C_2 are called the first and second radiation constants, respectively, with values $C_1 = 2\pi hc^2 = 3.742 \times 10^{-16}\ \mathrm{W \cdot m^2}$ and $C_2 = hc/k_B = 1.4388 \times 10^{-2}\ \mathrm{m \cdot K}$, where fundamental constants c (2.997×10^8 m/s), h (6.626×10^{-34} J s), and k_B (1.381×10^{-23} J/K) are the speed of light in vacuo, Planck's constant, and Boltzmann's constant, respectively.

In the process of radiometric calibration, it is necessary to keep laboratory conditions and thermal imager parameters unchanged. Nonuniformity correction (NUC) is often applied to make the response of each pixel identical. The procedure of radiometric calibration for an infrared thermal imager is as follows.

1. Turn off the auto gain function of the infrared thermal imager, and set focus to infinity;
2. Apply NUC to the blackbody source;
3. Save frames of image when the temperature of blackbody source is completely stable, then average the saved images to reduce random noises;
4. Collect DN value for the averaged images of different temperatures;

5. Calculate the radiance according to Equation (14);
6. Change the temperature of the blackbody and repeat the steps 3–5 until all temperatures are measured;
7. Fit DN value and the radiance in Equation (13) by the least square method.

The infrared thermal imager and the blackbody source used for radiometric calibration are shown in Figure 3.

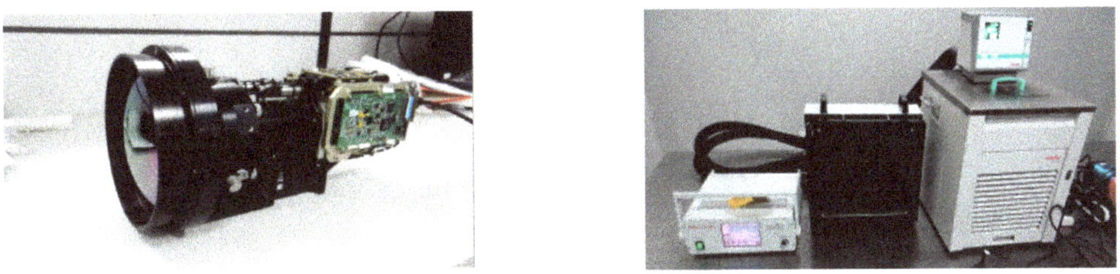

(a) (b)

Figure 3. The infrared thermal imager and the blackbody source. (**a**) Infrared thermal imager; (**b**) Blackbody source.

The properties of the infrared thermal imager and the blackbody source are listed in Tables 1 and 2, respectively.

Table 1. Properties of the infrared thermal imager.

Parameter	Value
Manufacturer	FLIR Systems
Model	MINICORE-600Z
Band Range	3.7–4.8 μm
Resolution	640 × 512 pixel
Pixel Size	15 μm × 15 μm
Output Bit	14 bit
F#	F4
Focus	30–600 mm continuous zooming
Cooling Type	Stirling cooling

Table 2. Properties of the blackbody source.

Parameter	Value
Manufacturer	CI Systems
Model	SR800-12LT
Size	305 mm × 305 mm
Temperature range	−40–150 °C
Emissivity	0.97 ± 0.02
Uniformity	±0.03 °C

2.4. The Proposed Blackbody-Based Method

Although MODTRAN software is widely used to estimate the transmittance, atmospheric radiation, and so on, it is only suitable for theoretical analysis not practical scenes, for MODTRAN is mainly based on the U.S atmospheric parameters which cannot represent the actual condition in the other places, especially in cities with heavy pollution. The proposed method can be considered as an alternative for practical scenes. The atmospheric transmittance and atmospheric radiation can both be derived from Equation (11) by setting two different temperatures denoted as T_{0H} and T_{0L} as follows:

$$G_{sH} = k\{\tau_{a\lambda}[\varepsilon_\lambda f(T_{0H}) + (1-\alpha_\lambda)f(T_u)] + \varepsilon_{a\lambda} f(T_a)\} + G_0 \tag{15}$$

$$G_{sL} = k\{\tau_{a\lambda}[\varepsilon_\lambda f(T_{0L}) + (1-\alpha_\lambda)f(T_u)] + \varepsilon_{a\lambda} f(T_a)\} + G_0 \tag{16}$$

Hence, the atmospheric transmittance and atmospheric radiation can be derived by Equation (15) minus Equation (16) and Equation (15) plus Equation (16) as follows:

$$\tau_{a\lambda} = \frac{G_{sH} - G_{sL}}{k[\varepsilon_\lambda f(T_{0H}) - \varepsilon_\lambda f(T_{0L})]} \tag{17}$$

$$\varepsilon_{a\lambda} f(T_a) = \frac{1}{2}\cdot\{\frac{G_{sH} + G_{sL} - 2G_0}{k} - \tau_{a\lambda}[\varepsilon_\lambda f(T_{0H}) + \varepsilon_\lambda f(T_{0L}) + 2(1-\alpha_\lambda)f(T_u)]\} \tag{18}$$

Figure 4 shows the workflow for correction of atmospheric transmission using the proposed approach.

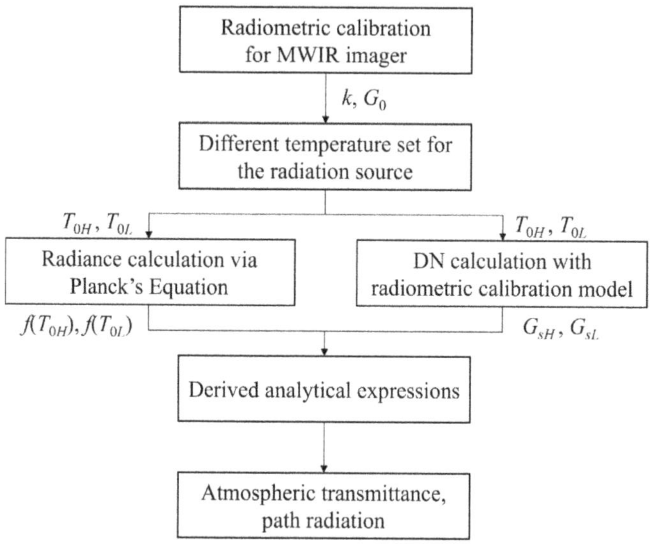

Figure 4. Workflow for the proposed method.

3. Results

3.1. Radiometric Calibration for the Infrared Thermal Imager

In the process of radiometric calibration for the infrared thermal imager, two-point NUC is applied to ensure the uniformity of images from the blackbody source. In addition, we use a DALSA X64-CL iPro image acquisition card to grab 100 image frames for each temperature, and set temperatures of the blackbody increasing from 35 to 115 degrees C at interval of 5 degrees C, with DN value of pixel (320,256) collected. The radiance calculated according to Planck' law of radiation and corresponding DN value are shown in Table 3.

Table 3. Radiance and DN value collected in the radiometric calibration.

Temperature/°C	Radiance/W·m^{-2}·str^{-1}	DN Value
35	2.4764	1986
40	2.9356	2257
45	3.4627	2584
50	4.0649	2979
55	4.7501	3399
60	5.5267	3900
65	6.4034	4491
70	7.3896	5162
75	8.4950	5904
80	9.7299	6722
85	11.1051	7668
90	12.6318	8717
95	14.3216	9880
100	16.1866	11,295
105	18.2395	12,658
110	20.4933	15,106
115	22.9614	15,114

Table 3 indicates that the DN value is of little increase during the temperature increasing from 110 degrees C to 115 degrees C. Considering the 14-bit output of the thermal imager, it can be concluded that the DN value has reached a saturation point around 110 degrees C. Therefore, data from 110 to 115 degrees C is removed for the least square, and only data from 35 to 105 degrees C is used. The result of the radiometric calibration under the condition of 2 ms integration time, 300 mm focal length, is given:

$$G_s = 679f(T_0) + 194 \tag{19}$$

with the fitted straight line using least square method shown in Figure 5.

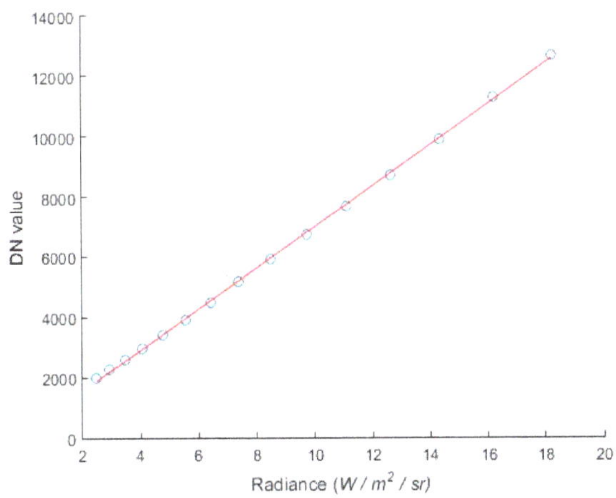

Figure 5. Fitted straight line with the least square method.

3.2. Experiments

We conduct an infrared radiation measuring experiment for the blackbody source in different temperatures. The blackbody source is settled horizontally 30 m away from the infrared thermal imager, with temperature increasing from 65 to 105 degrees C at intervals of 10 degrees C, as shown in Figure 6.

(a) (b) (c) (d) (e)

Figure 6. Blackbody source in field test with different temperatures in degree C: (**a**) 65; (**b**) 75; (**c**) 85; (**d**) 95; (**e**) 105.

The radiance calculated according to Planck' law of radiation and corresponding DN value are listed in Table 4.

Table 4. Radiance and DN value for the blackbody test.

Temperature/°C	Radiance/W·m^{-2}·str^{-1}	DN Value
65	6.4034	4072
75	8.4950	5298
85	11.1051	6764
95	14.3216	8605
105	18.2395	11,207

According to Equations (17) and (18) the atmospheric transmittance $\tau_{a\lambda}$ and atmospheric radiation $\varepsilon_{a\lambda} f(T_a)$ can be calculated through least square fitting in order to reduce error, values of which are approximately 88.8% and 0.0239 W/m^2/sr. In comparison with MODTRAN, the atmospheric parameters used for the blackbody test are listed in Table 5.

Table 5. Atmospheric parameters used in MODTRAN for the blackbody test.

Atmospheric Parameters	Value
Atmospheric profile	Mid-latitude summer
Altitude of observer (km)	0.216
Visibility (km)	13
Humidity (%)	54
Temperature (°C)	20
Atmospheric path	horizontal
Distance from observer to target (m)	30
CO_2 mixing ratio (ppmv)	370

Input these parameters listed in Table 5 into MODTRAN, the atmospheric transmittance and atmospheric radiation are 83.9% and 0.0352 W/m^2/sr, respectively. With all the parameters given above, the infrared radiation of targets can be calculated by Equation (12). The comparison of inversion results with the derivation from the analytical expression and with MODTRAN are given in Table 6. It is obvious that the inversion results derived by the expression in each row are closer to the actual radiance of the blackbody than that by MODTRAN, with the maximal error of 2.56% and 5.7% by the derivation from the analytical expression and by MODTRAN, respectively.

To extend the verification to other tests, we also carried out a field test for a fan-shaped infrared target. Due to the relatively small size of the blackbody, we cannot acquire enough pixels over a long distance in some cases. The target with a bigger fan-shaped infrared target is used in the experiment, which consists of 10 pieces of heating plates, each with an 18-degree central angle, the radius of 2.25 m, and the surface emissivity of 0.52. The fan-shaped target is shown in Figure 7, with the aluminum box of the temperature controller.

Table 6. Comparison of inversion results for the blackbody source.

Temp (°C)	DN Value	Actual Radiance (W/m²/sr)	Radiance with MODTRAN (W/m²/sr)	Error with MODTRAN (%)	Radiance with Derivation (W/m²/sr)	Error with Derivation (%)
65	4072	6.4034	6.7654	5.6	6.4348	0.48
75	5298	8.4950	8.9174	4.9	8.4696	0.29
85	6764	11.1051	11.4908	3.4	10.9010	1.83
95	8605	14.3216	14.7224	2.7	13.9552	2.55
105	11,207	18.2395	19.2899	5.7	18.2731	0.18

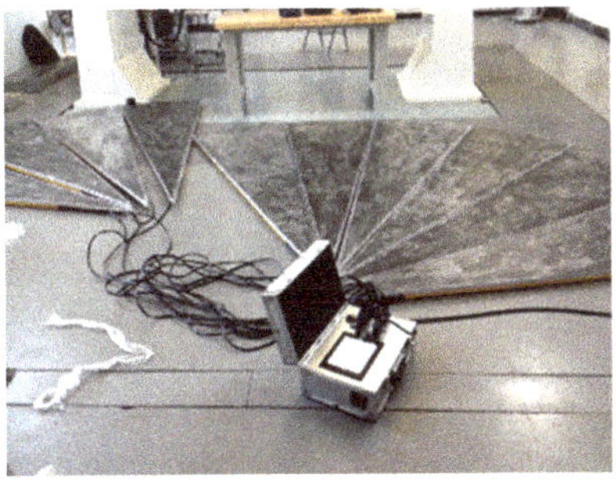

Figure 7. Heating plates used in the field test.

Two pieces of plates are chosen in the experiment, with the temperature increasing from 35 to 50 degrees C at intervals of 3 degrees C and with the ambient temperature of 28 degrees C, shown in Figure 8.

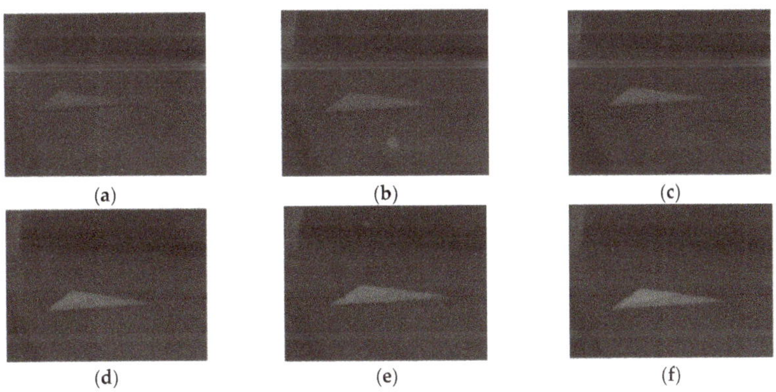

Figure 8. Heating plates with different temperatures in degree C: (**a**) 35; (**b**) 38; (**c**) 41; (**d**) 44; (**e**) 47; (**f**) 50.

The atmospheric parameters for the fan-shaped target test are listed in Table 7.

The transmittance and atmospheric radiation are calculated by Equation (12) with values of 73.3% and 1.17 W/m^2/sr, while by MODTRAN with values of 72.3% and 1.34 W/m^2/sr, respectively. Table 8 shows the comparison of inversion results for the two hearting plates, indicating that for each row, the radiance derived from the analytical expression is also closer to the actual radiance than that from MODTRAN over a relatively long distance, with the maximal error of 10.2% and 16.7% by the derivation from the analytical expression and by MODTRAN, respectively.

Table 7. Atmospheric parameters used in MODTRAN for the fan-shaped target test.

Atmospheric Parameters	Value
Atmospheric profile	Mid-latitude summer
Altitude of observer (km)	0.22
Visibility (km)	13
Humidity (%)	45
Temperature (°C)	28
atmospheric path	slant
distance from observer to target (m)	1560
CO_2 mixing ratio (ppmv)	370

Table 8. Comparison of inversion results for heating plates.

Temp (°C)	DN Value	Actual Radiance ($W/m^2/sr$)	Radiance with MODTRAN ($W/m^2/sr$)	Error with MODTRAN (%)	Radiance with Derivation ($W/m^2/sr$)	Error with Derivation (%)
35	1630	2.4764	2.0611	−16.7	2.4789	0.10
38	1701	2.7442	2.3392	−14.8	2.7532	0.33
41	1823	3.0354	2.8171	−7.2	3.2246	6.2
44	1944	3.3515	3.2911	−1.8	3.6922	10.2
47	1978	3.6941	3.4243	−7.3	3.8235	3.5
50	2041	4.0649	3.6711	−9.7	4.0669	0.05

4. Discussion

For long distance infrared signature of targets in the field, the inversion uncertainty is mainly affected by (1) the uncertainty of the radiometric calibration, (2) the uncertainty of the estimation for the atmospheric transmittance and atmospheric radiation, and (3) the uncertainty of the environmental condition changes in the field, such as temperature, wind, etc., other than that in the laboratory. Among these factors mentioned above, the uncertainty in the radiometric calibration, i.e., values of k and G_0 in Equation (13), under current measuring conditions is approximately 6% for a medium-wave infrared thermal imager [22]. The uncertainty in the estimation of the atmospheric transmittance and atmospheric radiation, depending on meteorological equipment and MODTRAN, is about 20–30%; and the environmental differences between the field and the laboratory is usually 10%. As a result, the uncertainty of the measurement for the infrared thermal imager can be calculated by the root mean square of the three factors listed above, approximately 23.2–32.2%. In our experiments, the inversion uncertainty derived from the analytical expression and by MODTRAN is 2.56%, 10.2% and 5.7%, 16.7%, respectively, both of which are acceptable according to the above analysis. It is obvious that the uncertainty derived from the analytical expressions is lower than that by MODTRAN for cooperative targets, i.e., targets that can be easily placed a blackbody or an object that served as a blackbody nearby. However, for non-cooperative targets such as an enemy vehicle, we can only employ MODTRAN to estimate the radiation of targets theoretically.

5. Conclusions

In this paper, a blackbody-based approach for estimating the actual radiation of measured cooperative target is proposed. Firstly, radiometric calibration is carried out in the laboratory for the infrared thermal imager to determine the slope and offset used in the linear regression. Then, the radiance of the blackbody and digital number value of images are calculated with a set of different temperatures. Finally, according to the analytical expressions derived the atmospheric transmittance and atmospheric radiation are determined, and actual radiance for the cooperative target is calculated. Lab and field tests demonstrate that the uncertainty of the actual radiance of measured cooperative target calculated via the proposed method is lower than that by MODTRAN, which can be considered as an alternative for practical scenes when measuring cooperative targets in engineering application. Future studies will mainly focus on lowering the uncertainty for non-cooperative targets over long distances, which aims to reduce the impact of the atmospheric transmission to improve furtherly the infrared radiation of measured targets.

Author Contributions: Conceptualization, M.Y.; methodology, M.Y.; software, M.Y.; validation, M.Y., L.X. and X.T.; formal analysis, M.Y.; investigation, H.S.; resources, H.S.; data curation, M.Y.; writing—original draft preparation, M.Y.; writing—review and editing, L.X. and X.T.; visualization, M.Y.; supervision, X.T.; project administration, L.X.; funding acquisition, L.X. All authors have read and agreed to the published version of the manuscript.

Funding: Science and Technology Resources Shared Services Platform Construction of JiLin Province (No.20230505009ZP).

Institutional Review Board Statement: Not applicable.

Informed Consent Statement: Not applicable.

Data Availability Statement: The data presented in this study are available from Tables 3–8.

Conflicts of Interest: The authors declare no conflict of interest.

References

1. Jacobs, A. *Thermal Infrared Characterization of Ground Targets and Backgrounds*; SPIE Optical Engineering Press: Washington, DC, USA, 1996; pp. 140–149.
2. Mahulikar, P.; Sonawane, R.; Arvind Rao, G. Infrared signature studies of aerospace vehicles. *Prog. Aerosp. Sci.* **2007**, *43*, 218–245. [CrossRef]
3. Zhao, B.; Xiao, S.; Lu, H.; Wu, D. Modeling and simulation of infrared signature of remote aerial targets. In Proceedings of the 2017 10th International Congress on Image and Signal Processing, BioMedical Engineering and Informatics (CISP-BMEI), Shanghai, China, 14–16 October 2017; pp. 1–5. [CrossRef]
4. Zhang, Y.-C.; Chen, Y.-M.; Fu, X.-B.; Luo, C. The research on the effect of atmospheric transmittance for the measuring accuracy of infrared thermal imager. *Infrared Phys. Technol.* **2016**, *77*, 375–381. [CrossRef]
5. Leslie, P.; Furxhi, O.; Short, R.; Grimming, R.; Driggers, R. Mid-Wave and Long-Wave Infrared Signature Model and Measurements of Power Lines Against Atmospheric Path Radiance. In Proceedings of the 2021 IEEE Research and Applications of Photonics in Defense Conference (RAPID), Miramar Beach, FL, USA, 2–4 August 2021; pp. 1–2. [CrossRef]
6. Talghader, J.J.; Gawarikar, A.S.; Shea, R.P. Spectral selectivity in infrared thermal detection. *Light. Sci. Appl.* **2012**, *1*, 6–16. [CrossRef]
7. Qiu, H.; Hu, L.; Zhang, Y.; Lu, D.; Qi, J. Absolute Radiometric Calibration of Earth Radiation Measurement on FY-3B and Its Comparison With CERES/Aqua Data. *IEEE Trans. Geosci. Remote Sens.* **2012**, *50*, 4965–4974. [CrossRef]
8. Poline, M.; Rebrov, O.; Larsson, M.; Zhaunerchyk, V. Theoretical studies of infrared signatures of proton-bound amino acid dimers with homochiral and heterochiral moieties. *Chirality* **2020**, *32*, 359–369. [CrossRef] [PubMed]
9. Liu, J.; Yue, H.; Lin, J.; Zhang, Y. A Simulation Method of Aircraft Infrared Signature Measurement with Subscale Models. *Procedia Comput. Sci.* **2019**, *147*, 2–16. [CrossRef]
10. Knežević, D.M.; Matavulj, P.S.; Nikolić, Z.M. Modeling of aircraft infrared signature based on comparative tracking. *Optik* **2021**, *225*, 165782. [CrossRef]
11. Zhang, D.; Bai, L.; Wang, Y.; Lv, Q.; Zhang, T. An improved SHDOM coupled with CFD for simulating infrared radiation signatures of rocket plumes. *Infrared Phys. Technol.* **2022**, *122*, 104054. [CrossRef]
12. Xu, M.; Bu, X.; Yu, J.; He, Z. Spinning projectile's attitude measurement with LW infrared radiation under sea-sky background. *Infrared Phys. Technol.* **2018**, *90*, 214–220. [CrossRef]
13. Haq, F.; Huang, J. Parametric design and IR signature study of exhaust plume from elliptical-shaped exhaust nozzles of a low flying UAV using CFD approach. *Results Eng.* **2022**, *13*, 100320. [CrossRef]
14. Cheng, W.; Wang, Z.; Zhou, L.; Shi, J.; Sun, X. Infrared signature of serpentine nozzle with engine swirl. *Aerosp. Sci. Technol.* **2019**, *86*, 794–804. [CrossRef]
15. Ata, Y.; Nakiboğlu, K.C. IR signature estimation of an object or a target by taking into account atmospheric effects. *Opt. Commun.* **2010**, *283*, 3901–3910. [CrossRef]
16. Berk, A.; Anderson, G.P.; Bernstein, L.S.; Acharya, P.K.; Dothe, H.; Matthew, M.W.; Adler-Golden, S.M.; Chetwynd, J.H., Jr.; Richtsmeier, S.C.; Pukall, B.; et al. MODTRAN4 radiative transfer modeling for atmospheric correction. In Proceedings of the SPIE's International Symposium on Optical Science, Engineering, and Instrumentation, Denver, CO, USA, 18–23 July 1999; pp. 348–353. [CrossRef]
17. Ni, N.; Zhang, K.; Hu, J.; Li, L.; Mi, S.; Zhang, Y.; Zhang, Y. Combined Use of Blackbody and Infrared Radiation for Accurate Measurement of Temperature Field of Aluminum Alloys. *Optik* **2022**, *268*, 169763. [CrossRef]
18. Shen, J.; Zhang, Y.; Xing, T. The study on the measurement accuracy of non-steady state temperature field under different emissivity using infrared thermal image. *Infrared Phys. Technol.* **2018**, *94*, 207–213. [CrossRef]

19. Zhang, Y.-C.; Chen, Y.-M.; Luo, C. A method for improving temperature measurement precision on the uncooled infrared thermal imager. *Measurement* **2015**, *74*, 64–69. [CrossRef]
20. Yang, G.; Yu, Y.; Sun, Z.; Li, Z.; Pang, X.; Zhang, T. Radiometric calibration algorithm for high dynamic range infrared imaging system. *Infrared Phys. Technol.* **2023**, *130*, 104607. [CrossRef]
21. Lin, D.; Cui, X.; Wang, Y.; Yang, B.; Tian, P. Pixel-wise radiometric calibration approach for infrared focal plane arrays using mul-tivariate polynomial correction. *Infrared Phys. Technol.* **2022**, *123*, 104110. [CrossRef]
22. Li, Z.; Yu, Y.; Tian, Q.-J.; Chang, S.-T.; He, F.-Y.; Yin, Y.-H.; Qiao, Y.-F. High-efficiency non-uniformity correction for wide dynamic linear infrared radiometry system. *Infrared Phys. Technol.* **2017**, *85*, 395–402. [CrossRef]

applied sciences

MDPI

Article

Two-Dimensional Space Turntable Pitch Axis Trajectory Prediction Method Based on Sun Vector and CNN-LSTM Model

Shuang Dai *, Ke-Fei Song, Yan-Long Wang and Pei-Jie Zhang

Changchun Institute of Optics, Precision Machinery and Physics, Chinese Academy of Sciences, Changchun 130033, China; songkf@ciomp.ac.cn (K.-F.S.); wangyl@ciomp.ac.cn (Y.-L.W.); zhangpj@ciomp.ac.cn (P.-J.Z.)

* Correspondence: dai-dai123@163.com

Abstract: A two-dimensional space turntable system has been used to ensure that the Solar X-ray and Extreme Ultraviolet Imager (X-EUVI) can track the Sun stably, and the prediction of the two-dimensional turntable trajectory is an important part of payload health management. Different from the dynamic model using traditional trajectory prediction, we propose a new method for predicting the pitch axis trajectory of the turntable based on the sun vector and a deep learning CNN-LSTM model. First, the ideal solar position of the pitch axis was calculated using the sun vector. Then, the ideal solar position was combined with the running turntable pitch axis motor speed, current, and solar position error signal as the CNN-LSTM model input data. The model parameters were trained and adjusted through test data simulation using Fengyun-3E satellite orbit data. Finally, the next position of the pitch axis was predicted. The test results showed that in the sun vector and CNN-LSTM model, the RMSE value was 0.623 and the MSE value was 0.388. It was better than the LSTM model or CNN model alone and could accurately predict the pitch axis position.

Keywords: two-dimensional turntable; pitch axis trajectory; sun vector; CNN-LSTM model; deep learning

Citation: Dai, S.; Song, K.-F.; Wang, Y.-L.; Zhang, P.-J. Two-Dimensional Space Turntable Pitch Axis Trajectory Prediction Method Based on Sun Vector and CNN-LSTM Model. *Appl. Sci.* **2023**, *13*, 4939. https://doi.org/10.3390/app13084939

Academic Editor: Eyad H. Abed

Received: 7 March 2023
Revised: 6 April 2023
Accepted: 11 April 2023
Published: 14 April 2023

1. Introduction

The Solar X-ray and Extreme Ultraviolet Imager (X-EUVI) is a payload of the Sun synchronous orbit (twilight) FY-3E satellite, which has an orbital altitude of 836 km and an orbital period of 102 min. This is the first space-based solar X-ray and extreme ultraviolet (EUV) imager for space weather and space physics in China [1]. FY-3E is a three-axis stabilized spacecraft with respect to the Earth and changes in the position of the Sun the X-EUVI coordinate system in real time. Therefore, the two-dimensional turntable system was developed to track the Sun [1]. It is shown in Figure 1.

When X-EUVI works in orbit, the two-dimensional turntable system first roughly points to the Sun according to the sun vector from the satellite. Then, X-EUVI precisely points to the Sun using a turntable lock-in control system based on Trace Guide Telescope (TGT) solar position data [1].

The two-dimensional turntable includes the pitch axis and the azimuth axis. The control system of the two-dimensional turntable realizes the accurate direction of the target by controlling the position, speed, and torque of the two brushless motors, namely the position axis and the pitch axis [2]. This paper takes the pitch axis trajectory as the research object. The two-dimensional turntable controls the trajectory depending on the sun vector data and the solar position data combined with the satellite attitude and other factors.

When the turntable controls the optical axis to point to the sun, the image obtained by the XEUV is in the center of the field of view as shown in Figure 2.

The main factors affecting the normal running of the turntable include the turntable motor current, motor voltage, target solar position, and solar position error signal. The prediction of the two-dimensional turntable trajectory can show the operation status of the turntable. It is not only important to understand the operation of the turntable itself but also that it has an important impact on the evaluation of the operation status of the satellite platform. The same applies to the trajectory prediction of the spaceborne integrated platform or intelligent load.

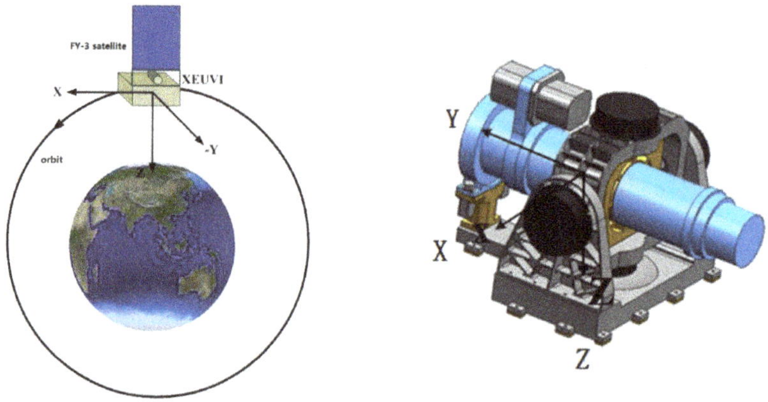

(**a**) (**b**)

Figure 1. Schematic diagram of the position of the sun, satellite, and turntable. (**a**) Schematic diagram of satellite orbit, and sun is in −Y direction. (**b**) Turntable optical axis coordinate system.

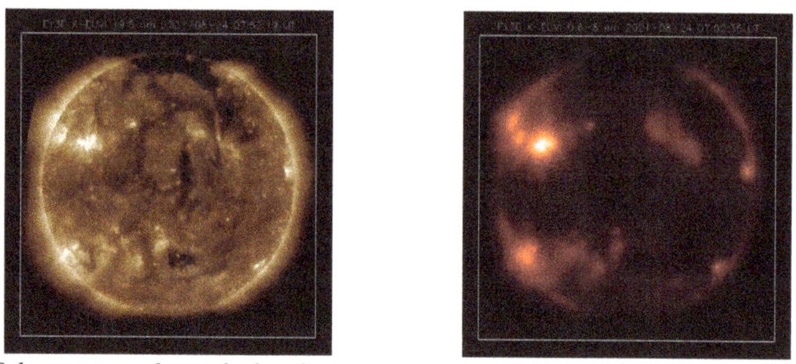

(**a**) Solar extreme ultraviolet band image (**b**) Solar X-band image

Figure 2. Solar images obtained by X-EUVI when the optical axis points to the sun (from the National Meteorological Satellite Center of China).

The existing models of trajectory prediction involve different algorithms in different fields, but research on trajectory predictions of a two-dimensional turntable in space is still lacking. By establishing the traditional motion model, the running track of the turntable can be predicted. However, due to many unknown and variable factors in the model, the error is difficult to measure, so the accuracy of the motion model is difficult to guarantee [3]. There are many different methods for trajectory prediction. They include the hidden Markov model (HMM), based on mathematical statistical methods [4], the Kalman filter [5], or neural networks and deep learning in machine learning methods [6–10]. The advantage of the hidden Markov model is that it has a relatively good prediction effect for tracks with variable states, and the disadvantage is that because of its memoryless nature, it cannot use the preorder information of track sequence. The hidden Markov model is often used for long-sequence prediction [11]. It is often used for pedestrian trajectory prediction. A Kalman filter relies on the information of the previous sequence point and the current position information to predict the next moment. The advantage is that the state estimation process is very stable, and because its calculation process is a continuous prediction and correction process, it is suitable for trajectory-prediction scenarios requiring strong real-time performance. However, the Kalman filter is extremely dependent on forecasting the next time based on the information of the last sequence point and the current position information.

Neural networks and deep learning have strong nonlinear mapping, self-learning, and adaptive abilities. The disadvantage is that they are very sensitive to the initial network's weight, and there is a local minimization problem. When the initial network weight is not uniform, the training results may be different, so this method is applicable to most trajectory prediction scenarios. Since there is no uniform standard for the structure of a neural network, the appropriate network structure should be selected according to the specific situation in practical application [12].

We proposed to use a sun vector and a one-dimensional convolutional neural network combined with a long short-term memory network (CNN-LSTM) hybrid neural network model as a method for predicting the trajectory of the pitch axis of the space turntable. First, according to the sun position calculation model, we calculated the ideal solar position value from the sun vector data and then used the ideal solar position value of the pitch axis, pitch axis motor speed, current, and position error signal data as the input data of the model. We then selected a specific step of the time sliding window and predicted the position of the pitch axis at the next time. It was built on a model of CNN-LSTM, the prediction sequence was set with an adaptive Adam optimizer, and the simulating telemetry data of a two-dimensional turntable was used for training. We used RMSE and MSE as performance evaluation indicators.

2. Materials and Methods

The pitch axis pointing model established a two-dimensional turntable to roughly point to the Sun from the sun vector broadcasted by the satellite platform.

2.1. Sun Vector Calculat Model

The sun vector in the orbital instant root broadcast was the unit vector in the orbital coordinate system. The sun vector was defined using the J2000 coordinate system [13,14]. After a series of coordinate transformations from the orbit coordinate system to the unit vector of the optical axis of the guide mirror, the transformation matrix of the satellite attitude had to be considered as shown in Figure 1. Many error factors were difficult to determine, setting all the installation errors was ignored, and the satellite attitude was ignored so the ideal value of the turntable motion could be calculated.

The ideal position value of the pitch axis can be calculated through the sun vector. It is shown in Equations (1) and (2):

Sun vector:

$$S_0 = [Xs(t)\ \ Ys(t)\ \ Zs(t)]^T \tag{1}$$

We calculated the ideal value of the pitch axis of the turntable as:

$$\theta_{pitcht} = \text{atan}\left[\frac{Zs(t)}{Ys(t)}\right] \times D_{pitcht} \tag{2}$$

where S_0 is the sun vector, θ_{pitcht} is the pitch axis angle, and D_{pitcht} is the error matrix, which is currently set as the unit matrix. According to the above formula, the initial position of the pitch angle of the turntable could be calculated in advance through the sun vector data broadcasted by the satellite platform. In the actual operation of the turntable, the platform attitude factor and the position error should also be considered. In addition, it was also affected by the operating speed and control current of the turntable itself.

2.2. One-Dimensional Convolution Neural Network (1D-CNN) Model

CNN is a successful deep learning framework first proposed by LeCun et al. [15]. In the study of deep learning, in 1D-CNN (also known as time-domain convolution), the convolution kernel is a vector with a length of N, which is used for neighborhood filtering of one-dimensional input signals and extracting local features. The kernel slides along a one-dimensional time axis. It is often used to process NLP and time series data. It is shown in Figure 3.

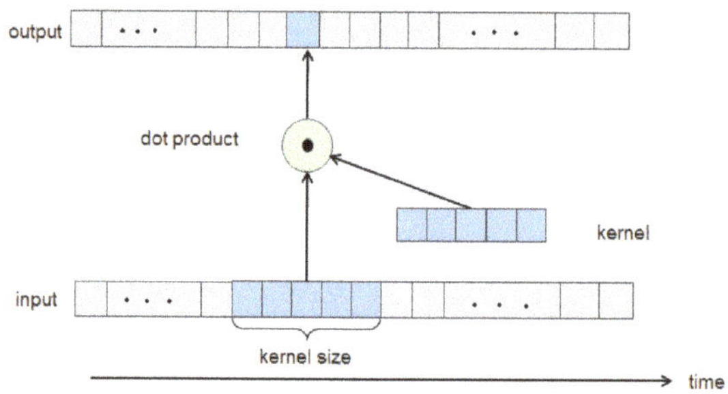

Figure 3. The 1D-CNN model structure, the blue color parts represents processing dot product with input slide window part and kernel.

In the convolution layer, the input data information needs to undergo convolution operation and activation function calculation before flowing to the next layer. The operation is shown in Equation (3):

$$h_t = \sigma_{cnn}(W_{cnn} * X_t + b_{cnn}) \tag{3}$$

where W_{cnn} represents the weight coefficient of the filter, namely the convolution kernel; X_t represents the data information of the time, while the input sample * represents the discrete convolution operation between X_t and W_{cnn}; b_{cnn} is a bias parameter, which will be obtained by learning when training the model; σ Cnn stands for the activation function; and h_t represents the output data after the convolution operation.

2.3. Long Short-Term Memory (LSTM) Network Model

A long short-term memory network (LSTM) is an improved cyclic neural network used to solve the problem that RNN networks cannot deal with long-distance dependence. Hochreiter proposed the LSTM algorithm [16], which can store data information in a longer time step. Regarding the problem of time series prediction and analysis, LSTM can predict future data characteristics through the data characteristics of the past period time. LSTM networks enable nodes to "remember" or "forget" data through a "gate" structure, which mainly includes three "gates": the forgetting gate, the information adding a gate, and the information output gate. Through these three "gates", the input of each cell state contains the output of the previous moment, and the input of the current moment also contains some information stored by the node itself. Therefore, LSTM has a better performance on longer sequences. It is shown in Figure 4.

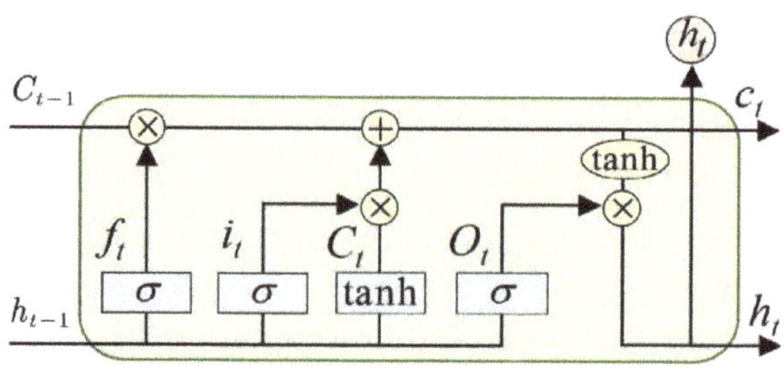

Figure 4. LSTM model structure.

The LSTM updates for time steps given inputs x_t, h_{t-1}, and C_{t-1}, The operation is shown in Equations (4)–(9):

$$f_t = \sigma\left(W_f \cdot [C_{t-1}, h_{t-1}, x_t] + b_f\right) \quad (4)$$

$$i_t = \sigma(W_i \cdot [C_{t-1}, h_{t-1}, x_t] + b_i) \quad (5)$$

$$\widetilde{C}_t = \tanh(W_C \cdot [h_{t-1}, x_t] + b_C) \quad (6)$$

$$C_t = f_t * C_{t-1} + i_t * \widetilde{C}_t \quad (7)$$

$$o_t = \sigma(W_o[h_{t-1}, x_t] + b_o) \quad (8)$$

$$h_t = o_t * \tanh(C_t) \quad (9)$$

where σ and tanh represent the sigmoid activation function and hyperbolic tangent activation function, respectively; W and b represent the weight matrix and bias parameters, respectively; x_t represents the input of the LSTM unit at time t; h_t represents the output of the unit corresponding to at time t; and C_t represents the state unit of the LSTM at time t. The whole LSTM unit includes three thresholds, namely forgetting gate f_t, input gate i_t, and output gate o_t.

2.4. CNN-LSTM Model

The CNN-LSTM model is a hybrid model of two neural network models. We first used CNN to extract data features and LSTM to further extract temporal features. The specific structure was as follows: the CNN model used a Conv1D layer and multiple input data as the time series; the kernel moved in one dimension along the time axis, then we input the data into LSTM layer and used the LSTM layer to obtain the long-term characteristics of the pitch axis data. Finally, it output the predicted value. It is shown in Figures 3 and 4.

According to the above, the main factors affecting the pitch axis position include the ideal position calculated from the sun vector, the pitch axis operation error, the motor current, and the motor speed. The pitch axis motor current (C; unit: A), initial position (I; unit: °), pitch axis motor speed (S; unit: °/s), and pitch axis operation error (E; unit: mV) were set as input data. The expected position of the pitch axis (Pt, unit: °) was output through two CNN network layers and one LSTM layer. The initial position value was equal to the ideal position value calculated using Formulas (1) and (2) as shown in Figures 5 and 6.

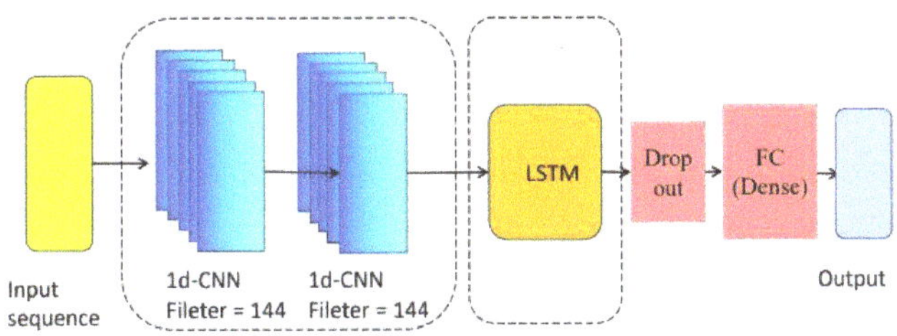

Figure 5. Proposed CNN-LSTM model architecture with input sequence, two CNNs layer, an LSTM layer, a dropout layer, dense layer, and an output.

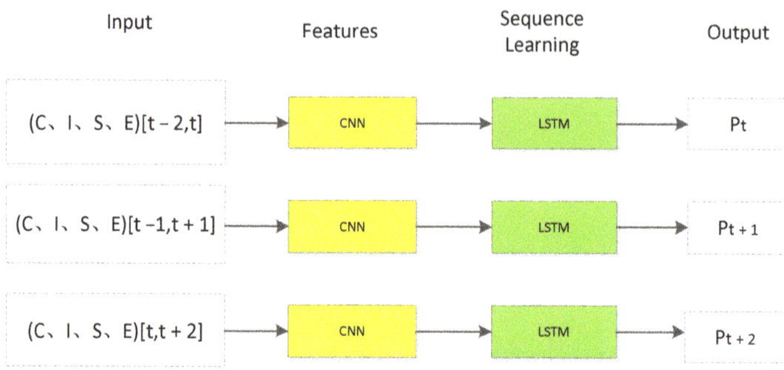

Figure 6. CNN-LSTM model operation diagram, the yellow parts represent the layer of CNN, and the green parts represent the layer of LSTM.

In Figure 6, *C* represents the pitch axis motor current, *I* represents the ideal sun position, *S* represents the pitch axis motor speed, *E* represents the pointing error, and P_t is the predicted position value of the output.

In the CNN network, we set the data input feature to 4. We conducted performance tests using input time sliding of 3, 5, or 10. Taking into account performance factors, the best performance was found for a time sliding of 3. The time sliding window was set at 3, the stride was set at 1, the kernel size was set at 1, and the activation function used RELU. The activation function of the LSTM layer was RELU. We used a grid search for hyperparameter optimization. We tested the performance of the SGD, Adagrad, and Adam optimizers [17] in this application. Finally, the optimizer selected Adam, the loss function selected MSE, and the drop layer was set to 0.35. The following network architecture design was sampled. It is shown in Figure 7 and Table 1.

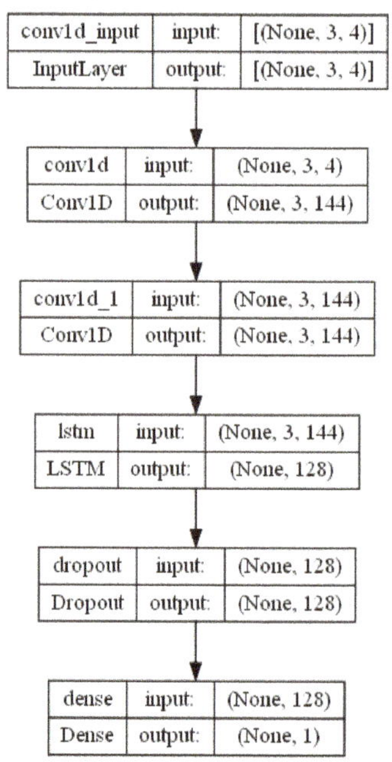

Figure 7. Proposed CNN-LSTM, model architecture parameter.

Table 1. Model parameter table.

Layer (Type)	Parameter [1]
conv1d (Conv1D)	720
conv1d_1 (Conv1D)	20,880
lstm (LSTM)	139,776
dropout (Dropout)	0
dense (Dense)	129

[1] Total params: 161,505; trainable params: 161,505; non-trainable params: 0.

3. Results

We constructed a data set based on orbit data simulated using Fengyun-3E satellite orbit data and the operation data of the two-dimensional turntable of the Solar X-ray and Extreme Ultraviolet Imager (X-EUVI).

The main steps to build the CNN-LSTM model were as follows:

(1) We set the time window size K and transformed the data set according to the time window size to transform the time series into a supervised sequence; that is, we used the past K values to predict the value of the next time and the original value of the next time as the supervised value.
(2) We divided the data set used into the training set and test set and converted the data format into the format required in the CNN-LSTM model, namely (samples, time steps, features).
(3) The parameters used in the model, including the number of iterations, the amount of data for each iteration, and the number of neurons, were determined through continuous attempts.
(4) We established a CNN-LSTM model. After the model for predicting the data in the data flow was built, the data could be predicted.

The specific process is shown in Figure 8.

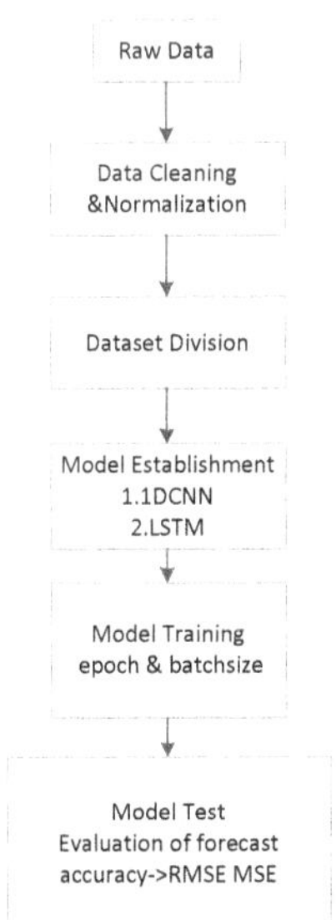

Figure 8. Prediction processing.

3.1. Data Feature Extraction and Data Set Establishment

The data sampling period was 32 s. A total of 34,559 sets of data were set up and sorted by time. The first 29,562 sets of data were used as training sets, and the last 4997 groups were used as test sets. The outliers in the data set were removed and normalized. The results are shown in Figures 9 and 10.

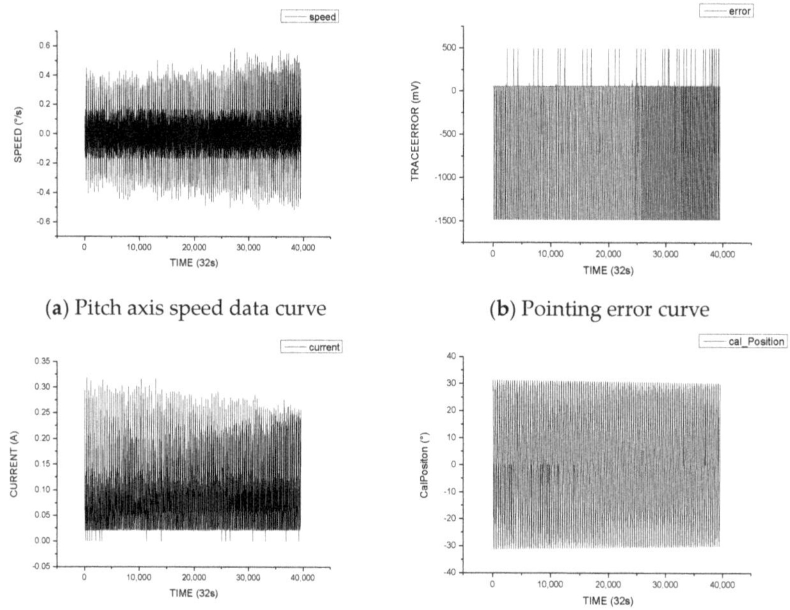

(**a**) Pitch axis speed data curve (**b**) Pointing error curve

(**c**) Pitch axis motor current curve (**d**) Initial position curve

Figure 9. Data feature curves of pitch axis.

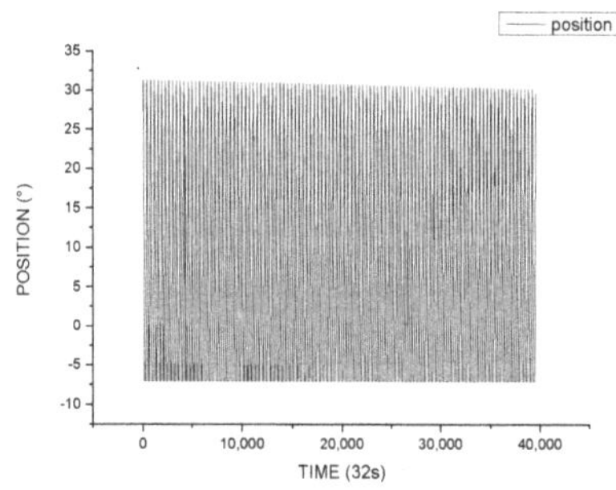

Figure 10. True position curve of pitch axis from data set.

3.2. Experiment Environment and Tools

The experimental environment of this research was an AMD FX (tm)—4100 Quad-Core Processor, 16 GB of memory, the Windows 10 operating system, and Python 3.9, and PyCharm as development tools. In PyCharm, we used the data packets keras and sklearn.

3.3. Experiment Result

We set epoch = 50 and batch_Size = 70. The change in the loss value with EPOCH is shown in Figure 11. The training set fluctuated in the early stage. With the increase in EPOCH, the loss value gradually converged. The predicted value and actual value are shown in Figure 12 below. It can be seen that the data prediction at some inflection points had more errors. It also can be seen that when using the CNN-LSTM model, the loss function of the training set converged better (see Figures 11 and 12).

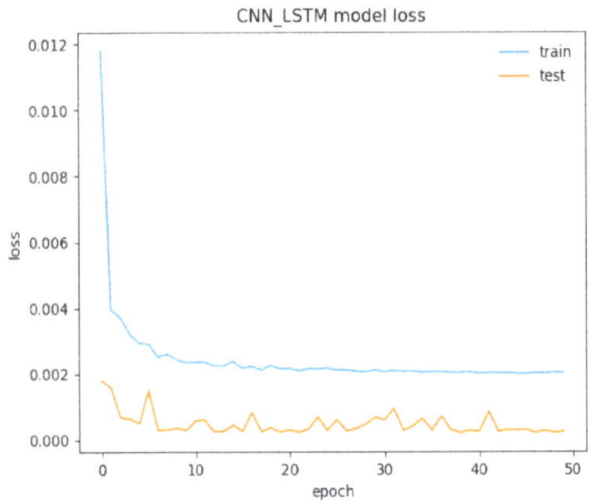

Figure 11. Loss curve for training set and test set.

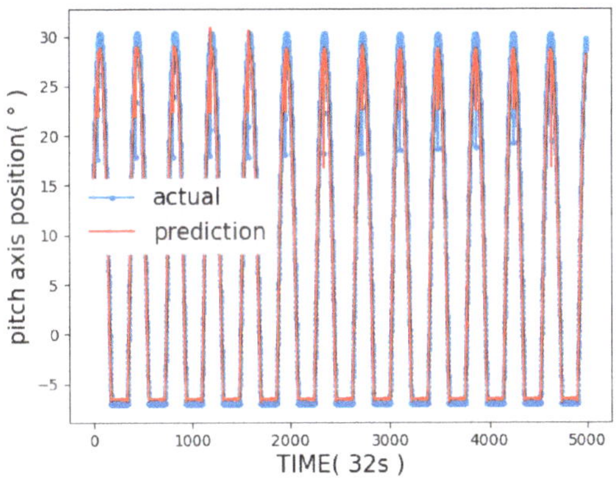

Figure 12. Forecast location map for test set.

Since we achieved good results using the CNN-LSTM model, we used the idea of ablation to evaluate the impact of each module on performance. In the CNN-LSTM model, we removed the CNN layers or LSTM layers, both of which had a large impact on the performance of the system (the number of CNN layers changes also affected the final result). We selected the CNN model and the LSTM model for the model comparison. The CNN model was set with three layers of 1D-CNN (Filter = 64), the LSTM model was set with two layers of LSTM, and the time sliding window was 3. The optimizer was consistent with the CNN-LSTM model, and the Adam optimizer was selected. The loss function was MSE. When EPOCH was set to 50, CNN-LMST had the lowest loss and the best effect. This is shown in Figure 13 for specific values and in Table 2. This model can be used for trajectory prediction of other motion units.

Table 2. Performance in test data sets for each model.

Model	RMSE	MSE
LSTM	0.678	0.459
CNN	0.632	0.399
CNN-LSTM	0.623	0.388

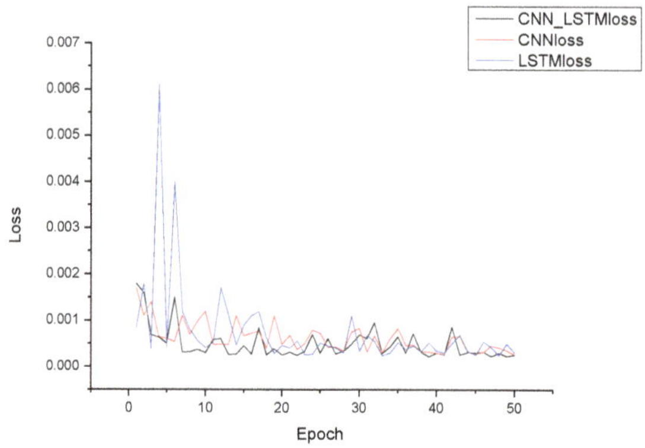

Figure 13. Loss diagram of each model for the test set.

4. Discussion

In this paper, we proposed using a sun vector to calculate the ideal turntable position value. Then we took the ideal position value and turntable speed, current, and solar position error signal as date features and used the CNN-LSTM model to realize predicting the trajectory of the pitch axis. The model could adjust parameters adaptively on the training set and had a better performance on the test set.

The premise of the conventional feedback method is the need to calibrate the optical axis in relation to the instrument body coordinate system and the satellite coordinate system. The position of the pitch axis is then derived by determining the coefficients of the controller. The disadvantage is that certain parameters may need to be adjusted after a long period of system operation. The advantage is the high reliability due to the rigorous model derivation. The advantage of the deep learning method for calculating the pitch axis position is that it can be used without coordinate system calibration, allowing adaptive adjustment of the parameters. However, due to its black-box nature, this research is still at a preliminary stage for on-orbit applications.

Compared with the trajectory prediction method mentioned above, the pitch axis trajectory prediction of a two-dimensional turntable is a typical time series prediction problem [18,19]. It is not only related to the previous running state of the turntable but also is affected by the running state of a period of the time window. From this point of view, HMM and an extended Kalman filter only predict the next state through the last state, and the results will affect the accuracy of the prediction [20]. We did not use the equation of motion either, thereby avoiding uncertainty error analysis. We used a deep learning model to predict the trajectory by extracting and learning data features. We used a sliding time window instead of just the last state, and our findings were in accordance with recent studies indicating that a prediction model based on deep learning can achieve satisfactory results [21–26]. The prediction model based on the LSTM model could effectively avoid gradient disappearance or gradient explosion. In the process of debugging parameters, we also found that the initial value of parameters affected the final performance of the model. The size of the dataset also limited our choice of models. This model had a better performance on a small data set. For the comparison of the performance indicators of the LSTM model and CNN model alone, we could see that the sun vector combined with CNN-LSTM model had a better performance.

The major limitation of the present study was that the model contained part of the black box, so the model lacked interpretability. At the present stage, we mainly used the calculated results to compare with the real values and then used the comparison results to reverse-extrapolate to achieve the purpose of optimizing the performance of the model. The generalization ability of the CNN-LSTM model has yet to be verified. In addition, when the sliding time window becomes larger, the calculation time of the model will become longer. It will affect the execution time of the model.

Despite its limitations, the model of CNN-LSTM had a significant short-term prediction effect on turntable trajectory prediction.

Future applications in orbit will be different from the ground simulation. For the hardware environment, we must choose the processor to meet the onboard radiation—a hardened and tolerant processor generally using an ARM and field-programmable gate array (FPGA) fabric for real-time processing [27]. Software algorithms will be transplanted and optimized accordingly, and the trained network structure needs to be arranged on the onboard platform to work.

5. Conclusions

With the introduction of intelligent load and spaceborne integration, increasing attention has been paid to the problems related to two-dimensional turntable motion [28].

In this paper, different from the motion model and equation using traditional trajectory prediction, we used the sun vector and the CNN-LSTM model to predict the pitch axis position of the two-dimensional turntable. It had the advantages of adaptive adjustment of parameters and easier establishment of models. We calculated the ideal sun position through the sun vector model and input it into the model as a feature to participate in the prediction. Through comparison of performance indicators, the CNN-LSTM model combined with a solar vector model was superior to the LSTM model or the CNN model. The test results showed that the RMSE value was 0.623 and the MSE value was 0.388. The CNN-LSTM model could accurately predict the two-dimensional turntable operation.

This can be applied not only to track prediction of a turntable but also extended to other track predictions; for example, vehicle trajectory prediction, navigation trajectory prediction, etc. It also has broad prospects in other applications such as fault detection by predicting the state [29–32].

Author Contributions: Project administration, K.-F.S.; data curation, Y.-L.W. and P.-J.Z.; S.D. wrote the manuscript. All authors have read and agreed to the published version of the manuscript.

Funding: This research received no external funding.

Institutional Review Board Statement: Not applicable.

Informed Consent Statement: Not applicable.

Data Availability Statement: Data are unavailable due to privacy or ethical restrictions.

Acknowledgments: Thanks to all the X-EUVI Project team members.

Conflicts of Interest: The authors declare no conflict of interest.

References

1. Chen, B.; Zhang, X.-X.; He, L.-P.; Song, K.-F.; Liu, S.-J.; Ding, G.-X.; Dun, J.-P.; Li, J.-W.; Li, Z.-H.; Guo, Q.-F.; et al. Solar X-ray and EUV imager on board the FY-3E satellite. *Light. Sci. Appl.* **2022**, *11*, 329. [CrossRef] [PubMed]
2. Wang, Z.; Huang, M.; Qian, L.; Zhao, B. Near-earth space two-dimension opto-electronic turntable design. *Optik* **2020**, *200*, 163387. [CrossRef]
3. Crassidis, J.L.; Alonso, R.; Junkins, J.L. Optimal Attitude and Position Determination from Line-of-Sight Measurements. *J. Astronaut. Sci.* **2000**, *48*, 391–408. [CrossRef]
4. Qiao, S.; Shen, D.; Wang, X.; Han, N.; Zhu, W. A Self-Adaptive Parameter Selection Trajectory Prediction Approach via Hidden Markov Models. *IEEE Trans. Intell. Transp. Syst.* **2014**, *16*, 284–296. [CrossRef]
5. Li, Q.; Li, R.; Ji, K.; Dai, W. Kalman filter and its application. In Proceedings of the 8th International Conference on Intelligent Networks and Intelligent Systems (ICINIS), Tianjin, China, 1–3 November 2015; IEEE: Piscataway, NJ, USA, 2015.
6. Zhou, Z.; Chen, J.; Shen, B.; Xiong, Z.; Shen, H.; Guo, F. A trajectory prediction method based on aircraft motion model and grey theory. In Proceedings of the 2016 IEEE Advanced Information Management, Communicates, Electronic and Automation Control Conference (IMCEC), Xi'an, China, 3–5 October 2016; IEEE: Piscataway, NJ, USA, 2016.
7. Wang, M.; Fu, W.; He, X.; Hao, S.; Wu, X. A survey on large-scale machine learning. *IEEE Trans. Knowl. Data Eng.* **2020**, *34*, 2574–2594. [CrossRef]

8. Zhou, H.; Chen, Y.; Zhang, S. Ship Trajectory Prediction Based on BP Neural Network. *J. Artif. Intell.* **2019**, *1*, 29–36. [CrossRef]
9. Zhou, H.; Zhang, S.; Peng, J.; Zhang, S.; Li, J.; Xiong, H.; Zhang, W. Informer: Beyond Efficient Transformer for Long Sequence Time-Series Forecasting. In Proceedings of the AAAI Conference on Artificial Intelligence, No. 23, Online, 2–9 February 2021; Volume 35.
10. Kuang, D.; Xu, B. Predicting kinetic triplets using a 1d convolutional neural network. *Thermochim. Acta* **2018**, *669*, 8–15. [CrossRef]
11. Yusoff, M.I.M.; Mohamed, I.; Bakar, M.R.A. Hidden Markov models: An insight. In Proceedings of the 6th International Conference on Information Technology and Multimedia, Putrajaya, Malaysia, 18–20 November 2014; IEEE: Piscataway, NJ, USA, 2014.
12. Qiao, S.J.; Wu, L.C.; Han, N.; Huang, F.L.; Mao, R.; Yuan, C.A.; Gutierrez, L.A. Multiple-motion-pattern Trajectory Prediction of Moving Objects with Context Awareness: A Survey. *J. Softw.* **2021**, *34*, 312–333.
13. Yu, F.X.; Zheng, Y.M.; Xie, C.X.; Jin, J.J.; Jin, Z.H. Error analysis of pico-satellite attitude angle measurement based on magnetometer. *Jilin Daxue Xuebao* **2007**, *37*, 1460.
14. Wertz, J.R. *Spacecraft Attitude Determination and Control*; Springer Science & Business Media: Berlin/Heidelberg, Germany, 2012; Volume 73.
15. Lecun, Y.; Bottou, L.; Bengio, Y.; Haffner, P. Gradient-based learning applied to document recognition. *Proc. IEEE* **1998**, *86*, 2278–2324. [CrossRef]
16. Sepp, H.; Schmidhuber, J. Long short-term memory. *Neural Comput.* **1997**, *9*, 1735–1780.
17. Kingma, D.P.; Ba, J. Adam: A method for stochastic optimization. *arXiv* **2014**, arXiv:1412.6980.
18. Montgomery, D.C.; Jennings, C.L.; Kulahci, M. *Introduction to Time series Analysis and Forecasting*; John Wiley & Sons: Hoboken, NJ, USA, 2015.
19. Horvatic, D.; Stanley, H.E.; Podobnik, B. Detrended cross-correlation analysis for non-stationary time series with periodic trends. *Europhys. Lett.* **2011**, *94*, 18007. [CrossRef]
20. He, W.; Williard, N.; Chen, C.; Pecht, M. State of charge estimation for electric vehicle batteries using unscented kalman filtering. *Microelectron. Reliab.* **2013**, *53*, 840–847. [CrossRef]
21. Chemali, E.; Kollmeyer, P.J.; Preindl, M.; Emadi, A. State-of-charge estimation of Li-ion batteries using deep neural networks: A machine learning approach. *J. Power Sources* **2018**, *400*, 242–255. [CrossRef]
22. Lee, J.-Y.; Jo, B.-U.; Moon, G.-H.; Tahk, M.-J.; Ahn, J. Intercept Point Prediction of Ballistic Missile Defense Using Neural Network Learning. *Int. J. Aeronaut. Space Sci.* **2020**, *21*, 1092–1104. [CrossRef]
23. Baccouche, M.; Mamalet, F.; Wolf, C.; Garcia, C.; Baskurt, A. Sequential deep learning for human action recognition. In *Human Behavior Understanding, Proceedings of the Second International Workshop, HBU 2011, Amsterdam, The Netherlands, 16 November 2011; Proceedings 2*; Springer: Berlin/Heidelberg, Germany, 2011.
24. Sutskever; Ilya; Vinyals, O.; Le, Q.V. Sequence to sequence learning with neural networks. In Proceedings of the Advances in Neural Information Processing Systems 27: Annual Conference on Neural Information Processing Systems 2014, Montreal, QC, Canada, 8–13 December 2014; Volume 27.
25. Guo, L.; Lei, Y.; Xing, S.; Yan, T.; Li, N. Deep Convolutional Transfer Learning Network: A New Method for Intelligent Fault Diagnosis of Machines with Unlabeled Data. *IEEE Trans. Ind. Electron.* **2019**, *66*, 7316–7325. [CrossRef]
26. Tang, Y.; Dou, L.; Zhang, R.; Zhang, X.; Liu, W. Deep Transfer Learning-based Fault Diagnosis of Spacecraft Attitude System. In Proceedings of the 39th Chinese Control Conference (CCC), Shenyang, China, 27–29 July 2020.
27. Mandl, D.J. Real-Time Data Products and the Intelligent Payload Module. In Proceedings of the HyspIRI and Surface Biology and Geology Science and Applications Workshop, Washington, DC, USA, 15–17 August 2018. No. GSFC-E-DAA-TN60324.
28. You, Z.; Wang, C.; Xing, F.; Sun, T. Key technologies of smart optical payload in space remote sensing. *Spacecr. Recovery Remote Sens.* **2013**, *34*, 35–43.
29. Liu, B.; Yan, S.; Li, J.; Qu, G.; Li, Y.; Lang, J.; Gu, R. A Sequence-to-Sequence Air Quality Predictor Based on the n-Step Recurrent Prediction. *IEEE Access* **2019**, *7*, 43331–43345. [CrossRef]
30. Zhou, Y.; Dong, M.; Wu, J. Hyperparameter Optimization for SOC Estimation by LSTM with Internal Resistance. In Proceedings of the 2021 International Conference on Computer Network, Electronic and Automation (ICCNEA), Xi'an, China, 24–26 September 2021; IEEE: Piscataway, NJ, USA, 2021.
31. Zhou, X.; Shi, J.; Gong, K.; Zhu, C.; Hua, J.; Xu, J. A Novel Quench Detection Method Based on CNN-LSTM Model. *IEEE Trans. Appl. Supercond.* **2021**, *31*, 4702105. [CrossRef]
32. Fu, J.; Sun, C.; Yu, Z.; Liu, L. A hybrid CNN-LSTM model based actuator fault diagnosis for six-rotor UAVs. In Proceedings of the 2019 Chinese Control and Decision Conference (CCDC), Nanchang, China, 3–5 June 2019; pp. 410–414. [CrossRef]

applied sciences

MDPI

Article

Hyperspectral Image Classification Based on Fusion of Convolutional Neural Network and Graph Network

Luyao Gao [1,2], Shulin Xiao [1,2], Changhong Hu [1,*] and Yang Yan [3]

1 Changchun Institute of Optics, Fine Mechanics and Physics, Chinese Academy of Sciences, Changchun 130033, China; gaoluyao20@mails.ucas.ac.cn (L.G.); xiaoshulin20@mails.ucas.ac.cn (S.X.)
2 University of Chinese Academy of Sciences, Beijing 100049, China
3 College of Computer Science and Technology, Changchun Normal University, Changchun 130033, China; yanyang2016@hotmail.com
* Correspondence: changhonghu@ciomp.ac.cn; Tel.: +86-181-8688-9840

Abstract: Convolutional neural networks (CNNs) have attracted significant attention as a commonly used method for hyperspectral image (HSI) classification in recent years; however, CNNs can only be applied to Euclidean data and have difficulties in dealing with relationships due to their limitations of local feature extraction. Each pixel of a hyperspectral image contains a set of spectral bands that are correlated and interact with each other, and the methods used to process Euclidean data cannot effectively obtain these correlations. In contrast, the graph convolutional network (GCN) can be used in non-Euclidean data but usually leads to over-smoothing and ignores local detail features due to the need for superpixel segmentation processing to reduce computational effort. To overcome the above problems, we constructed a fusion network based on the GCN and CNN which contains two branches: a graph convolutional network based on superpixel segmentation and a convolutional network with an added attention mechanism. The graph convolutional branch can extract the structural features and capture the relationships between the nodes, and the convolutional branch can extract detailed features in the local fine region. Owing to the fact that the features extracted from the two branches are different, the classification performance can be improved by fusing the complementary features extracted from the two branches. To validate the proposed algorithm, experiments were conducted on three widely used datasets, namely Indian Pines, Pavia University, and Salinas. An overall accuracy of 98.78% was obtained in the Indian Pines dataset, and overall accuracies of 98.99% and 98.69% were obtained in the other two datasets. The results show that the proposed fusion network can obtain richer features and achieve a high classification accuracy.

Keywords: hyperspectral images; convolutional neural networks; graph convolutional networks; feature fusion

Citation: Gao, L.; Xiao, S.; Hu, C.; Yan, Y. Hyperspectral Image Classification Based on Fusion of Convolutional Neural Network and Graph Network. *Appl. Sci.* **2023**, *13*, 7143. https://doi.org/10.3390/app13127143

Academic Editor: Juan A. Gómez-Pulido

Received: 20 April 2023
Revised: 11 June 2023
Accepted: 12 June 2023
Published: 14 June 2023

1. Introduction

Hyperspectral imaging technology combines imaging technology with spectral technology and has achieved wide application in recent years. With the advancement of hyperspectral imaging technology, hyperspectral imaging systems can simultaneously acquire abundant spectral information and two-dimensional spatial information of a feature and then form a hyperspectral image (HSI) [1–3]. Therefore, hyperspectral imaging technology has become a hotspot for research due to its rich spectral and spatial information. An HSI provides from tens to hundreds of continuous spectral bands [4]. The abundance of spectral information greatly enhances the ability to distinguish objects. Therefore, an HSI is commonly used in disaster monitoring, vegetation classification, fine agriculture, and medical diagnosis due to its extremely high spectral resolution [1,2,5].

As the focus of the field of hyperspectral image analysis, the HSI classification task has always received significant attention from scholars. Hyperspectral image classification aims to classify each pixel point in the image [6]. In the early days, most HSI classification methods mainly relied on some traditional machine learning algorithms [7] which were mainly divided into two processes: traditional manual feature engineering and classifier classification [8]. Feature engineering aims to process data based on knowledge so that the processed features can be better used in subsequent classification algorithms. Commonly used feature engineering methods include principal component analysis (PCA), independent component analysis (ICA), and other dimensionality reduction methods.

Typical classification algorithms include the support vector machine (SVM) [9], random forest (RF) [10], and k-nearest neighbor (KNN) [11], etc. [12,13]. The above machine learning approaches only focus on the spectral information of an HSI. It is inaccurate to use the spectral information only for the classification task, thus limiting the improvement in the classification accuracy and the gradual elimination of spectral information.

As a result of the triumph of deep learning in areas such as computer vision, many approaches based on deep learning have also been adopted for hyperspectral image classification [14]. Among the deep learning methods, convolutional neural networks (CNNs) [15] have become a popular method for hyperspectral image classification due to their excellent performance. Deep-learning-based methods represented by CNNs have replaced traditional machine-learning-based HSI classification methods and have become a research hotspot [16].

Deep learning methods of 1D-CNN [17] and 2D-CNN were first applied to hyperspectral image classification, and the performance surpassed machine learning methods. However, the above methods suffer from the underutilization of spatial and spectral information. Therefore, the 3D-CNN model [16] was proposed, which can extract spatial–spectral features simultaneously and therefore obtain better classification results, but the model has a large computational burden. To extract richer features, some scholars have proposed a hybrid spectral CNN (HybridSN) [18] which combines 3D-CNN and 2D-CNN to exploit the spatial–spectral features of an HSI with less computational burden than 3D-CNN.

With the purpose of finding correlations between data, highlighting important features, and ignoring irrelevant noise information, an attention mechanism has been proposed. Li et al. proposed a two-branch double attention network (DBDA) [19] which contained two branches to extract spatial and spectral features and added an attention mechanism to obtain better classification results. In order to capture richer features, deeper network layers are needed, but the deeper network layers will lead to computational complexity and make the model training difficult. Zhong et al. introduced a residual structure based on the 3D-CNN model [20], constructed a spectral residual module and a spatial residual module, and achieved more satisfactory classification results.

Although the classification results achieved by CNN-based classification methods have been good, there are still some limitations. First, the CNN is designed for Euclidean data, and the traditional CNN model can only convolve regular rectangular regions, so it is difficult to obtain complex topological information. Second, CNNs cannot capture and utilize the relationship between different pixels or regions in hyperspectral images; they can only extract detailed features in the local fine region, but the structure features and dependency relationship between the nodes may provide useful information for the classification process [21,22].

In order to obtain the relationship between objects, graph convolutional networks (GCNs) have been developed rapidly in recent years [23]. GCNs are designed to process graph-structured data. CNNs are used for processing Euclidean data such as images, which are a regular matrix. Therefore, no matter where the convolution kernel of a CNN is located in the image, the consistency of the result of the operation is guaranteed (translational invariance). However, the graph-structured data are non-Euclidean data, and the graph structure is irregular, so it is impossible to apply the CNN on graph data. The graph convolution is designed to resolve this situation. The most important innovation of the GCN is to overcome the inapplicability of translation invariance on non-Euclidean data, so it can be applied to extract the features of the graph structure.

Kipf et al. proposed the GCN model [24] which is able to operate on non-Euclidean data and extract the structural relationship between different nodes [21]. Some scholars have attempted to apply the GCN to hyperspectral classification tasks [25], and various studies have shown that the classification results are not only affected by spectral information but are also related to the spatial structure information of the pixels [22,26]. By treating each pixel or superpixel in the HSI as a graph node, the hyperspectral image can be converted into graph-structured data, and then the GCN can be used to obtain the spatial structure information in the image and provide a more effective information for the classification. Hong et al. [22] proposed the MiniGCN method and constructed an end-to-end fusion network which was able to sample images in small batches, classify images as subgraphs, and achieve good classification results. Wan et al. proposed MDGCN [27], which is different from the commonly used GCN. Working on a fixed graph model, MDGCN is able to make the graph structure update dynamically so that the two steps benefit each other. In addition, we cannot consider each pixel of an HSI as a graph node due to the limitation of computational complexity, so hyperspectral images are usually preprocessed as superpixels. The superpixel segmentation technique is applied to the construction of the graph structure, which reduces the complexity of model training significantly. However, the superpixel segmentation technique leads to another problem. Superpixel segmentation often leads to smooth edges of the classification map and a lack of local detailed information of the features. This problem restricts the improvement of the classification performance and has an impact on the analysis of the results.

To obtain the relational features of an HSI and to solve the problem of missing details due to superpixel segmentation, inspired by [28], we designed a feature fusion of the CNN and GCN (FCGN). The algorithm consisted of two branches: the GCN branch and CNN branch. We applied the superpixel segmentation technique in the GCN branch. The superpixel segmentation technique can aggregate similar pixels into a superpixel. Then, we treated these superpixels as graph nodes. Graph convolution processes the data by aggregating the features of each node as well as its neighboring nodes. This approach can capture the structure features and dependency relationship between the nodes and thus better represent the features of the nodes. Compared with the CNN branch, the GCN branch based on superpixel segmentation can acquire structure information over a longer distance, while the CNN branch can obtain the pixel-level features of the HSI and perform a fine classification of local regions. Finally, the different features acquired by the two branches were fused to obtain richer image features by complementing their strengths. In addition, the attention mechanism and depth-wise separable convolution algorithm [29] were applied to further optimize the classification results and network parameters.

2. Methodology

This section presents the proposed FCGN for HSI classification, which includes the overall structure of the FCGN and the function of each module in the network.

2.1. General Framework

To solve the problem of missing local details in classification maps due to superpixel segmentation, we proposed a feature fusion of the CNN and GCN, as shown in Figure 1. The proposed network framework contained a spectral dimension reduction module (see Section 2.2 for details), a graph convolutional branch (see Section 2.3 for details), a convolutional branch (see Section 2.4 for details), a feature fusion module, and a Softmax classifier. It should be noted that the features extracted from convolutional neural networks were different from those of graph convolutional networks. Feature fusion methods can utilize different features of an image to complement each other's strengths, thus obtaining more robust and accurate results. Because of that, it is possible to obtain better classification results than a single branch by fusing features from two branches.

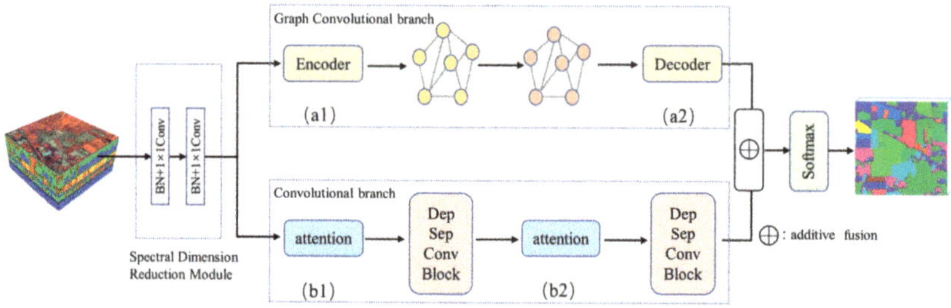

Figure 1. The framework of the FCGN algorithm. (**a1**,**a2**) Feature conversion module; (**b1**,**b2**) SE attention module.

The original HSI was handled by the spectral dimensionality reduction module first, which was used for spectral information transformation and feature dimensionality reduction. Then, we used convolutional neural networks to extract the detailed features in a local fine region. Considering the problem that the CNN-based method may induce overfitting with too many parameters and an insufficient number of training samples, we used a depth-wise separable convolution to reduce the parameters and enhance the robustness. To further improve the model, we added attention modules to the convolution branch. We used the SE attention module to optimize the proposed network [30]. The SE module can obtain the weight matrix of different channels. Then, the weight values of each channel calculated by the SE module were multiplied with the two-dimensional matrix of the corresponding channel of the original feature map. We used graph convolutional networks to extract the superpixel-level contextual features. In this branch, we applied a graph encoder and a graph decoder to implement the transformation of pixel features and superpixel-level features (see Section 2.5 for details). Next, the different features acquired by the two branches were fused to obtain richer image features by complementing their strengths. Finally, after the processing of the Softmax classifier, we obtained the label of each pixel. The role of Softmax is to assign a probability value to each output classification, indicating the probability of belonging to each class.

2.2. Spectral Dimension Reduction Module

There is a significant amount of redundant information in the original hyperspectral image. Using dimension reduction modules, it is possible to significantly reduce the computational cost without significant performance loss. The 1×1 convolutional layer has the ability to remove useless spectral information and increase nonlinear characteristics. Moreover, it is usually used as a dimension reduction module to remove computational cost, as shown in Figure 2. In the FCGN network, hyperspectral images are first processed using two 1×1 convolutional blocks. Specifically, each 1×1 convolutional block contains a BN layer, a 1×1 convolution layer, and an activation function layer. The role of the BN layer is to accelerate the convergence of the network, and the activation function layer can significantly increase the network's nonlinearity to achieve better expressiveness. The activation function in this module adopts Leaky ReLU.

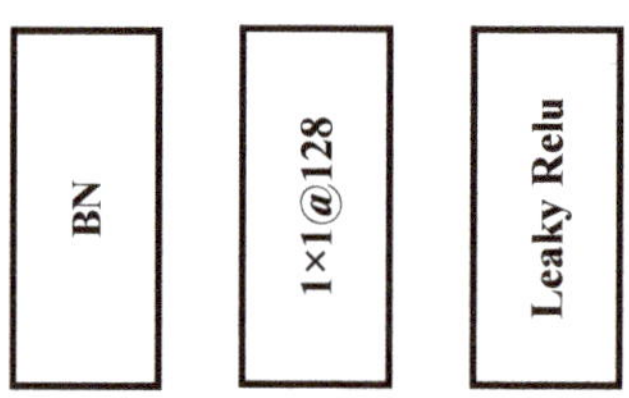

Figure 2. Dimension reduction module. Each 1×1 convolution block contains the above three parts.

We have:

$$\mathbf{X}^*_{\mathrm{h,w,n}} = \sigma\left(\sum_{\mathrm{x,y,b}} \mathbf{W}_{\mathrm{x,y,b,n}} \widetilde{\mathbf{X}}_{\mathrm{h,w,b}} + \mathbf{B}_{\mathrm{x,y,b,n}}\right) \tag{1}$$

where $\mathbf{X}_{\mathrm{h,w,b}}$ denotes the input feature map, $\widetilde{\mathbf{X}}_{\mathrm{h,w,b}}$ denotes the batch-normalized input feature map, $\mathbf{X}^*_{\mathrm{h,w,n}}$ denotes the output feature map, $\mathbf{W}_{\mathrm{x,y,b,n}}$ denotes the convolution kernel of the input feature map in row x and column y, $\mathbf{B}_{\mathrm{x,y,b,n}}$ denotes bias, and n is the number of convolution kernels. σ represents the Leaky ReLU activation function.

2.3. Graph Convolution Branch

Numerous studies have shown that the classification accuracy can be effectively improved by combining the different features of images. Traditional CNN models can only convolve images in regular image regions using convolution kernels of a fixed size and weight, resulting in an inability to obtain global features and structural features of images. Therefore, it is often required to deepen the network layer to alleviate this problem. However, as the number of network layers deepens, the chance of overfitting increases subsequently, especially when processing data with a small amount of training samples such as HSIs. Such a result is unacceptable to us.

Therefore, a GCN branch based on superpixel segmentation was constructed to obtain the structural features. Different from the CNN, the GCN is a method used for the graph structure. The GCN branch can extract the structure features and dependency relationship between the nodes from images. These features are different from the neighborhood spatial features in a local fine region extracted by the CNN branch. Finally, the property of the network can be enhanced by fusing the different features extracted from the two branches. The graph structure is a non-Euclidean structure that can be defined as $\mathrm{G} = (\mathrm{V}, \mathrm{E})$, where V is the set of nodes and E is the set of edges. V and E are usually encoded into a degree matrix $\mathbf{D}$ and node matrix $\mathbf{A}$, where $\mathbf{D}$ records the relationship between each pixel of the hyperspectral image and $\mathbf{A}$ denotes the number of edges associated with each node.

Because the degree of each graph node in the graph structure is not the same, the GCN cannot directly use the same-size local graph convolution kernel for all nodes similar to the CNN. Considering that the convolution in the spatial domain is equivalent to the product in the frequency domain, researchers hope to implement the convolution operation on topological graphs with the help of the theory of graph spectra, and they have proposed the frequency domain graph convolution method [31]. The Laplacian matrix of the graph structure is defined as $\mathbf{L} = \mathbf{D} - \mathbf{A}$. The symmetric normalized Laplacian matrix is defined as:

$$\mathbf{L} = \mathbf{I}_N - \mathbf{D}^{-1/2}\mathbf{A}\mathbf{D}^{-1/2} \tag{2}$$

The graph convolution operation can be expressed by Equation (3).

$$g_\theta \bullet x = \mathbf{U} g_\theta(\Lambda) \mathbf{U}^{\mathrm{T}} x \tag{3}$$

where $\mathbf{U}$ is the orthogonal matrix composed of the feature vectors of the Laplacian matrix L by column, and $g_\theta(\Lambda)$ is a diagonal matrix consisting of parameter θ, representing the parameter to be learned. The above is the general form of graph convolution, but Equation (3) is computationally intensive because the complexity of the eigenvector matrix $\mathbf{U}$ is $O(N^2)$. Therefore, Hammond et al. [32] showed that this process can be obtained by fitting a Chebyshev polynomial, as in Equation (4).

$$g_\theta \bullet x = \sum_{k=0}^{K} \theta_k T_k\left(\widetilde{L}\right) x \tag{4}$$

where $\widetilde{L} = \frac{2}{\lambda_{\max}} L - I_N$ and $\lambda_{\max}$ are the largest eigenvalues of L. θ_k is the vector of the Chebyshev coefficients. In order to reduce the computational effort, the literature [33] only calculates up to K = 1. $\lambda_{\max}$ is approximated as two; then, we have:

$$g_\theta \bullet x \approx \theta\left(\mathbf{I}_N + \mathbf{D}^{-1/2}\mathbf{A}\mathbf{D}^{-1/2}\right)x \tag{5}$$

In addition, self-normalization is introduced:

$$\mathbf{I}_N + \mathbf{D}^{-1/2}\mathbf{A}\mathbf{D}^{-1/2} \rightarrow \widetilde{\mathbf{D}}^{-1/2}\widetilde{\mathbf{A}}\widetilde{\mathbf{D}}^{-1/2} \tag{6}$$

where $\widetilde{\mathbf{A}} = \mathbf{A} + \mathbf{I}_N$,$\widetilde{\mathbf{D}}_{ii} = \sum_j \widetilde{\mathbf{A}}_{ij}$. Finally, the graph convolution is:

$$\mathbf{H}^{l+1} = \sigma\left(\widetilde{\mathbf{D}}^{-1/2}\widetilde{\mathbf{A}}\widetilde{\mathbf{D}}^{-1/2}\mathbf{H}^l\mathbf{W}^l\right) \tag{7}$$

2.4. SE Attention Mechanism

The attention mechanism can filter key information from the input images and enhance the accuracy of the model with a limited computational capability. Therefore, we applied the attention mechanism to the convolutional branch. For simplicity, we chose the SE attention mechanism. The purpose of the SE module is to obtain more important feature information by a weight matrix that provides different weights to different positions of the image from the perspective of the channel domain. The SE module consists of three steps. First, the compression operation performs feature compression from the spatial dimension to turn the feature of H × W × B into a 1 × 1 × B feature. Second, the excitation operation generates weights for each feature channel by introducing the w parameter. Finally, the weight outputs from the excitation block are considered as the importance of each feature channel after selection, and the weights of each channel calculated by the SE module are multiplied with the two-dimensional matrix of the corresponding channel of the original feature map to complete the rescaling of the original features in the channel dimension to highlight the important features. The SE module is shown in Figure 3.

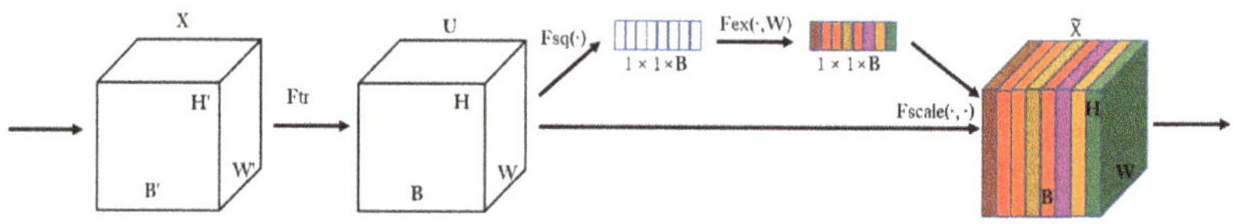

Figure 3. Schematic diagram of SE attention mechanism mainly divided into two operations of squeeze and excitation.

2.5. Superpixel Segmentation and Feature Conversion Module

The GCN can only be applied on graph-structured data, and in order to apply the GCN to hyperspectral images, the hyperspectral image needs to be constructed as a graph structure first. The simplest method is to consider each pixel of the image as each node of the graph structure, but this method leads to a huge computational cost. Therefore, it is common to first apply superpixel segmentation to the HSI.

Currently, common superpixel segmentation algorithms include SLIC [34], Quick-Shift [35], and Mean-Shift [36]. Among them, the SLIC algorithm assigns image pixels to the nearest clustering centers to form superpixels based on the distance and color difference between pixels. This method is computationally simple and has excellent results compared with other segmentation methods.

In general, the SLIC algorithm has only one parameter: the number of superpixels K. Suppose an image with M pixel is expected to be partitioned into K superpixel blocks; then, each superpixel block contains M/K pixels. Under the assumption that the length and width of each superpixel block are uniformly distributed, the length and width of each superpixel block can be defined as S, S = sqrt (M/K).

Second, in order to avoid the seed points falling on noisy points or line edges of the image and thus affecting the segmentation results, the positions of the seed points are also adjusted by recalculating the gradient values of the pixel points in the 3 × 3 neighborhood of each seed point and setting the new seed point to the minimum gradient in that neighborhood.

Finally, the new clustering centers are calculated iteratively by clustering. The pixel points in the 2S × 2S region around the centroid of each superpixel block are traversed. After that, each pixel is divided into the superpixel blocks closest to it; thus, an iteration is completed. The coordinates of the centroid of each superpixel block are recalculated and iterated, and convergence is usually completed in 10 iterations. Figure 4 represents the diagram of different number of superpixels in a image.

Figure 4. Schematic diagram of different number of superpixels for one image.

In this paper, the number of superpixel was not the same in each dataset but rather varied according to the total number of pixels in the dataset, for which the number of superpixels is specified as $K = (H \times W)/\beta$, where H and W are the length and width of the dataset, and β is a segmentation factor to control the number of superpixels, which is 100 in this paper.

It is worth noting that since each superpixel had a different number of pixels, and because the data structures of the two branches were different, the CNN branch and the GCN branch could not be fused directly. Inspired by [28], we applied a data transformation module that allowed the features obtained from the GCN branch to be fused with the features from the CNN branch, as shown in Figure 5.

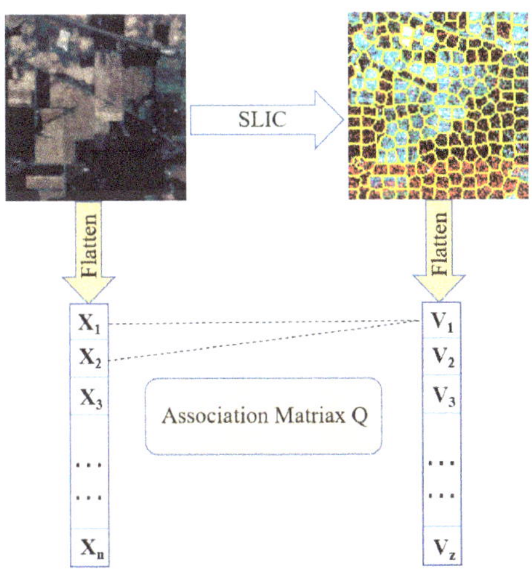

Figure 5. Pixel and superpixel data conversion module. This module allows features to be propagated between pixels and superpixels.

X_i denotes the i-th pixel in the flattened HSI and V_j denotes the average radiance of the pixels contained in the superpixels S_j. Let $\mathbf{Q} \in \mathbf{R}^{HW \times Z}$ be the association matrix between pixels and superpixels, where Z denotes the number of superpixels; then, we have:

$$\mathbf{Q}_{i,j} = \begin{cases} 1, \text{ if } \bar{\mathbf{X}}_i \in S_j \\ 0, \text{ if } \bar{\mathbf{X}}_i \notin S_j \end{cases}, \tag{8}$$

where $\bar{\mathbf{X}} = \text{Flatten}(\mathbf{X})$, $\mathbf{Q}_{i,j}$ denotes the value of $\mathbf{Q}$ at the association matrix, and $\bar{\mathbf{X}}_i$ denotes the i-th pixel in $\bar{\mathbf{X}}$. Finally, the feature conversion process can be represented by:

$$\mathbf{V} = Encoder(\mathbf{X};\mathbf{Q}) = \hat{\mathbf{Q}}^{\text{T}}\text{flatten}(\mathbf{X}), \tag{9}$$

$$\widehat{\mathbf{X}} = Decoder(\mathbf{V};\mathbf{Q}) = \text{reshape}(\mathbf{QV}), \tag{10}$$

where $\hat{\mathbf{Q}}$ denotes the normalized $\mathbf{Q}$ by column, and reshape$(\mathbf{QV})$ denotes restoring the spatial dimension of the flattened data. $\mathbf{V}$ denotes the nodes composed of superpixels and $\widehat{\mathbf{X}}$ denotes the feature converted back to Euclidean domains. In summary, features can be projected from the image space to the graph space using the graph encoder. Accordingly, the graph decoder can assign node features to pixels.

3. Experiments and Discussion

The overall accuracy (OA), average accuracy (AA), and kappa coefficient (kappa) were employed as the evaluation indices of the classification performance. The AA is equal to the sum of the number of correctly classified samples divided by the total number of samples. AA represents the average value of each accuracy for each category. The kappa coefficient is a reference metric that enables the calculation of overall consistency and classification consistency.

3.1. Experimental Datasets

To evaluate the effectiveness of the model, three commonly used hyperspectral datasets—Indian Pines (IP), Pavia University (PU), and Salinas (SA)—were used to evaluate the FCGN algorithm in this paper.

The IP dataset was acquired by the Airborne Visible/Infrared Imaging Spectrometer (AVIRIS) sensor over the northwestern Indian region. This dataset contains 145 × 145 pixels with 220 spectral bands ranging from 0.4 to 2.5 μm. After removing 20 water absorption and noisy spectral bands, 200 bands were reserved for the experiment. The land cover scene consists of 16 classes with 10,366 labeled pixels. The dataset was divided into training, validation, and test sets. For this dataset, the sample size was relatively small, and the number of samples of each class was extremely unbalanced. Overall, 5%, 1%, and 94% of samples per class were randomly selected for training, validation, and testing, respectively, as presented in Table 1.

The PU dataset was captured by the reflective optics system imaging spectrometer (ROSIS) sensor at Pavia University. This dataset contains 610 × 340 pixels with 125 spectral bands ranging from 0.43 to 0.86 μm. In total, 103 bands were utilized after discarding noisy bands. There are nine land cover categories in this dataset. Overall, 0.5%, 0.5%, and 99% of samples per class were selected for training, validation, and testing, respectively, as listed in Table 2.

The SA dataset is another commonly used dataset for hyperspectral image classification. It was recorded by the AVIRIS sensor over the Salinas Valley. This dataset contains 512 × 217 pixels with 224 spectral bands, and 204 bands were utilized. There are 16 land cover categories in this dataset. Because this dataset has a larger number of samples compared with Indian Pines, 0.5% of the labeled samples were selected as the training set and the validation set, and 99% of samples per class were randomly selected for testing, as listed in Table 3.

Table 1. The dataset division for each class of the IP dataset.

NO.	CLASS	Train	Val	Test
1	Alfalfa	3	1	50
2	Corn-notill	72	14	1348
3	Corn-mintil	42	8	784
4	Corn	12	2	220
5	Pasture	25	4	468
6	Trees/Grass	37	7	703
7	Pasture-mowed	2	1	23
8	Hay-windrowed	24	4	461
9	Oats	1	1	18
10	Soybeans-notill	48	9	911
11	Soybeans-mintill	123	24	2321
12	Soybeans-cleantill	31	6	577
13	Wheat	11	2	199
14	Woods	65	12	1217
15	Building-Grass	19	3	358
16	Stone-steelTowers	5	1	89
	Total	520	99	9747

Table 2. The dataset division for each class of the PU dataset.

NO.	CLASS	Train	Val	Test
1	Asphalt	34	34	6563
2	Meadows	94	94	18,461
3	Gravel	11	11	2077
4	Trees/Grass	16	16	3032
5	Metalsheets	7	7	1331
6	Baresoil	26	26	4977
7	Bitumen	7	7	1316
8	Bricks	19	19	3644
9	Shadows	5	5	937
	Total	219	219	42,338

Table 3. The dataset division for each class of the SA dataset.

NO.	CLASS	Train	Val	Test
1	Brocoli_green_weds_1	11	11	1987
2	Brocoli_green_weds_2	19	19	3688
3	Fallow	10	10	1956
4	Fallow_rough_plow	7	7	1380
5	Fallow_smooth	14	14	2650
6	Stubble	20	20	3919
7	Celery	18	18	3543
8	Grapes_untrained	57	57	11,157
9	Soil_vinyard_develop	32	32	6139
10	Corn_senesced_green_weeds	17	17	3244
11	Lettuce_romaine_4wk	6	6	1056
12	Lettuce_romaine_5wk	10	10	1907
13	Lettuce_romaine_6wk	5	5	906
14	Lettuce_romaine_7wk	6	6	1058
15	Vinyard_untrained	37	37	7194
16	Vinyard_vertical_trellis	10	10	1787
	Total	279	279	53,571

3.2. Experimental Settings

The proposed architecture consisted of three modules. The number of layers in the spectral dimension reduction module, graph convolution branch, and convolution branch were all set to two. The spectral dimension reduction modules started with two 1×1 convolution layers (128 filters and 128 filters). The size of the convolution kernels in the CNN branch was set to 3×3 (128 and 64 filters), and the sample output dimensions in the GCN branch were set to 128 and 64. We used the Adam optimizer to train our model with a learning rate of 0.001, and the training epoch was set to 500. The number of superpixels for each dataset was set to $1/100$ of the number of pixels.

The proposed algorithm is implemented in Python 3.8.12 and Pytorch1.1.0 using Python language. The hardware used for training is an i7-10750H CPU and a NVIDIA GeForce RTX 2060s GPU.

3.3. Classification Results

To verify the performance of the model, several advanced HSI classification methods were selected for comparison with this model, including 3D-CNN [37], GCN [24], MiniGCN [22], HybirdSN [18], DBDA [19], and MDGCN [27]. The number of training samples and test samples selected for each method were the same, and the hyperparameters were the same as in the original paper. The classification accuracies of the different methods on each dataset are shown in Tables 4–6 the best results in each class were bolded and the classification maps obtained by these methods are illustrated in Figures 6–8; the experimental results are the average of five experiments. It is worth noting that although we have minimized the risk of data leakage, the issue may still exist and affect the classification results.

Table 4 shows the classification results of the different models on the IP dataset. The lack of training samples on the IP dataset and the imbalance in the number of samples from different categories made classification challenging, but our classification method obtained the best classification results. It can be observed that the classification accuracy obtained by the 3D-CNN was lower than other methods, which might have been due to the fact that the 3D-CNN had more parameters, but the number of training samples was small in this experiment. In addition, it did not take full advantage of the relationship information contained in the samples, which eventually led to poor classification results. HybirdSN combines 3D-CNN layers with 2D-CNN layers, which has a stronger feature representation capability by combining spatial and spectral information and a lower number of parameters, but the accuracy was still lower in the case of a small number of samples. DBDA contains two branches to obtain spatial–spectral features, respectively, and introduces the attention mechanism and eventually achieved better classification results than HybirdSN. The GCN-based classification method can generally obtain better classification results with a smaller number of samples. MiniGCN adds a convolutional branch and adopts a small batch strategy compared with GCN. MiniGCN achieved better classification results but did not take into account the different importance of different features. In contrast, the FCGN obtained the best classification results, which was greatly due to the design of two branches to obtain complementary features. The graph convolution branch based on superpixel segmentation can obtain large-scale irregular features of the image and the relationship between different nodes, reducing the classification error caused by noise. The convolutional neural network with the added attention mechanism can acquire regular image features at a small image scale and generate detailed edge features, which complements the smooth features acquired by the superpixel segmentation-based graph convolution branch to obtain better classification results on both large and small scales. The convolution branch was able to process the local fine area to obtain the detailed features of the image; due to the misclassification of pixels, the classification result of the convolution branch contained more noise. By fusing the features of the two branches, the influence of noise on the classification results was greatly reduced. In terms of running time, the FCGN had a medium running time compared to the other comparison algorithms due to the use of the depth-separable convolution algorithm.

Table 5 shows the classification results of the different models on the Pavia University dataset. It can be observed that the classification results of each algorithm slightly improved relative to those of the IP dataset, which may be because of the fewer sample classes in the PU dataset and because the number of samples in each class was similar. It is remarkable that the DBDA obtained better classification results than HybirdSN, which may have resulted from the two-branch structure of HybirdSN and the attention mechanism. The FCGN performed better than the compared methods, with an OA of 98.99%, because the FCGN could fully exploit the features of the samples. Moreover, the addition of the attention mechanism also improved the classification results. The runtime of the FCGN algorithm slightly increased compared to some comparison algorithms, but considering the competitive classification results of this algorithm and the short testing time, the increase in the runtime is worth it.

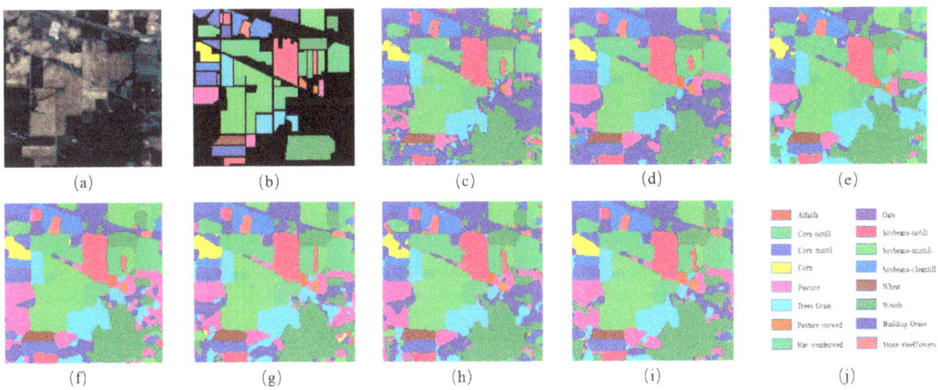

Figure 6. The classification maps for Indian Pines. (**a**) False-color image; (**b**) ground truth; (**c**) 3DCNN; (**d**) GCN; (**e**) MiniGCN; (**f**) HybirdSN; (**g**) DBDA; (**h**) MDGCN; (**i**) FCGN; (**j**) Figure legend.

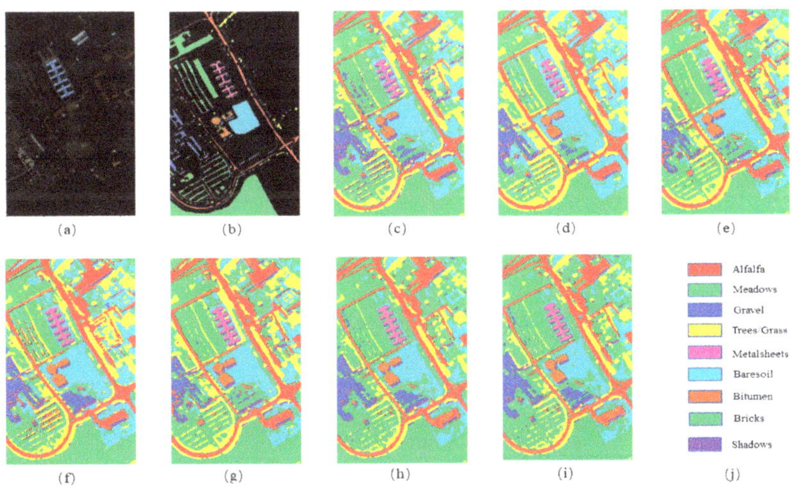

Figure 7. The classification maps for Pavia University. (**a**) False-color image; (**b**) ground truth; (**c**) 3DCNN; (**d**) GCN; (**e**) MiniGCN; (**f**) HybirdSN; (**g**) DBDA; (**h**) MDGCN; (**i**) FCGN; (**j**) Figure legend.

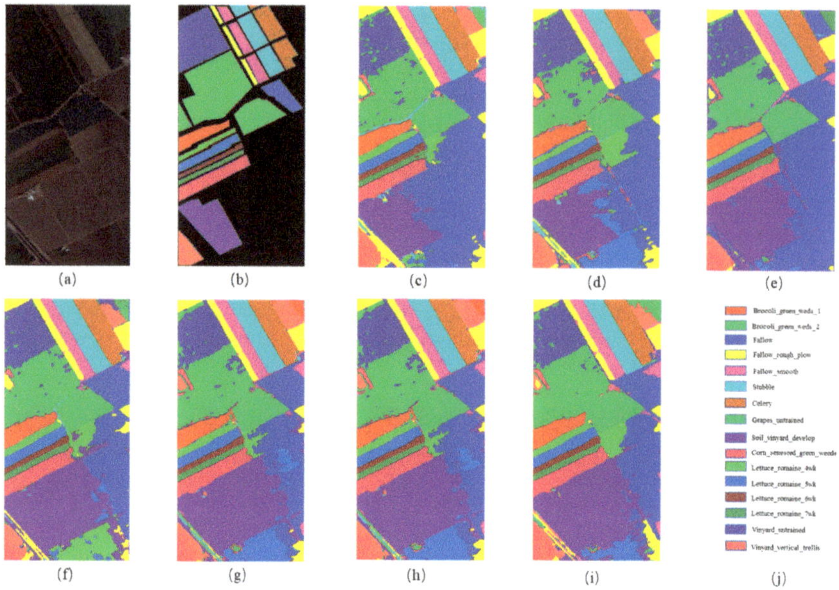

Figure 8. The classification maps for Salinas. (**a**) False-color image; (**b**) ground truth; (**c**) 3DCNN; (**d**) GCN; (**e**) MiniGCN; (**f**) HybirdSN; (**g**) DBDA; (**h**) MDGCN; (**i**) FCGN; (**j**) Figure legend.

Table 4. Classification results of the IP dataset.

Class	3DCNN	GCN	MiniGCN	HybirdSN	DBDA	MDGCN	FCGN
1	80.00	94.66	93.73	83.44	99.51	88.88	**100.00**
2	90.16	70.37	72.54	80.38	93.55	96.16	**96.51**
3	80.05	66.89	76.83	82.21	94.11	94.17	**98.46**
4	86.54	86.88	98.16	**99.19**	96.42	92.59	97.36
5	84.43	89.27	93.40	96.47	97.64	**96.79**	97.13
6	79.12	93.38	92.64	98.81	96.23	99.50	**100.00**
7	67.22	92.91	88.48	86.89	96.66	**96.82**	92.89
8	90.87	96.14	97.59	98.04	91.35	95.72	**98.68**
9	70.00	**100.00**	99.73	73.11	89.37	99.98	**100.00**
10	79.11	85.37	75.98	90.41	70.30	85.70	**97.30**
11	90.81	68.45	79.42	74.23	90.32	96.06	**98.63**
12	72.50	78.90	79.51	91.00	97.38	**98.88**	95.17
13	70.88	**99.84**	98.93	71.88	97.99	97.21	99.59
14	85.83	85.12	87.88	98.33	98.34	**99.93**	99.33
15	92.30	82.67	89.82	94.48	**96.25**	96.17	95.71
16	69.22	97.41	**100.00**	70.22	86.66	94.82	97.75
OA(%)	78.47	86.67	88.67	87.99	94.55	97.62	**98.78**
AA(%)	80.57	86.77	89.04	86.82	93.26	95.58	**97.80**
Kappa	80.70	84.38	88.39	87.44	94.01	95.49	**97.99**
Train time(s)	250.11	59.02	342.55	220.99	298.32	1204.15	204.50
Test time(s)	15.04	5.60	15.38	14.51	21.77	20.33	5.92

Table 5. Classification results of the PU dataset.

Class	3DCNN	GCN	MiniGCN	HybirdSN	DBDA	MDGCN	FCGN
1	80.85	77.26	86.22	94.10	96.36	**99.00**	98.41
2	80.49	76.97	92.21	94.36	99.11	98.21	**99.91**

Table 5. *Cont.*

Class	3DCNN	GCN	MiniGCN	HybirdSN	DBDA	MDGCN	FCGN
3	69.77	69.19	86.13	82.40	90.32	86.81	**97.35**
4	95.99	90.88	92.06	95.27	**97.99**	94.55	97.96
5	91.30	94.27	95.11	95.77	99.01	**99.76**	99.73
6	90.57	92.98	90.34	92.44	97.55	**99.80**	99.00
7	80.21	82.81	88.99	89.06	94.37	98.07	**99.89**
8	89.73	86.91	82.77	80.04	88.94	96.92	**98.68**
9	91.12	95.55	92.87	**99.11**	98.39	98.38	98.41
OA(%)	86.33	85.41	89.67	92.99	97.22	98.22	**98.99**
AA(%)	85.56	85.20	89.63	92.81	95.78	96.83	**98.81**
Kappa	85.21	80.37	87.09	89.98	96.72	97.27	**97.90**
Train time(s)	131.44	251.59	1058.37	122.61	145.88	3265.31	1283.37
Test time(s)	88.21	17.33	50.15	65.48	118.37	57.29	38.94

Table 6. Classification results of the SA dataset.

Class	3DCNN	GCN	MiniGCN	HybirdSN	DBDA	MDGCN	FCGN
1	88.31	98.64	96.19	99.34	99.62	**100.00**	99.74
2	88.35	98.99	99.02	98.61	99.14	**100.00**	**100.00**
3	82.01	73.84	86.32	94.38	97.45	99.16	**100.00**
4	84.02	99.49	98.32	97.04	94.77	**100.00**	99.82
5	87.76	**99.66**	96.35	98.24	98.02	94.32	97.71
6	91.42	99.97	99.55	99.03	**99.99**	99.98	98.98
7	90.94	93.54	98.54	96.89	97.62	98.85	**99.99**
8	80.07	**94.70**	91.40	86.55	87.35	86.18	94.25
9	94.88	**100.00**	99.74	99.12	89.37	99.97	99.97
10	88.76	70.82	84.25	89.89	89.57	93.84	**96.68**
11	83.62	80.85	83.51	91.23	90.32	98.29	**99.01**
12	87.99	95.05	94.99	97.92	97.38	94.98	**99.99**
13	72.15	94.94	89.47	**99.46**	98.99	97.00	99.36
14	73.05	97.82	98.94	97.66	95.69	97.12	**99.10**
15	91.34	54.25	67.39	81.47	86.77	**95.92**	94.56
16	92.96	65.60	64.61	**99.28**	96.34	98.65	98.67
OA(%)	86.30	91.47	91.76	96.25	92.55	96.80	**98.69**
AA(%)	86.10	90.92	90.53	95.38	94.90	97.14	**98.61**
Kappa	85.09	88.01	88.39	92.09	93.37	95.34	**97.18**
Train time(s)	153.09	269.04	1094.67	146.96	176.55	3377.41	1357.15
Test time(s)	93.37	23.02	57.47	72.72	120.17	65.33	42.46

Table 6 shows the classification results of the different models on the Salinas dataset. We can see that the FCGN was superior to other methods in terms of the OA, AA, and Kappa coefficient, proving the effectiveness of the FCGN algorithm again. By observing the classification results of Grapes untrained and Vineyard untrained ground features in the Salinas dataset, the classification accuracy is relatively low; this is largely because the two ground features are mixed together. We can see from Figure 8 that the FCGN method had fewer misclassified pixels than other methods and was more accurate for classifying large-scale regions.

4. Discussion

4.1. Influence of Label Ratio

To evaluate the generalizability and robustness of the proposed FCGN and other methods, we set the number of training samples per class from 5 to 25 with an interval of 5. Figure 9 shows the OA obtained by the different methods on the three datasets. It can be observed that the proposed FCGN achieved a better classification accuracy than other methods, and the classification accuracy of each method improved as the number of training samples increased, which proves the great robustness of the proposed FCGN.

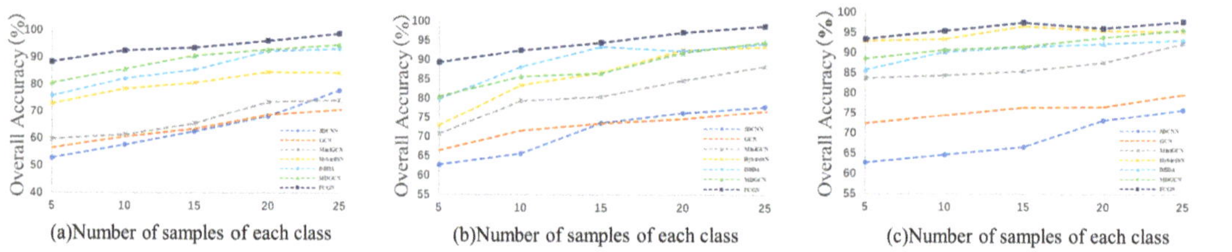

Figure 9. The classification performance of each method with different training set ratios. (**a**) India Pines; (**b**) Pavia University; (**c**) Salinas.

4.2. Influence of Segmentation Factor

The larger the segmentation factor, the smaller the number of superpixels; therefore, the larger the size of the superpixels, the more they can preserve larger objects and suppress more noise. Conversely, the smaller the segmentation factor, the larger the number of superpixels; therefore, the smaller the size of the segmentation map obtained, the more smaller objects which can be preserved and noise which can be contained.

In order to investigate the influence of the number of superpixel blocks on the performance of the FCGN, in this section, the segmentation coefficients were set to 50, 100, 150, and 200, and the influence of different segmentation factors on the classification accuracy of the FCGN was tested on the three datasets, as shown in Figure 10. It can be seen that the classification accuracy of the FCGN on the IP dataset decreased with the increase in the segmentation coefficients, which was due to the fact that the samples in the IP dataset were of a smaller scale. The size of the superpixel was too large, which missed the sample detail information. The OA of the PU dataset was similar when the segmentation factor was 50 and 100, and the highest accuracy was achieved when the factor was 100. The sample scale on the SA dataset was much larger, so as the size of the superpixels increased, the classification results did not decrease. Instead, more noise effects were removed, increasing the accuracy. However, when the segmentation factor reached a certain size, the classification accuracy was bound to decrease gradually. In order to prevent the classification map from being smooth and missing too much detailed information, the segmentation factor was set to 100 in this chapter.

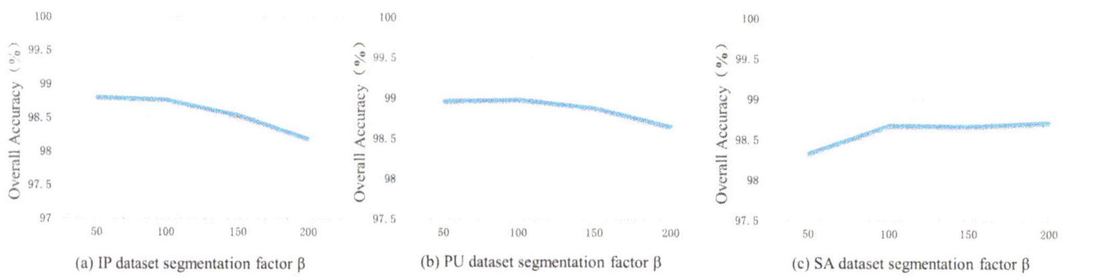

Figure 10. The classification performance with different segmentation factors. (**a**) India Pines; (**b**) Pavia University; (**c**) Salinas.

4.3. Ablation Study

The proposed FCGN mainly contains a graph convolutional branch based on superpixel segmentation and a convolutional branch with an added attention mechanism. To further validate the contribution of the two branches, we tested the OA of the two branches on three datasets separately. In addition, we tested the impact of the attention mechanism. We can intuitively see from Table 7 that the overall classification accuracy decreased when any branch was missing, which proves that the complementary features obtained by combining the graph convolution branch and the convolution branch can indeed improve the classification performance. We can also observe that the addition of attention mechanisms resulted in some improvement in the classification results, which indicates that by adding appropriate attention mechanisms to the network, we can obtain the importance of different features and capture long-range features and high-level features to improve the classification results.

Table 7. OA(%) indices of the ablation experiment of the FCGN.

Dataset	Branch1	Branch2	Without SE	FCGN
Indian Pines	93.58	94.69	96.44	98.78
Pavia University	95.65	93.41	97.98	98.99
Salinas	97.33	94.50	97.88	98.69

5. Conclusions

To reduce the complexity of graph structure construction, superpixel segmentation is often performed on an HSI first; however, superpixel segmentation processing leads to similar features within each superpixel node, resulting in a lack of local details in the classification map. To solve the above problems, a new hyperspectral image classification algorithm, the FCGN, was proposed in this paper, in which a graph convolutional network based on superpixel segmentation was fused with an attentional convolutional network for feature fusion, a GCN based on superpixel segmentation was used to extract superpixel-level features, an attentional convolutional network was used to extract local detail features, and, finally, the obtained complementary features were used to improve the classification results. In order to verify the effectiveness of the algorithm, experiments were conducted on three datasets and compared with some excellent algorithms. The experimental results show that the FCGN achieved a better classification performance. Although the FCGN achieved better classification results, there are still some shortcomings. In particular, this paper did not consider the variability of different neighbor nodes during the construction of the graph structure which may limit the ability of the model. In addition, only a simple feature splicing fusion method was used in this paper, so the construction of the graph structure and new fusion mechanism will be further explored in subsequent research.

Author Contributions: Conceptualization, L.G.; methodology, L.G.; software, L.G.; validation, L.G., S.X. and C.H.; formal analysis, L.G.; investigation, L.G.; resources, L.G., S.X. and C.H.; data curation, L.G.; writing—original draft preparation, L.G.; writing—review and editing, L.G. and Y.Y.; visualization, L.G. and S.X.; supervision, C.H. and Y.Y.; project administration, C.H.; funding acquisition, C.H. and Y.Y.; All authors have read and agreed to the published version of the manuscript.

Funding: This research was funded by Advanced technology industrialization of science and technology cooperation between Jilin Province and Chinese Academy of Sciences (grant number 2020SYHZ0028).

Data Availability Statement: Indian Pines dataset, Pavia University dataset, Salinas dataset (https://www.ehu.eus/ccwintco/index.php/Hyperspectral_Remote_Sensing_Scenes, accessed on 11 June 2022).

Conflicts of Interest: The authors declare no conflict of interest.

References

1. Paoletti, M.E.; Haut, J.M.; Plaza, J.; Plaza, A. Deep Learning Classifiers for Hyperspectral Imaging: A Review. *ISPRS J. Photogramm. Remote Sens.* **2019**, *158*, 279–317. [CrossRef]
2. Fauvel, M.; Tarabalka, Y.; Benediktsson, J.A.; Chanussot, J.; Tilton, J.C. Advances in Spectral-Spatial Classification of Hyperspectral Images. *Proc. IEEE* **2013**, *101*, 652–675. [CrossRef]
3. Ma, Z.; Jiang, Z.; Zhang, H. Hyperspectral Image Classification Using Feature Fusion Hypergraph Convolution Neural Network. *IEEE Trans. Geosci. Remote Sens.* **2021**, *60*, 5517314. [CrossRef]

4. Pan, B.; Xu, X.; Shi, Z.; Zhang, N.; Luo, H.; Lan, X. DSSNet: A Simple Dilated Semantic Segmentation Network for Hyperspectral Imagery Classification. *IEEE Geosci. Remote Sens. Lett.* **2020**, *17*, 1968–1972. [CrossRef]
5. Tan, Y.; Lu, L.; Bruzzone, L.; Guan, R.; Chang, Z.; Yang, C. Hyperspectral Band Selection for Lithologic Discrimination and Geological Mapping. *IEEE J. Sel. Top. Appl. Earth Obs. Remote Sens.* **2020**, *13*, 471–486. [CrossRef]
6. Dong, Y.; Liang, T.; Zhang, Y.; Du, B. Spectral–Spatial Weighted Kernel Manifold Embedded Distribution Alignment for Remote Sensing Image Classification. *IEEE Trans. Cybern.* **2021**, *51*, 3185–3197. [CrossRef]
7. Wu, P.; Cui, Z.; Gan, Z.; Liu, F. Three-Dimensional ResNeXt Network Using Feature Fusion and Label Smoothing for Hyperspectral Image Classification. *Sensors* **2020**, *20*, 1652. [CrossRef]
8. Farooque, G.; Xiao, L.; Yang, J.; Sargano, A.B. Hyperspectral Image Classification via a Novel Spectral–Spatial 3D ConvLSTM-CNN. *Remote Sens.* **2021**, *13*, 4348. [CrossRef]
9. Peng, J.; Zhou, Y.; Chen, C.L.P. Region-Kernel-Based Support Vector Machines for Hyperspectral Image Classification. *IEEE Trans. Geosci. Remote Sens.* **2015**, *53*, 4810–4824. [CrossRef]
10. Belgiu, M.; Drăguţ, L. Random Forest in Remote Sensing: A Review of Applications and Future Directions. *ISPRS J. Photogramm. Remote Sens.* **2016**, *114*, 24–31. [CrossRef]
11. Zhang, M.-L.; Zhou, Z.-H. ML-KNN: A Lazy Learning Approach to Multi-Label Learning. *Pattern Recognit.* **2007**, *40*, 2038–2048. [CrossRef]
12. Rish, I. An Empirical Study of the Naive Bayes Classifier. In Proceedings of the IJCAI 2001 Workshop on Empirical Methods in Artificial Intelligence, Seattle, WA, USA, 4 August 2001; Volume 3, pp. 41–46.
13. Wan, A.; Dunlap, L.; Ho, D.; Yin, J.; Lee, S.; Jin, H.; Petryk, S.; Bargal, S.A.; Gonzalez, J.E. NBDT: Neural-Backed Decision Trees. *arXiv* **2020**. [CrossRef]
14. Audebert, N.; Le Saux, B.; Lefevre, S. Deep Learning for Classification of Hyperspectral Data: A Comparative Review. *IEEE Geosci. Remote Sens. Mag.* **2019**, *7*, 159–173. [CrossRef]
15. Paoletti, M.E.; Haut, J.M.; Plaza, J.; Plaza, A. A New Deep Convolutional Neural Network for Fast Hyperspectral Image Classification. *ISPRS J. Photogramm. Remote Sens.* **2018**, *145*, 120–147. [CrossRef]
16. Acción, Á.; Argüello, F.; Heras, D.B. Dual-Window Superpixel Data Augmentation for Hyperspectral Image Classification. *Appl. Sci.* **2020**, *10*, 8833. [CrossRef]
17. Hu, W.; Huang, Y.; Wei, L.; Zhang, F.; Li, H. Deep Convolutional Neural Networks for Hyperspectral Image Classification. *J. Sens.* **2015**, *2015*, e258619. [CrossRef]
18. Roy, S.K.; Krishna, G.; Dubey, S.R.; Chaudhuri, B.B. HybridSN: Exploring 3-D–2-D CNN Feature Hierarchy for Hyperspectral Image Classification. *IEEE Geosci. Remote Sens. Lett.* **2020**, *17*, 277–281. [CrossRef]
19. Li, R.; Zheng, S.; Duan, C.; Yang, Y.; Wang, X. Classification of Hyperspectral Image Based on Double-Branch Dual-Attention Mechanism Network. *Remote Sens.* **2020**, *12*, 582. [CrossRef]
20. Zhong, Z.; Li, J.; Luo, Z.; Chapman, M. Spectral–Spatial Residual Network for Hyperspectral Image Classification: A 3-D Deep Learning Framework. *IEEE Trans. Geosci. Remote Sens.* **2018**, *56*, 847–858. [CrossRef]
21. Wang, L.; Wang, X. Dual-Coupled CNN-GCN-Based Classification for Hyperspectral and LiDAR Data. *Sensors* **2022**, *22*, 5735. [CrossRef]
22. Hong, D.; Gao, L.; Yao, J.; Zhang, B.; Plaza, A.; Chanussot, J. Graph Convolutional Networks for Hyperspectral Image Classification. *IEEE Trans. Geosci. Remote Sens.* **2021**, *59*, 5966–5978. [CrossRef]
23. Wu, Z.; Pan, S.; Chen, F.; Long, G.; Zhang, C.; Philip, S.Y. A Comprehensive Survey on Graph Neural Networks. *IEEE Trans. Neural Netw. Learn. Syst.* **2020**, *32*, 4–24. [CrossRef]
24. Kipf, T.; Welling, M. Semi-Supervised Classification with Graph Convolutional Networks. *arXiv* **2017**. [CrossRef]
25. Full Article: Graph Inductive Learning Method for Small Sample Classification of Hyperspectral Remote Sensing Images. Available online: https://www.tandfonline.com/doi/full/10.1080/22797254.2021.1901064 (accessed on 28 January 2023).
26. Yu, W.; Wan, S.; Li, G.; Yang, J.; Gong, C. Hyperspectral Image Classification With Contrastive Graph Convolutional Network. *IEEE Trans. Geosci. Remote Sens.* **2023**, *61*, 5503015. [CrossRef]
27. Wan, S.; Gong, C.; Zhong, P.; Du, B.; Zhang, L.; Yang, J. Multiscale Dynamic Graph Convolutional Network for Hyperspectral Image Classification. *IEEE Trans. Geosci. Remote Sens.* **2020**, *58*, 3162–3177. [CrossRef]
28. Liu, Q.; Xiao, L.; Yang, J.; Wei, Z. CNN-Enhanced Graph Convolutional Network With Pixel- and Superpixel-Level Feature Fusion for Hyperspectral Image Classification. *IEEE Trans. Geosci. Remote Sens.* **2021**, *59*, 8657–8671. [CrossRef]
29. Chollet, F. Xception: Deep Learning with Depthwise Separable Convolutions. In Proceedings of the 2017 IEEE Conference on Computer Vision and Pattern Recognition (CVPR), Honolulu, HI, USA, 21–26 July 2017; pp. 1800–1807.
30. Hu, J.; Shen, L.; Sun, G. Squeeze-and-Excitation Networks. In Proceedings of the 2018 IEEE/CVF Conference on Computer Vision and Pattern Recognition, Salt Lake City, UT, USA, 18–23 June 2018; pp. 7132–7141.
31. Shuman, D.I.; Narang, S.K.; Frossard, P.; Ortega, A.; Vandergheynst, P. The Emerging Field of Signal Processing on Graphs: Extending High-Dimensional Data Analysis to Networks and Other Irregular Domains. *IEEE Signal Process. Mag.* **2013**, *30*, 83–98. [CrossRef]
32. Hammond, D.K.; Vandergheynst, P.; Gribonval, R. Wavelets on Graphs via Spectral Graph Theory. *Appl. Comput. Harmon. Anal.* **2011**, *30*, 129–150. [CrossRef]

33. Defferrard, M.; Bresson, X.; Vandergheynst, P. Convolutional Neural Networks on Graphs with Fast Localized Spectral Filtering. *arXiv* **2017**. [CrossRef]
34. Achanta, R.; Shaji, A.; Smith, K.; Lucchi, A.; Fua, P.; Süsstrunk, S. SLIC Superpixels Compared to State-of-the-Art Superpixel Methods. *IEEE Trans. Pattern Anal. Mach. Intell.* **2012**, *34*, 2274–2282. [CrossRef]
35. Zhang, S.; Ma, Z.; Zhang, G.; Lei, T.; Zhang, R.; Cui, Y. Semantic Image Segmentation with Deep Convolutional Neural Networks and Quick Shift. *Symmetry* **2020**, *12*, 427. [CrossRef]
36. Comaniciu, D.; Meer, P. Mean Shift Analysis and Applications. In Proceedings of the the Seventh IEEE International Conference on Computer Vision, Kerkyra, Greece, 20–27 September 1999; Volume 2, pp. 1197–1203.
37. Chen, Y.; Jiang, H.; Li, C.; Jia, X.; Ghamisi, P. Deep Feature Extraction and Classification of Hyperspectral Images Based on Convolutional Neural Networks. *IEEE Trans. Geosci. Remote Sens.* **2016**, *54*, 6232–6251. [CrossRef]

applied sciences

MDPI

Article

A Novel Constant Damping and High Stiffness Control Method for Flexible Space Manipulators Using Luenberger State Observer

Tao Yang [1,2,3,4,5,*], Fang Xu [1,2], Si Zeng [4,5], Shoujun Zhao [4,5], Yuwang Liu [1,2] and Yanbo Wang [4,5]

1 State Key Laboratory of Robotics, Shenyang Institute of Automation, Chinese Academy of Sciences, Shenyang 110016, China
2 Institutes for Robotics and Intelligent Manufacturing, Chinese Academy of Sciences, Shenyang 110169, China
3 University of Chinese Academy of Sciences, Beijing 100049, China
4 Beijing Institute of Precision Mechatronics and Controls, Beijing 100076, China
5 Laboratory of Aerospace Servo Actuation and Transmission, Beijing 100076, China
* Correspondence: yangtao@sia.cn; Tel.: +86-010-88537922

Abstract: This paper presents a novel control strategy for transferring large inertia loads using flexible space manipulators in orbit. The proposed strategy employs a Luenberger state observer and damping-stiffness controller to address issues of large tracking error and vibration. A comprehensive joint dynamics model is developed to identify the main sources of disturbance, and a Luenberger state observer is designed to estimate unmeasurable transmission deformation. Transmission stiffness and load inertia perturbations are identified based on the estimated results. By adjusting velocity damping and the gain of the forward channel, perturbations are suppressed to maintain optimal system damping and stiffness. Simulation and physical experiments demonstrate the effectiveness of the algorithm, with simulation experiments showing smoother joint output characteristics and minimal vibration under large load inertia changes, and a 97% reduction in internal deformation. Physical experiments demonstrate improved joint dynamic command tracking performance, with an 88% reduction in position tracking error. The algorithm provides a practical and efficient approach for transferring large inertia scientific payloads in space.

Keywords: space manipulator; flexible joint; dynamic; Luenberger observer; automatic control

Citation: Yang, T.; Xu, F.; Zeng, S.; Zhao, S.; Liu, Y.; Wang, Y. A Novel Constant Damping and High Stiffness Control Method for Flexible Space Manipulators Using Luenberger State Observer. *Appl. Sci.* **2023**, *13*, 7954. https://doi.org/10.3390/app13137954

Academic Editor: Subhas Mukhopadhyay

Received: 3 May 2023
Revised: 6 June 2023
Accepted: 9 June 2023
Published: 7 July 2023

1. Introduction

The space manipulator is a powerful tool for on-orbit space services due to its large working range, high positioning accuracy, and strong load capacity [1]. However, the joint—a key component of the space manipulator—shows obvious flexibility characteristics due to the application of harmonic reducers that increase output torque. During the transfer of a target object for a science experiment in orbit, nonlinear changes in the transmission stiffness of the joint harmonic reducer and the load inertia can cause perturbations in the dynamic model, leading to out-of-tolerance joint trajectory tracking and flexible vibration, as shown in Figure 1. Reducing motion speed and modal analysis have been used to avoid violent flexible vibration of the space manipulator. In addition, the dynamic control of flexible joints is a point of difficulty in the control of exoskeleton robots, medical robots, and other equipment that are in direct contact with the human body. Therefore, the high-performance control of flexible joint robots has always been a research hotspot in the industry [2].

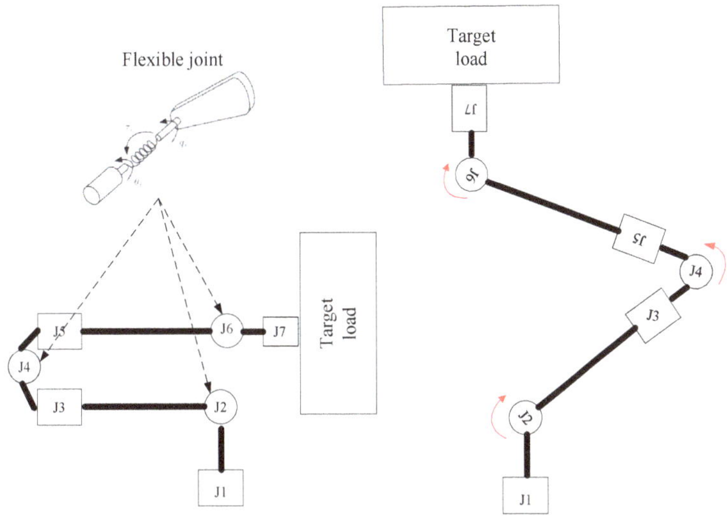

Figure 1. Schematic of a space manipulator with flexible joints.

1.1. Related Works

Scholars have proposed various control strategies, such as full-state feedback control (FSFB), backstep control, compensation control based on state observer, synovial membrane control, and neural network control, to address the motion control of flexible joints with nonlinear time-varying load inertia [3–5]. The backstep method compensates for the interference of the contained severe uncertainty by building a state feedback controller [6]. The state observer is capable of compensating for model disturbances and thereby reducing their impact on joint control. Jian Li, for a class of uncertain flexible joint robots with variable stiffness actuators (VSAs), explicitly constructed a state feedback controller through backstepping and disturbance learning mechanism [7], and designed a key switching mechanism to adjust online parameters in the controller, which achieved good control effect. However, when the order of the controlled system is high, the calculation of the partial derivative of the virtual control quantity will have the disadvantage of differential explosion [8], which is not suitable for space occasions with limited computing resources [9]. The FSFB method has a simple structure and a small calculation amount; however, due to the general effect on the flexible time-varying inertial control system, the flexible vibration phenomenon occurs easily when the gain is large [10–12]. Feedback linearization control is an effective means to solve the nonlinear links in the control system; however, this method relies on high-precision mathematical models, and the control effect is greatly reduced once the model is biased [12]. The synovial membrane controller has the characteristics of fast response and strong robustness to disturbances. However, when the state trajectory reaches the sliding modal plane, it does not easily slide strictly along the sliding mode towards the equilibrium point. Rather, it crosses back and forth on both sides to approach the equilibrium point, resulting in jitter [13]. The intelligent algorithm based on a neural network also has the problem of excessive computation and is difficult to apply in space engineering [14].

The state observer can compensate the model to reduce the impact of model disturbance on the control system, and there is no differential explosion problem. Z. Bowen et al. propose an accurate estimation of the uncertainty of the velocity state of the manipulator and the stiffness of the joint by the Extended State Observer (ESO) [15]. M. J. Kim et al. use the inverse method to generate adaptive controllers, in which dynamic uncertainty is compensated for by a radial basis function neural network (RBFNN) and joint-stiffness uncertainty is eliminated by ESO estimation. The articulated motor side interference observer is proposed and fed back to the proportion differential controller, which, in turn, introduces a nominally stable controller input [16]. A. Fagiolini et al. achieve linear state observation of the plastic deformation of the flexible joint reducer by measuring the joint and motor position [17]. L. Dou et al. propose a terminal sliding mode control (TSMC) strategy based on fuzzy interference observer to target liquid shaking interference of a flexible liquid-filled spacecraft [18]. C. Pan et al. design a state observer for single-degree-of-freedom flexible joint manipulator disturbance to compensate for interference, turning the nonlinear system into a linear integral sequence system [19]. Y. Lin et al. propose a linear expansion state observer for industrial robotic arms to estimate joint flexibility and torque [20]. Yong Nie developed two low-order extended state observers to handle the external load force and the impact of the pressure compensator in a high-order nonlinear hydraulic system. Additionally, backstepping methods were designed to ensure robust stability of the system [21]. Due to the difficulty of modeling nonlinear and flexible environments, some scholars propose the use of artificial neural networks to control the flexible joints [22–25], and W. Quanwei et al. propose a new neural network based on a disturbance observer-based integrated sliding mode controller that is only numerically simulated [26]. However, the state observer method can only realize the estimation of the internal state of the model, and must be combined with other closed-loop control methods to build a stable control system. Therefore, it is very important to choose a suitable observer and control strategy.

1.2. Motivations and Contributions

This paper investigates and proposes a new high-performance tracking control method for flexible space manipulator joints with a simple model structure, moderate calculation requirements, and practical applications. This method maintains the damping and stiffness of the control system at the desired state, despite simultaneous disturbances from external load inertia and internal transmission stiffness in the joint. To achieve this, the paper designs a Luenberger state observer to estimate the internal flexible deformation of the joint reducer and observe real-time transmission stiffness and load inertia. By simplifying the model, the paper analyzes the influence of parameter changes on the control characteristics, improving on traditional three-loop feedback control algorithms that are sensitive to changes in model parameters. The paper introduces speed dampers and feed-forward gain regulators to keep the damping and stiffness of the flexible joint control system at the desired state, Henceforth, this strategy is referred to as Constant Damping and High Stiffness (CDHS) algorithm. Thereby improving the joint control system's robustness and the ability of the flexible space manipulator to transfer large-inertia scientific payloads in orbit. The paper's contributions to the control of flexible space manipulators include:

- A comprehensive state-space dynamic modeling method for flexible joints is proposed, and the perturbation parameters of the flexible joint model are comprehensively analyzed.
- A Luenberger state observer [27] based on the joint dynamics model is constructed, which enables the accurate estimation of the internal flexible deformation of a joint that cannot be directly measured.
- The dynamic identification of joint transmission stiffness and external inertia is realized according to the estimated value of joint flexible deformation and joint physical motion characteristics.
- A control algorithm for flexible joints with constant damping and high stiffness is proposed, based on a dual-adjustment mechanism of speed damping and feed-forward gain. The algorithm is proposed and its effectiveness is verified by simulation experiments and physical experiments.

The remainder of this paper is organized as follows: In Section 2, we present the problem formulation and the comprehensive joint dynamics model. Section 3 describes the design of the CDHS controller with a Luenberger observer. In Section 4, we compare and discuss the simulation results and physical experiment results. Finally, Section 5 contains the conclusions of this study.

2. Dynamic Modeling of Flexible Joints

Due to the limitations of space transportation and the usage environment, space manipulators are longer but lighter in weight compared to industrial manipulators, resulting in significantly lower structural rigidity [28]. The flexible space manipulator studied in this paper is illustrated in Figure 2a. It has an arm length of approximately 3300 mm, with a weight of only 60 kg. The joint structure of the manipulator, as shown in Figure 2b, is driven by a brushless DC motor with a harmonic reducer as the transmission part. Both the motor and the output shafts of the reducer are equipped with position sensors. The reducer has the lowest stiffness among the manipulator's components. Hence, this paper only considers the flexibility of the space manipulator caused by the joint harmonic reducer while disregarding other factors.

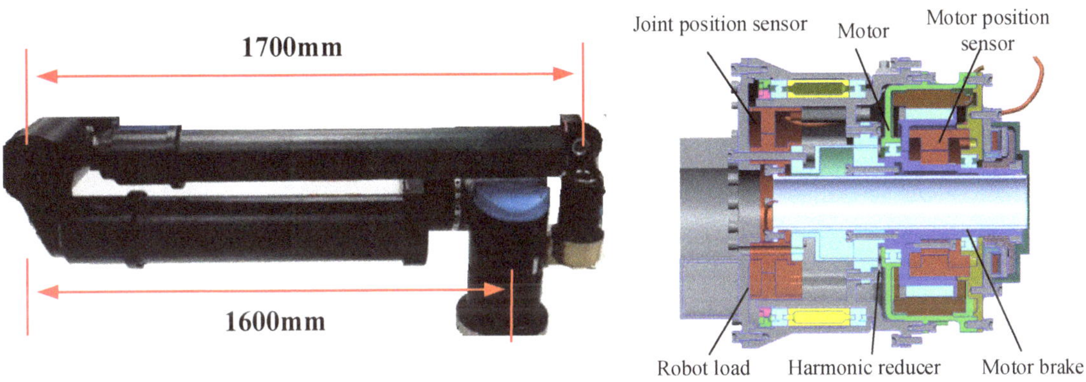

(**a**) (**b**)

Figure 2. Developed space manipulator. (**a**) Prototype; (**b**) Model of the flexible joint.

Full-Elements Joint Dynamics Model

The flexible joint of the space manipulator is composed of several components, including the motor, motor position sensor, joint position sensor, and harmonic reducer, as shown in Figure 2b. To simplify the engineering analysis, the model is divided into three parts: the motor unit, transmission unit, and output unit, as depicted in Figure 3. The inertia and damping of the motor position sensor and its associated structure are included in the motor unit, while the inertia and damping of the output position sensor and its subsidiary structure are included in the output unit. The equivalent inertias and damping of the joint unit, transmission unit, and output unit are denoted by I_m, I_g, I_l, and d_m, d_g, d_l, respectively. The rotation angles are q_m, q_g, q_l, and the moments are τ_m, τ_g, τ_l. Additionally, to account for the stiffness and damping characteristics of the output shaft and reducer, an ideal torsion spring-damping link is introduced between the motor unit and the transmission unit and between the transmission unit and the output unit. The equivalent stiffness and damping between the motor unit and the transmission unit are denoted by k_g and d_{mg}, respectively, while the equivalent stiffness and damping between the transmission unit and the output unit are denoted by k_{gl} and d_{gl}, respectively.

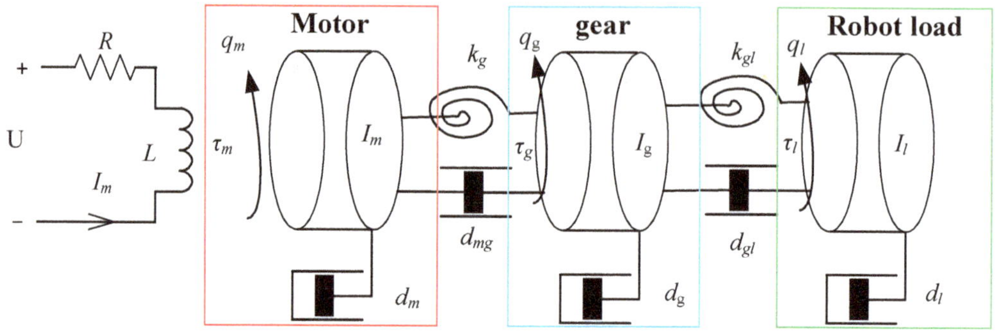

Figure 3. Full-elements dynamic model of flexible joint.

The dynamics model of the flexible joint consists of two parts: an electrical model and a mechanical model. The electrical model describes the characteristics of motor current input and torque output, while the mechanical model describes the characteristics of motor torque input and joint load torque output, which is further divided into linear and nonlinear parts. The dynamic model of the flexible joint can be described as follows:

$$\tau = \mathbf{Kq} + \mathbf{D\dot{q}} + \mathbf{I\ddot{q}} \tag{1}$$

$$q = \left[q_m, q_g, q_l\right] \tag{2}$$

The joint stiffness matrix **K**, the inertia matrix **I**, and the damping matrix **D** are defined as follows:

$$\mathbf{K} = \begin{bmatrix} k_g & -k_g & 0 \\ -k_g & k_g + k_b & -k_b \\ 0 & -k_b & k_b \end{bmatrix} \tag{3}$$

$$\mathbf{I} = \begin{bmatrix} I_m & 0 & 0 \\ 0 & I_g & 0 \\ 0 & 0 & I_l \end{bmatrix} \tag{4}$$

$$\mathbf{D} = \begin{bmatrix} d_m + d_{mg} & -d_{mg} & 0 \\ -d_{mg} & d_g + d_{mg} + d_{gl} & -d_{gl} \\ 0 & -d_{gl} & d_l + d_{gl} \end{bmatrix} \tag{5}$$

The nonlinear part of joint dynamics mainly includes joint viscous friction f_{vis} and coulomb friction f_{cou}.

$$\mathbf{f_{vis}} = \left[0, 0, 0, d_{mn} - d_m, d_{gn} - d_g, d_{ln} - d_l\right] \boldsymbol{q}^T \tag{6}$$

$$\mathbf{f}_{cou} = \frac{e^{\mathbf{k_d}} - e^{-\mathbf{k_d}}}{e^{\mathbf{k_d}} + e^{-\mathbf{k_d}}} \mathrm{X}^T \tag{7}$$

$$\mathbf{k_d} = \left[0, 0, 0, d_m, d_g, d_l\right] \tag{8}$$

According to the dynamic model, the flexible joint of a space manipulator is a dynamic system of multi-variable inputs and outputs, and the joint dynamics model shown in Figure 4 is established by the state space method. In order to clarify the physical meaning of the model, the position and velocity of the joint motor unit, transmission unit, and output unit are selected as the state vector **X**.

$$\begin{cases} \dot{\boldsymbol{X}}_J = \boldsymbol{A}_J \boldsymbol{X}_J + \boldsymbol{B}_J \boldsymbol{U}_J \\ \boldsymbol{Y}_J = \boldsymbol{C}_J \boldsymbol{X}_J \end{cases} \tag{9}$$

$$\boldsymbol{X}_J = \left[q_m, q_g, q_l, \dot{q}_m, \dot{q}_g, \dot{q}_l\right]^T \tag{10}$$

$$\dot{\boldsymbol{X}}_J = \left[\dot{q}_m, \dot{q}_g, \dot{q}_l, \ddot{q}_m, \ddot{q}_g, \ddot{q}_l\right]^T \tag{11}$$

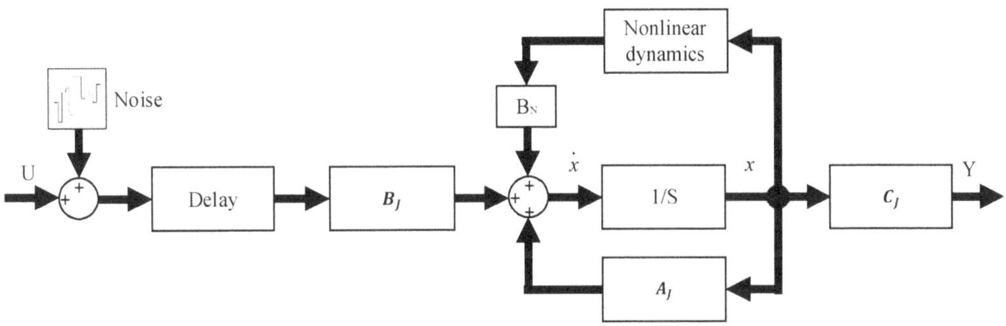

Figure 4. State space model of flexible joint.

The input vectors of the state-space model are the motor torque and the total output torque.

$$\boldsymbol{U}_J = [\tau_m, \tau_l]^T \tag{12}$$

The joint model state transformation matrix **A** $\boldsymbol{A}_J$, input matrix $\boldsymbol{B}_J$, and output matrix $\boldsymbol{C}_J$ are, respectively:

$$\boldsymbol{A}_J = \begin{bmatrix} 0 & 0 & 0 & 1 & 0 & 0 \\ 0 & 0 & 0 & 0 & 1 & 0 \\ 0 & 0 & 0 & 0 & 0 & 1 \\ \frac{k_g}{I_m} & -\frac{k_g}{I_m} & 0 & -\frac{d_{mg}+d_m}{I_m} & \frac{d_{mg}}{I_m} & 0 \\ -\frac{k_g}{I_g} & \frac{k_g+k_b}{I_g} & -\frac{k_b}{I_g} & \frac{d_{mg}}{I_g} & -\frac{d_{mg}+d_g+d_{gl}}{I_g} & \frac{d_{gl}}{I_l} \\ 0 & -\frac{k_b}{I_l} & \frac{k_b}{I_l} & 0 & \frac{d_{gl}}{I_l} & -\frac{d_{gl}+d_l}{I_l} \end{bmatrix} \tag{13}$$

$$\boldsymbol{B}_J = \begin{bmatrix} 0 & 0 & 0 & \frac{1}{I_m} & 0 & 0 \\ 0 & 0 & 0 & 0 & 0 & \frac{1}{I_l} \end{bmatrix} \tag{14}$$

$$\boldsymbol{C}_J = \begin{bmatrix} 1 & 0 & 0 & 0 & 0 & 0 \\ 0 & 1 & 0 & 0 & 0 & 0 \\ 0 & 0 & 1 & 0 & 0 & 0 \\ 0 & 0 & 0 & 1 & 0 & 0 \\ 0 & 0 & 0 & 0 & 1 & 0 \\ 0 & 0 & 0 & 0 & 0 & 1 \end{bmatrix} \tag{15}$$

3. CDHS Controller with Luenberger Observer

3.1. Structure of CDHS

Based on the analysis of Equation (9), the disturbances of the dynamic model of the flexible joint of the space manipulator are primarily caused by changes in internal parameters, such as the load inertia disturbance I_l caused by manipulator configuration changes, the stiffness disturbances k_g and k_{gl} resulting from harmonic reducer stiffness changes, and the friction coefficient. Additionally, there is a nonlinear damping disturbance caused by changes in d_{mg} and d_{gl}. While the position–speed–current (torque) three-loop feedback control algorithm utilized in transmission flexible joint control can adapt to nonlinear disturbances caused by friction changes, significant changes in load inertia and transmission stiffness can result in significantly worse control characteristics, larger tracking errors of motion trajectory, and flexible joint vibrations [29].

The constant damping and high stiffness control algorithm proposed in this paper is based on the three-loop feedback control algorithm, and addresses the limitation that the latter cannot handle time-varying loads with large inertia and flexibility [30,31]. To overcome this, we incorporate a system-stiffness controller and a system-damping controller to mitigate perturbations caused by changes in joint inertia and transmission stiffness, thus ensuring that the damping and stiffness parameters of the flexible joint dynamics control system are always maintained in the desired state. The schematic diagram of the constant damping–high stiffness controller is presented in Figure 5 to illustrate its design and implementation for achieving high quality control performance.

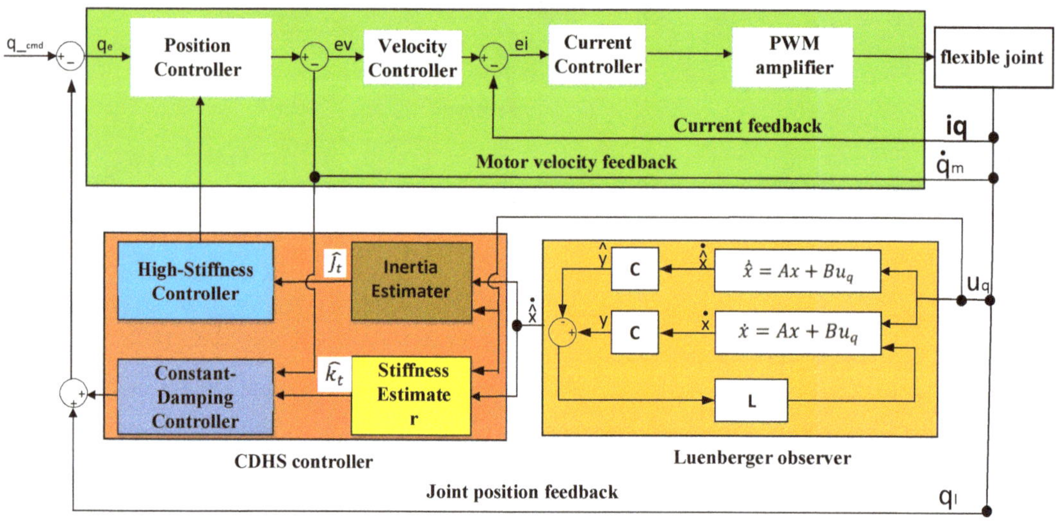

Figure 5. The structure of CDHS controller.

3.2. The Luenberger Observer

The flexible deformation of the joint harmonic reducer is the most sensitive component that reacts to changes in joint stiffness and external load inertia. However, due to the reliability requirements of the space environment, a sensor for detecting harmonic deformation cannot be installed inside the joint. Instead, the working state of the internal transmission mechanism of the joint can be estimated through a state observer [32]. In this paper, a Luenberger state observer is constructed based on the joint dynamics model. The specific calculation process is as follows:

Take the reducer position q_g and the reducer speed $\dot{q}_g$ as the state observation vector x, and take the motor position q_m, motor speed $\dot{q}_m$, motor acceleration $\ddot{q}_m$, joint output position q_l, and joint output speed $\dot{q}_l$ that can be directly measured as the input vector $\boldsymbol{U}$.

$$\mathbf{X} = \left[q_g, \dot{q}_g\right]^T \tag{16}$$

$$\mathbf{U} = \left[q_m, q_l, \dot{q}_m, \dot{q}_l, \ddot{q}_m\right] \tag{17}$$

According to the joint dynamics state space model Formula (5), the following formula is obtained.

$$\dot{x}_1 = x_2 \tag{18}$$

$$\dot{x}_2 = -\frac{k_g + k_b}{I_g}x_1 - \frac{d_{mg} + d_{gl} + d_g}{I_g}x_2 + \mathbf{BU} \tag{19}$$

$$\mathbf{Y} = -k_g x_1 - d_{mg}x_2 + \mathbf{DU} \tag{20}$$

where **B** and **D** are the coefficients corresponding to the state observer.

$$\mathbf{B} = \begin{bmatrix} 0 & 0 & 0 & 0 & 0 \\ \frac{k_g}{I_g} & \frac{k_b}{I_g} & \frac{d_{mg}}{I_g} & \frac{d_{gb}}{I_g} & 0 \end{bmatrix} \tag{21}$$

$$\mathbf{D} = \begin{bmatrix} k_g & d_{mg} & d_m & 0 & I_m \end{bmatrix} \tag{22}$$

Define a Luenberger state observer based on the above calculation.

$$\begin{cases} \dot{\mathbf{X}} = \mathbf{AX} + \mathbf{Bu_q} + \mathbf{L}(\mathbf{y} - \mathbf{Cx}) \\ \mathbf{Y} = \mathbf{CX} + \mathbf{DU} \end{cases} \tag{23}$$

$$\mathbf{A} = \begin{bmatrix} 0 & 1 \\ -\frac{k_g + k_b}{I_g} & -\frac{d_{mg} + d_{gl} + d_g}{I_g} \end{bmatrix} \tag{24}$$

$$\mathbf{C} = \begin{bmatrix} -k_g & -d_{mg} \end{bmatrix} \tag{25}$$

Let the estimation error be $e = x - \dot{x}$. Then, the observation error dynamic equation is as follows:

$$\dot{e} = (A - LC)e \tag{26}$$

The characteristic state equation of the system is as follows:

$$|\lambda I - (A - LC)| = 0 \tag{27}$$

$$\mathbf{L} = \begin{bmatrix} \lambda_1 \\ \lambda_2 \end{bmatrix} \tag{28}$$

$$\lambda^2 + \left[k_g + \frac{(1 + I_g)d_{mg} + d_g + d_{gl}}{I_g}\right]\lambda + \frac{\left(1 + I_g + d_g + d_{gl}\right)k_g + k_b(1 - d_{mg})d_g}{I_g} = 0 \tag{29}$$

The characteristic equation of the state observer is calculated to obtain λ_1 and λ_2, and then the estimated value $\hat{\mathbf{x}}$ of the state observation vector **x** is calculated in real-time according to Equation (29).

3.3. CDHS Control Layer

3.3.1. Realtime Stiffness and Load Inertia Identification

The real-time load inertia and real-time stiffness of the joint are identified according to the estimated value of the joint's flexible deformation by the observer. The equivalent stiffness k_g on the input side of the motor is related to the position q_m of the motor and the position q_g of the reducer, and the equivalent stiffness k_{gl} on the output side of the joint is related to the output position q_l of the joint and the position q_g of the reducer. The corresponding relationship can be obtained through calibration of harmonics [33].

$$\begin{aligned} \hat{k}_g &= f_1(q_m - \hat{x}_1) \\ \hat{k}_{gl} &= f_2(\hat{x}_1 - q_l) \end{aligned} \tag{30}$$

The joint real-time stiffness k_t is determined by the series connection of two elastic links, and the estimated value of the stiffness $\hat{k}_t$ is calculated as follows:

$$\hat{k}_t = \frac{\hat{k}_g \times \hat{k}_{gl}}{\hat{k}_g + \hat{k}_{gl}} \tag{31}$$

The real-time load inertia of the joint is determined by the quotient of the joint output load moment and the load acceleration. The calculation formula is as follows:

$$\hat{J}_t = \frac{k_{gl}(\hat{x}_1 - U_2) + d_{gl}(\hat{x}_2 - U_4)}{\ddot{q}_l} \tag{32}$$

3.3.2. CDHS Controller

Due to the limited computing power of the processor used in a space manipulator, a simplified model must be used. Given that the stiffness of the joint's output unit is significantly greater than that of the transmission unit, the stiffness and damping of the output and transmission units can be combined. Assuming constant joint transmission stiffness and only considering viscous friction, the joint's dynamic equation can be simplified to a transfer function model, where the motor input angle is θ_i, the joint output angle is θ_o, the joint equivalent stiffness is k_t, the equivalent inertia is J_t, and the viscous friction system is f. The transfer function model is as follows:

$$\frac{\theta_o(s)}{\theta_i(s)} = \frac{k_t}{J_t s^2 + fs + k_t} \tag{33}$$

The undamped oscillation frequency ω_n and damping ratio ξ of the system can be calculated using the following equations:

$$\omega_n = \sqrt{\frac{k_t}{J_t}} \xi = \frac{f}{2\sqrt{J_t k_t}} \tag{34}$$

According to the above formula, it can be observed that an increase in load inertia leads to a decrease in system damping and a decrease in undamped oscillation frequency, which may result in joint motion overshoot or instability. To address this issue, a closed-loop damping control system is introduced via speed damping feedback, and a position variable gain controller is added to adjust the closed-loop stiffness of the system. Here, k_1 represents the position control gain, k_s represents the damping control gain, and the transfer function of the joint closed-loop control system is as follows:

$$\frac{\theta_o(s)}{\theta_i(s)} = \frac{k_t k_1}{J_t s^2 + (f + k_t k_1 k_s)s + k_t k_1} \tag{35}$$

The new system undamped oscillation frequency ω_n and system damping ξ are calculated as follows:

$$\omega_n = \sqrt{\frac{k_t k_1}{J_t}} \xi = \frac{f}{2\sqrt{J_t k_t k_1}} + \frac{k_s}{2}\sqrt{\frac{k_t k_1}{J_t}} \tag{36}$$

The above formula shows that the speed damping feedback introduces an additional term of $\frac{k_s}{2}\sqrt{\frac{k_t k_1}{J_t}}$ compared to the original system. By setting a desired reference inertia

J_{exp}, for the joint, the speed damping feedback coefficient k_s can be adjusted to maintain a constant closed-loop damping of the joint.

$$k_s = \frac{f}{\hat{k}_t k_1}\left(\sqrt{\frac{\hat{J}_t}{J_{exp}}}-1\right) \tag{37}$$

In the same way, changing the position gain k_1 can make the system undamped oscillation frequency ω_n stable at the expected value ω_{nexp}.

$$k_1 = \frac{\hat{J}_t \times {\omega_{nexp}}^2}{\hat{k}_t} \tag{38}$$

4. Experiments and Analysis

The flexible joint used in the experiment is the self-developed manipulator joint module with a rated output torque of 40 N.m and a self-weight of 3.5 kg. The motor of the joint has six pairs of poles, and the coil resistance r, coil inductance L_q, and Torque constant k_t are shown in Table 1. The joint is equipped with various sensors, including a motor resolver, joint angular position sensor, joint angular velocity sensor, and harmonic end resolver (output position sensor). The main characteristic parameters of the joint are shown in Table 1.

Table 1. Main parameters of space flexible joint.

m_m [kg]	I_m [kg·m^2]	I_g [kg·m^2]	d_m [N·s/m]	d_g [N·s/m]	d_l [N·s/m]	k_g [N·s/rad]
3.5	0.23	0.0717	0.0036	0.002	0.0003	1000
d_{mg} [N·s/m]	d_{gl} [N·s/m]	P [pairs]	k_b [Nm/rad]	k_t [Nm/A]	L_q [H]	R [Ω]
0.001	0.002	6	16,000	0.78	0.002	15.2

4.1. The Simulation Experiment

To verify the effectiveness of the high stiffness constant damping control algorithm, a comparative experiment was conducted between the traditional three-loop feedback controller and the CDHS controller using a typical excitation signal for fast track applications in orbit. The joint servo system was set to a sinusoidal signal with a frequency of 0.05 Hz and an amplitude of 0.1 rad, while the load inertia signal was set to a sinusoidal signal with a frequency of 0.01 Hz and an amplitude of 100 kg·m^2.

The position command tracking curve and position tracking error curve obtained using the traditional three-loop feedback controller are presented in Figures 6 and 7, respectively. The curves indicate the occurrence of flexible vibration when the algorithm is used to control the flexible joint to track the sinusoidal position command signal, particularly at the peak and trough of the signal. Notably, the algorithm fails to adapt to changes in joint stiffness. Additionally, the position tracking error curve reveals that the error in joint position at different times, such as 30 s and 50 s, is inconsistent. This indicates that the command tracking error is not solely dependent on the joint position but is also influenced by the load inertia, with the maximum error being $\pm 1.5°$.

The position command tracking curve and position error curve of the CDHS algorithm are presented in Figures 8 and 9, respectively. As shown in the command tracking curve, the CDHS algorithm effectively eliminates flexible vibrations, indicating its ability to adapt to changes in joint stiffness. Furthermore, the position tracking error curve indicates that the command tracking error is consistent for the same joint position at different times, indicating that the algorithm is not affected by load inertia. The maximum error is only $\pm 0.5°$, highlighting the improved tracking performance achieved by the CDHS algorithm. Figure 8 shows that the position feedback and the position command basically overlap when the joint tracks the position sinusoidal excitation signal. Figure 9 shows that the position tracking error is also similar to a sinusoidal waveform, and the tracking error peak appears at the peak and trough commutation moments of the position command.

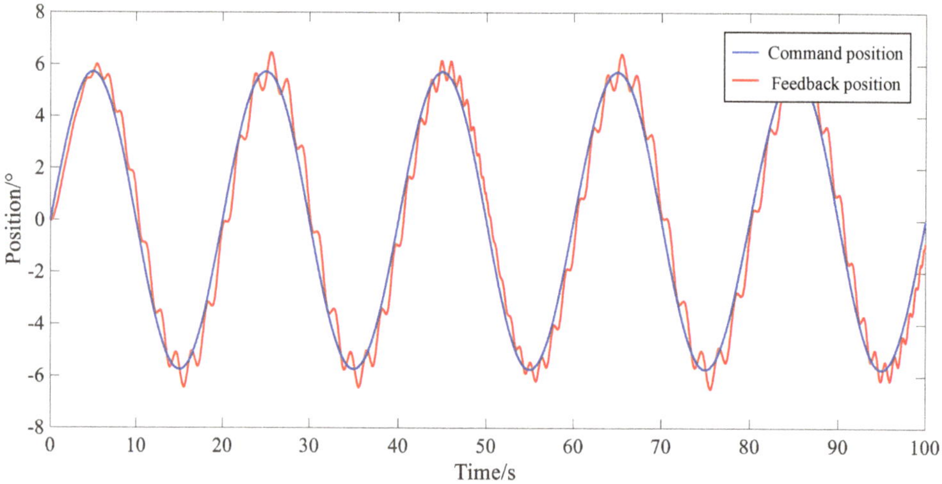

Figure 6. Joint position command tracking curve using traditional three-loop feedback controller.

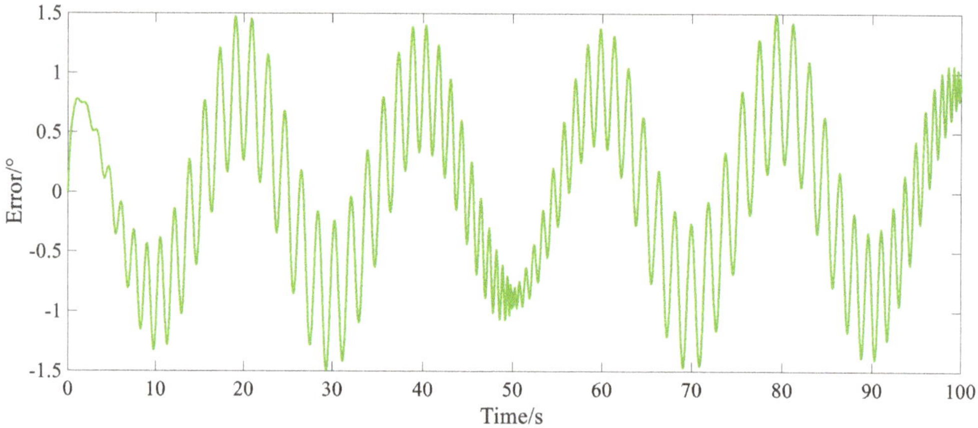

Figure 7. Joint position tracking error curve using traditional three-loop feedback controller.

Figures 10 and 11 depict the flexible deformation curves of the joint reducer under the traditional three-loop feedback controller and CDHS controller, respectively. The curves indicate that the joint harmonic reducer is significantly affected by changes in load inertia under the three-loop feedback controller, resulting in noticeable fluctuations and a maximum deformation of $\pm 0.05°$. Conversely, the maximum deformation of the joint harmonic reducer under the CDHS controller is only $\pm 0.004°$, which is not comparable to the three-loop feedback controller; rather, it is reduced by an order of magnitude. This demonstrates the CDHS controller's ability to achieve smooth transmission of the reducer under variable load conditions.

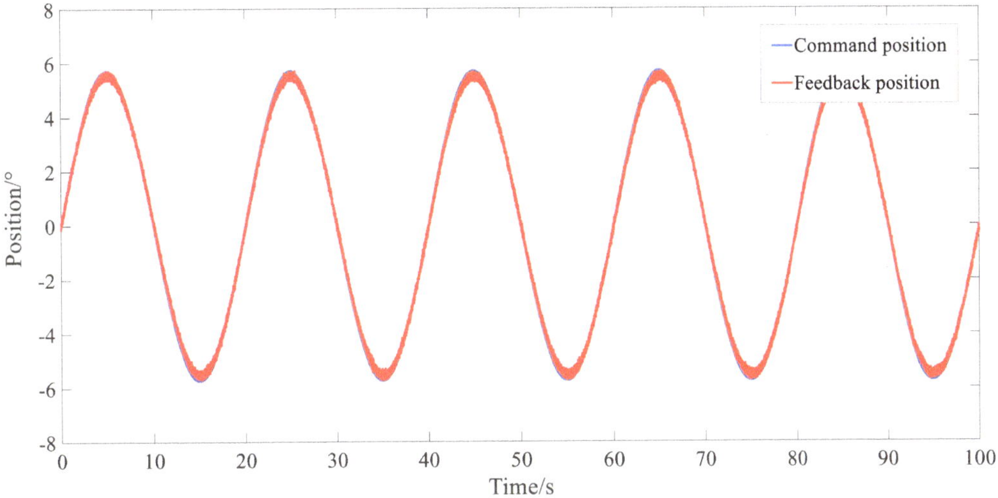

Figure 8. Joint position command tracking curve using CDHS controller.

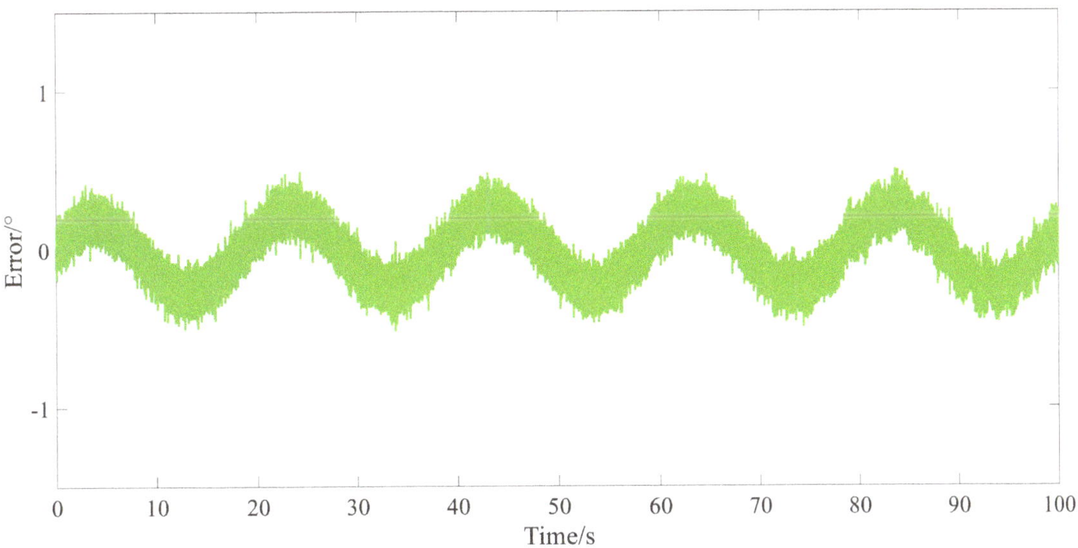

Figure 9. Joint position tracking error curve using CDHS controller.

Table 2 shows the statistics on the joint position tracking error and flexible deformation of the reducer as key indicators for evaluating the performance of the controller. The results show a big improvement in both indicators with the CDHS controller compared to the three-loop feedback controller. Specifically, the joint tracking error is reduced by 73.1% and the flexible deformation of the reducer is reduced by 97.9%. These results indicate that the CDHS controller improves the control accuracy and robustness of the joint system, particularly under variable load conditions.

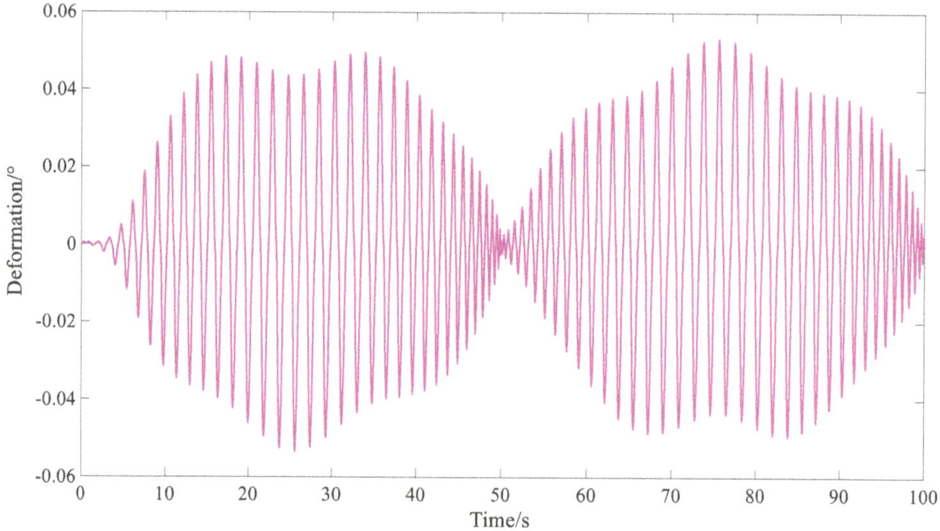

Figure 10. The flexible deformation of harmonic reducer using three-loop feedback controller.

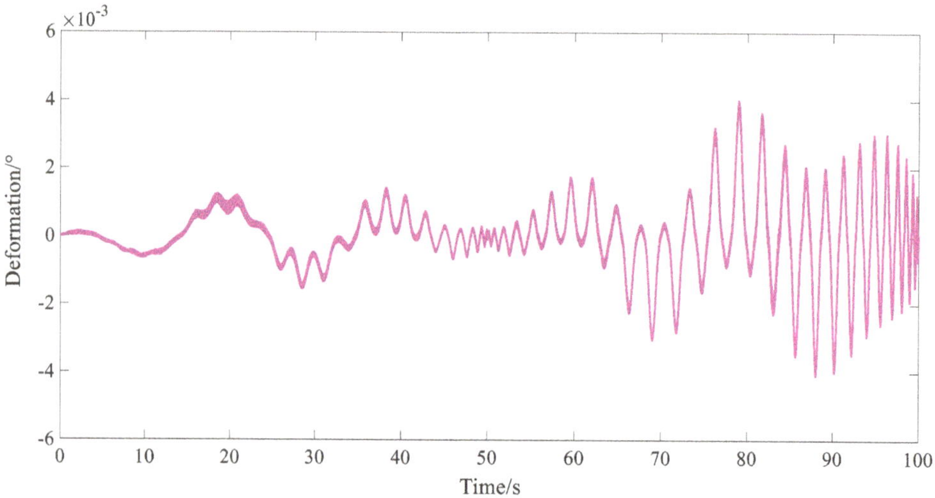

Figure 11. The flexible deformation of harmonic reducer using CDHS controller.

Table 2. Comparison of key indicators resulting from three-loop and CDHS controllers.

Controller Method	Maximum Error/°	Average Error/°	Error Reduction Rate	Maximum Deformation /°	Average Deformation /°	Deformation Reduction Rate	Flexible Vibration
Three-loop	±1.5	0.606	1	±0.05	0.019	1	Exist
CDHS	±0.5	0.163	73.1%	±0.004	0.0004	97.9%	None

4.2. Physical Experiment

In order to further validate the effectiveness of the control strategy, a physical test platform was constructed, as shown in Figure 12. The platform was used to simulate flexible joints controlling the displacement of large inertia target loads, with an inertia of about 100 kg·m^2. The joint driver utilized Altera's FPGA A3PE300 as the core computing unit, which allowed for efficient operation of the algorithm through the parallel computing capability of the FPGA. The host computer sent continuous position commands with a period of 16 ms to the joint driver through the CAN bus. The joint driver then controlled the compliant joint to drive the simulated load to track the position command. The photoelectric encoder located between the joint and the load recorded the position response curve of the joint output.

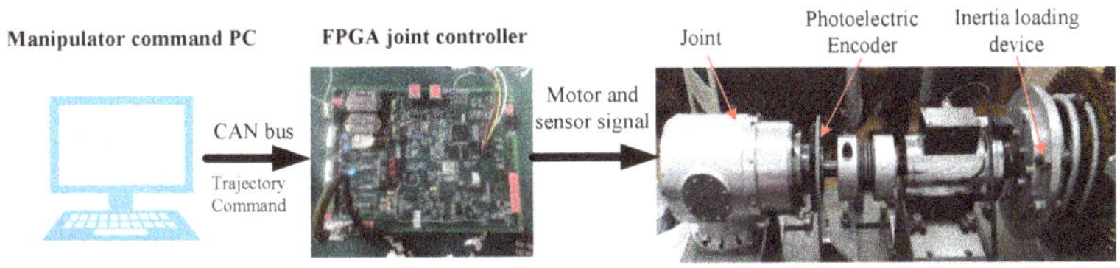

Figure 12. Flexible joint experiment platform.

The joint position control performance of the three-loop feedback controller and the CDHS controller was compared on a physical test platform. The controllers were configured with the same control parameters and tested on the joint's range of motion from $-135°$ to $-125°$, with varying speeds. The highest speed was reached at 40 s, followed by a gradual reduction in speed until the joint stopped. The joint driver recorded the position tracking curves and error curves for both controllers. Figures 13–16 show the results for different controller and speed configurations.

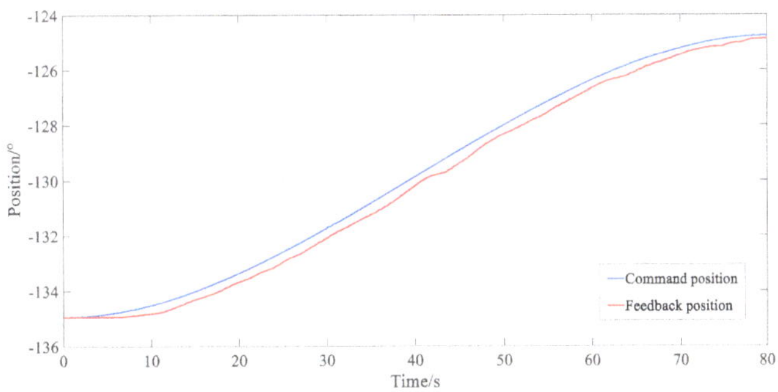

Figure 13. Physical joint position command tracking curve using three-loop feedback controller.

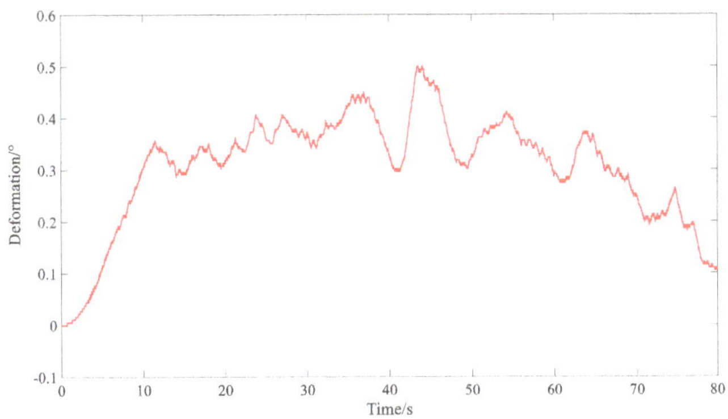

Figure 14. Physical joint position tracking error curve using three-loop feedback controller.

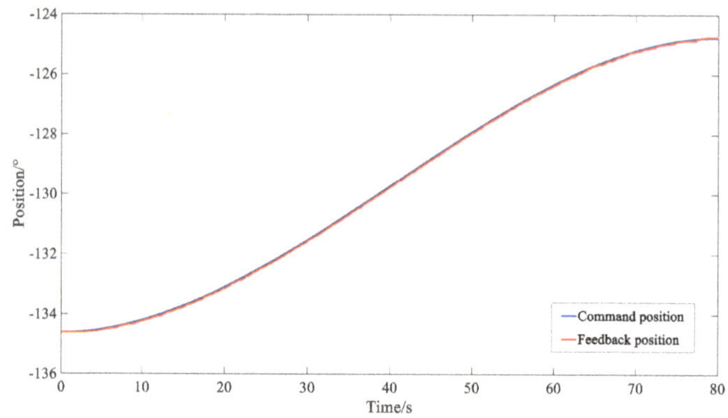

Figure 15. Physical joint position command tracking curve using CDHS controller.

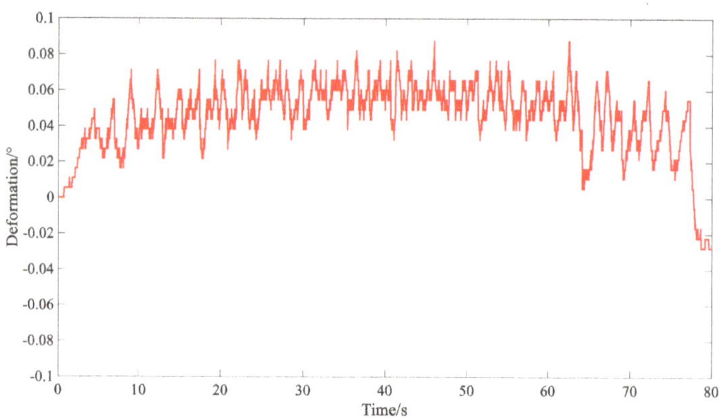

Figure 16. Physical joint position tracking error curve using CDHS controller.

The position tracking curve indicates that the tracking error of the three-loop feedback controller is larger than that of the CDHS controller. When the three-loop feedback controller is in operation, the average errors in the high-speed and low-speed sections are 0.7° and 0.5°, respectively. In contrast, the CDHS controller operating in the high-speed and low-speed sections has average errors of 0.08° and 0.05°, respectively. The position command tracking errors of the CDHS controller in the high-speed section and low-speed section are reduced by 88.6% and 90%, respectively, compared to the three-loop feedback controller.

5. Conclusions and Future Work

During space scientific load position transfer control, the joint load inertia and transmission stiffness change nonlinearly due to the manipulator arm shape, leading to flexible deformation of the joint harmonic reducer and nonlinear variation in joint transmission stiffness. These changes perturb the joint dynamics model and affect the control system's key parameters, resulting in decreased damping and stiffness, which increase joint position tracking error and flexible vibration. To address this issue, we first established a full-element flexible joint dynamic model and used a Luenberger state observer to estimate joint flexible deformation, which cannot be directly measured, and to identify joint load inertia disturbance and transmission stiffness disturbance. Then, we designed a disturbance compensation control strategy based on a simplified model that is convenient for engineering application, which maintains the joint control system's desired damping and stiffness under internal and external disturbances and improves control quality. Finally, we compared the new algorithm with the traditional three-loop feedback controller through simulation and physical experiments. The results show that the CDHS algorithm has more stable motion characteristics than the three-loop feedback control algorithm, reducing the joint's flexible deformation by 97% and improving joint dynamic tracking performance, with an 88% reduction in position tracking error. These results are consistent with the theoretical analysis.

In the future, experiments will be conducted on an air-bearing platform to validate the effectiveness of the algorithm in controlling the real load displacement of space manipulators in microgravity, especially in the presence of multi-joint coupling disturbances. The research and experiments on the constant damping and high stiffness control method of flexible space manipulators based on the Luenberger state observer have positive implications for the future of space manipulators. These implications include the ability to capture, transfer, and maintain large spacecraft in orbit, and the capability to perform on-orbit servicing missions with high precision and reliability. The findings of this research will serve as a valuable guide for the development of advanced space manipulator technologies and contribute to the progress of space exploration and technology.

Author Contributions: Conceptualization, T.Y.; Methodology, F.X.; Validation, S.Z. (Shoujun Zhao); Formal analysis, Y.L.; Resources, T.Y. and Y.W.; Writing—original draft, T.Y.; Project administration, S.Z. (Si Zeng). All authors have read and agreed to the published version of the manuscript.

Funding: This research was funded by National Natural Science Foundation of China, grant number 1975566.

Institutional Review Board Statement: Not applicable.

Informed Consent Statement: Not applicable.

Data Availability Statement: Not applicable.

Conflicts of Interest: The authors declare no conflict of interest.

References

1. Zhang, W.; Li, F.; Li, J.; Cheng, Q. Review of on-Orbit Robotic Arm Active Debris Capture Removal Methods. *Aerospace* **2023**, *10*, 13. [CrossRef]
2. Malzahn, J.; Roozing, W.; Tsagarakis, N.G. The Compliant Joint Toolbox for MATLAB: An Introduction with Examples. *IEEE Robot. Autom. Mag.* **2019**, *26*, 52–63. [CrossRef]
3. Zhang, X.; Jiang, Z.; Zhao, Z.; He, Y.; Xu, Z.; Liu, Y. Intelligent Control of a Space Manipulator Ground Unfold Experiment System with Lagging Compensation. *Appl. Sci.* **2023**, *13*, 5508. [CrossRef]
4. Carabis, D.S.; Wen, J.T. Trajectory Generation for Flexible-Joint Space Manipulators. *Front. Robot. AI* **2022**, *9*, 687595. [CrossRef]
5. Zhang, W.; Wu, W.; Li, Z.; Du, X.; Yan, Z. Three-Dimensional Trajectory Tracking of AUV Based on Nonsingular Terminal Sliding Mode and Active Disturbance Rejection Decoupling Control. *J. Mar. Sci. Eng.* **2023**, *11*, 959. [CrossRef]
6. Abdul-Adheem, W.R.; Ibraheem, I.K.; Humaidi, A.J.; Azar, A.T. Model-free active input–output feedback linearization of a single-link flexible joint manipulator: An Improved Active Disturbance Rejection Control Approach. *Meas. Control.* **2020**, *54*, 002029402091717. [CrossRef]
7. Jian, L.; KaiFa, M.; ZhaoJing, W. Tracking control via switching and learning for a class of uncertain flexible joint robots with variable stiffness actuators. *Neurocomputing* **2022**, *469*, 130–137.
8. Yong, N.; Jiajia, L.; Lao, Z.; Chen, Z. Modeling and Extended State Observer-Based Backstepping Control of Underwater Electro Hydrostatic Actuator with Pressure Compensator and External Load. *Electronics* **2022**, *11*, 1286.
9. Yang, K.; Zhao, L. Command-filter-based Backstepping Control for Flexible Joint Manipulator Systems with Full-state Constrains. *Int. J. Control. Autom. Syst.* **2022**, *20*, 2231–2238. [CrossRef]
10. Zhang, Q.; Liu, X.; Cai, G. Dynamics and Control of a Flexible-Link Flexible-Joint Space Robot with Joint Friction. *Int. J. Aeronaut. Space Sci.* **2021**, *22*, 415–432. [CrossRef]
11. Meng, D.; She, Y.; Xu, W.; Lu, W.; Liang, B. Dynamic modeling and vibration characteristics analysis of flexible-link and flexible-joint space manipulator. *Multibody Syst. Dyn.* **2018**, *43*, 321–347. [CrossRef]
12. Rajgade, S.C.; Shendge, P.D.; Londhe, P.S. Feedback linearization-based robust control for flexible joint robotic system using proportional integral observer. *Int. J. Dyn. Control.* **2023**, *11*, 396–410. [CrossRef]
13. Long, C.; Hai, W.; Yunzhi, H.; Ping, Z.; Yu, M.; Zheng, X.; Ye, M.; Hu, Y. Robust hierarchical sliding mode control of a two-wheeled self-balancing vehicle using perturbation estimation. *Mech. Syst. Signal Process.* **2020**, *139*, 106584.

14. Gao, M.M.; Jin, X.Z.; Ding, L.J. Robust adaptive backstepping INTSM control for robotic manipulators based on ELM. *Neural Comput. Appl.* **2022**, *34*, 5029–5039. [CrossRef]
15. Bowen, Z.; Minghe, J.; Jian, L. Extended-state-observer-based adaptive control of flexible-joint space manipulators with system uncertainties. *Adv. Space Res.* **2022**, *69*, 3088–3102.
16. Kim, M.J.; Chung, W.K. Disturbance-Observer-Based PD Control of Flexible Joint Robots for Asymptotic Convergence. *IEEE Trans. Robot.* **2015**, *31*, 1508–1516. [CrossRef]
17. Fagiolini, A.; Trumić, M.; Jovanović, K. An Input Observer-Based Stiffness Estimation Approach for Flexible Robot Joints. *IEEE Robot. Autom. Lett.* **2020**, *5*, 1843–1850. [CrossRef]
18. Dou, L.; Du, M.; Zhang, X.; Du, H.; Liu, W. Fuzzy Disturbance Observer-Based Sliding Mode Control for Liquid-Filled Spacecraft with Flexible Structure under Control Saturation. *IEEE Access* **2019**, *7*, 149810–149819. [CrossRef]
19. Pan, C.; Luo, J.; Zhou, L.; Xiong, P. Linear Active Disturbance Rejection Control Design for a Single-Link Flexible-Joint Manipulator with Uncertainties. In Proceedings of the 2020 Chinese Automation Congress (CAC), Shanghai, China, 6–8 November 2020; pp. 2446–2451.
20. Lin, Y.; Zhao, H.; Ding, H. External Force Estimation for Industrial Robots with Flexible Joints. *IEEE Robot. Autom. Lett.* **2020**, *5*, 1311–1318. [CrossRef]
21. Li, Y.-R.; Chen, C.-C.; Peng, C.-C. Integral Backstepping Control Algorithm for a Quadrotor Positioning Flight Task: A Design Issue Discussion. *Algorithms* **2023**, *16*, 122. [CrossRef]
22. Zheng, D.; Li, P.; Xie, W.; Li, D. Identification and Control of Flexible Joint Robot Using Multi-Time-Scale Neural Network. *J. Shanghai Jiaotong Univ.* **2020**, *25*, 553–560. [CrossRef]
23. Yan, G.; Zhao, B. Analytical Solution and Shaking Table Test on Tunnels through Soft-Hard Stratum with a Transition Tunnel and Flexible Joints. *Appl. Sci.* **2022**, *12*, 3151. [CrossRef]
24. Yen, V.T.; Nan, W.Y.; Van, C.P. Robust Adaptive Sliding Mode Neural Networks Control for Industrial Robot Manipulators. *Int. J. Control. Autom. Syst.* **2019**, *17*, 783–792. [CrossRef]
25. Wu, Y.; Fang, H.; Xu, T.; Wan, F. Adaptive Neural Fixed-time Sliding Mode Control of Uncertain Robotic Manipulators with Input Saturation and Prescribed Constraints. *Neural Process. Lett.* **2022**, *54*, 3829–3849. [CrossRef]
26. Wen, Q.; Yang, X.; Huang, C.; Zeng, J.; Yuan, Z.; Liu, P.X. Disturbance Observer-based Neural Network Integral Sliding Mode Control for a Constrained Flexible Joint Robotic Manipulator. *Int. J. Control. Autom. Syst.* **2023**, *21*, 1243–1257. [CrossRef]
27. Cheng, Y.; Li, C. Luenberger observer-based microgrid control strategy for mixed load conditions. *Energies* **2020**, *15*, 3655. [CrossRef]
28. Zhao, Y.; Yang, T.; Yue, H.; Yang, X.; Bai, D.; Yang, F. Design and Analysis of a New Deployable Docking Mechanism for Microsatellites. *Remote Sens.* **2022**, *14*, 5002. [CrossRef]
29. Azar, A.T.; Abed, A.M.; Abdul-Majeed, F.A.; Hameed, I.A.; Jawad, A.J.M.; Abdul-Adheem, W.R.; Ibraheem, I.K.; Kamal, N.A. Design and Stability Analysis of Sliding Mode Controller for Non-Holonomic Differential Drive Mobile Robots. *Machines* **2023**, *11*, 470. [CrossRef]
30. Wu, H.; Zhang, X.; Song, L.; Zhang, Y.; Wang, C.; Zhao, X.; Gu, L. Parallel Network-Based Sliding Mode Tracking Control for Robotic Manipulators with Uncertain Dynamics. *Actuators* **2023**, *12*, 187. [CrossRef]
31. Chang, P.-H.; Park, K.C.; Lee, S. An extension to operational space for kinematically redundant manipulators: Kinematics and dynamics. *IEEE Trans. Robot. Autom.* **2000**, *16*, 592–596. [CrossRef]
32. Doroudchi, A.; Khodambashi, R.; Sharifzadeh, M.; Li, D.; Berman, S.; Aukes, D.M. Tracking Control of a Miniature 2-DOF Manipulator with Hydrogel Actuators. *IEEE Robot. Autom. Lett.* **2021**, *6*, 4774–4781. [CrossRef]
33. Malarczyk, M.; Zychlewicz, M.; Stanislawski, R.; Kaminski, M. Speed Control Based on State Vector Applied for Electrical Drive with Elastic Connection. *Automation* **2022**, *3*, 337–363. [CrossRef]

applied sciences

MDPI

Article

An Image Denoising Method for a Visible Light Camera in a Complex Sky-Based Background

Zelong Ma, Qinglei Zhao *, Xin Che *, Xinda Qi, Wenxian Li and Shuxin Wang

Changchun Institute of Optics, Fine Mechanics and Physics, Chinese Academy of Sciences, Changchun 130033, China; mazelong@ciomp.ac.cn (Z.M.); qixinda@ciomp.ac.cn (X.Q.); liwenxian@ciomp.ac.cn (W.L.); wangsx@ciomp.ac.cn (S.W.)

* Correspondence: zhaoql_ciomp@163.com (Q.Z.); chexin@ciomp.ac.cn (X.C.)

Abstract: For space target images captured by a sky-based visible light camera, various conditions are influenced by the imaging system itself and the observation environment; these conditions include uneven image background intensity, complex noise, stray light composition, and diverse target forms. A mean wavelet transform algorithm is proposed. This algorithm initially performs mean filtering and wavelet transform filtering on the noise-containing space target images and then performs a wavelet inverse transform on the filtered results to generate the final image. The experimental results show that our algorithm has good denoising performance and can effectively maintain the image details.

Keywords: sky-based visible light camera; space target images; denoising; wavelet transform; mean filtering

check for updates

Citation: Ma, Z.; Zhao, Q.; Che, X.; Qi, X.; Li, W.; Wang, S. An Image Denoising Method for a Visible Light Camera in a Complex Sky-Based Background. *Appl. Sci.* **2023**, *13*, 8484. https://doi.org/10.3390/app13148484

Academic Editor: Emanuel Guariglia

Received: 30 May 2023
Revised: 18 July 2023
Accepted: 20 July 2023
Published: 22 July 2023

1. Introduction

Studies on sky-based image denoising techniques have focused on extracting space targets by processing the image prior to object detection. The objective is to suppress or eliminate inhomogeneous background noise.

Methods for suppressing or eliminating image noise can be categorized into two groups: transform domain methods [1–4] and spatial domain methods [5–13]. The transform domain approach involves removing the noise in the transformed domain of the image and then inverting the transform to achieve image denoising. Spatial domain methods, on the other hand, involve directly operating on the original image, taking advantage of the grayscale values in the neighborhood of the pixels to perform relevant operations on the pixels.

Unlike conventional images, in space target images, the captured images not only contain Gaussian noise, salt-and-pepper noise, Poisson noise, speckle noise, etc., but also the imaging size, shape, and gray value of space targets are highly similar to noise, and the background strength is unevenly distributed due to the scattered light. These factors cause difficulty for traditional image denoising algorithms [14–19] to be applied to the denoising of star maps, easily causing loss of target information or false targets. New requirements for image processing technology are needed to suppress or eliminate the complex background interference in star maps while preserving space target information.

This paper proposes a mean curvelet transform method. Curvelet transform is a multiscale and directionally sensitive transformation method that extracts detailed features from an image. On the other hand, mean processing is a basic image smoothing method that reduces noise by calculating the average value of pixel neighborhoods. By integrating these two methods, one can first utilize the curvelet transform to extract detailed features from the image and then use mean processing to smooth the extracted details, thereby reducing noise. The effectiveness of the proposed algorithm is verified through simulation experiments.

2. Materials and Methods

2.1. Wavelet Transform and Curvelet Transform

2.1.1. Wavelet Transform

In the 1980s, J. Morlet first proposed the concept of wavelets [20]. In the same period, mathematicians Y. Meyer and S. Mallat further studied the Mallat algorithm based on previous work, and the wavelet transform was officially created [21,22]. From a theoretical perspective, wavelet transform theory evolved from Fourier analysis. Fourier analysis can be used to analyze the characteristics of data separately using the time axis and frequency axis transformations, but it cannot reflect the characteristics of signals changing simultaneously along both the time and frequency axes. It can only consider one of the two axes, allowing a global understanding of the signal's characteristics, but is unable to describe the local characteristics of a specific time-frequency region of the signal. In reality, signals are often nonstationary, and it is difficult to maintain stability. Most signals are nonstationary, and the local time-frequency characteristics precisely describe the distinctive features of nonstationary signals. While scaling the signal by shrinking or expanding the variable multiplier and transforming the signal through smooth distance shifts, the wavelet transform can decompose the signal into multiple levels, surpassing the limitations of the Fourier transform in this aspect.

The traditional approach of the wavelet transform is to separate the high-frequency and low-frequency signals for processing. The low-frequency signal, which undergoes minimal changes, is expanded to analyze the fine details of the rapidly changing parts in the high-frequency signal. In the time dimension, the signal is further segmented, while in the low-frequency signal, the frequency is divided more finely, adjusting the time-frequency characteristics analysis based on the specific high and low frequencies of the signal.

The continuous wavelet transform involves extensive data operations, so it needs to be discretized during implementation to simplify the form. In the expression of continuous wavelets, the binary number scale parameter a and the translation parameter b remain continuous, and only the parameters in the semi-discrete continuous wavelet formula are discretized while the variable t remains unchanged. The discrete form of the wavelet transform is:

$$(W_{\Psi}f)\left(\frac{1}{2^j},b\right) = \int_{-\infty}^{\infty} f(t)\left\{2^{j/2}\Psi\left(2^j(x-b)\right)\right\}dx \quad (1)$$

This transformation is also known as the binary wavelet transform, which meets the condition:

$$A \le \sum_{j\in z}\left(2^{-j}\omega\right)\Big|^2 \le B \quad (2)$$

Here, $0 < A < B < \infty$, where A and B are constants. For a certain level of decomposition, the binary wavelet transform can be considered a function of the variable b in Equation (2), and b is not discretized. Equation (2) can be transformed into a convolution operation expression, which is shown below:

$$W_s f(x) = f^{*}\Psi_s(x) \quad (3)$$

By scaling the wavelet mother function $\Psi(x)$ with a factor of s (where $s = 2j$), we obtain the wavelet function as follows:

$$\Psi_s(x) = \frac{1}{s}\Psi_s\left(\frac{x}{s}\right) \quad (4)$$

Let us assume $f_\tau(x) = f(x-\tau)$. Then,

$$(W_{2j}f_\tau)(x) = W_{2j}[f(x)] \quad (5)$$

If we first shift the function f along the scale and then perform the binary wavelet transform, it is equivalent to first applying the binary wavelet transform and then shifting it along the scale. In other words, $W_{2j}f$ possesses the properties of f, which is the meaning of translation invariance.

Although the binary wavelet transform is in a discrete form, the variable b is not discretized along the scale, and the transformation of the wavelet transform at different scales is not discrete. This translation invariance exhibits a zooming characteristic. By adjusting b to enlarge or reduce the scaling factor, the wavelet transform acts like a microscope, enabling the study of finer or coarser details in the signal.

2.1.2. Limitations of the Wavelet Transform and Introduction of the Ridgelet Transform

Although the wavelet transform, known as the "mathematical microscope", can express the details of image waveforms and has significant application value and advantages, it has a limitation due to the lack of anisotropy in wavelet bases. The wavelet transform can only reflect discontinuities or points where the derivative is zero in 1D images, and it cannot capture the edge features of 2D spatial images. Therefore, at the end of the 20th century, Candes developed the ridgelet transform algorithm [23], which expresses line singularities and can provide a good approximation of line singularities in multivariable functions. The expression of the ridgelet transform is given by:

$$\Psi_\gamma(x) = a^{-\frac{1}{2}}\Psi\left(\frac{[u,x]-b}{a}\right) \tag{6}$$

where Ψ_γ is called a ridgelet generated from Ψ under specific admissibility conditions. a represents the scale parameter, "u" represents the direction parameter, and b represents the position parameter. Ψ_γ has rapid decay properties that satisfy:

$$\int \Psi(t)dt = 0 \tag{7}$$

For functions with several variables, the ridgelet transform is defined as:

$$R(f)(r) = \langle f, \Psi_r \rangle \tag{8}$$

The ridgelet transform represents the direction by adding an additional parameter to the wavelet basis function. In the cross-section, the basis of the ridge is the wavelet. The continuous ridgelet transform in 2D space is expressed in Equation (9).

$$CRT_f(a, b_x, b_y) = \int_{-\infty}^{\infty}\int_{-\infty}^{\infty} f(x,y)\Psi a, b_x, b_y(x,y)dxdy \tag{9}$$

When comparing the expressions of the 2D wavelet transform, as shown in Equation (10), with the ridgelet transform, we can observe the following contrast: in the wavelet transform formula, the parameter is a point, whereas in the ridgelet transform, the parameter is a line.

$$w_f(a, b_x, b_y) = \int_{-\infty}^{\infty}\int_{-\infty}^{\infty} f(x,y)\Psi a, b_x, b_y(x,y)dxdy \tag{10}$$

The introduction of the ridgelet transform solves the problem of the wavelet transform's inability to describe 2D edge singularities. However, the ridgelet transform still has limitations when dealing with functions with curve singularities. Moreover, in practical images, edges are rarely straight lines, which limits the widespread application of the ridgelet transform. Additionally, the ridgelet transform suffers from high computational redundancy. Therefore, in the last year of the 20th century, Candes and Donoho developed the curvelet transform and formulated its theory. The curvelet transform effectively captures the singular points and characteristics of curves in images. The basis of the curvelet transform exhibits directionality and anisotropy, allowing for the optimal approximation and description of nonlinear parts of 2D images. Thus, curvelet analysis combines the properties of wavelet multiresolution and locality with directionality, providing an optimal approach for representing images with these three features.

2.1.3. Curvelet Transform

The curvelet transform, introduced by E.J. Cande and David L. Donoho at the end of the 20th century, is a method aimed at addressing the limitations of the wavelet transform in expressing the anisotropy of image boundaries and line features [24]. While the wavelet transform is suitable for representing isotropic objects, it falls short in capturing the directional variations in image boundaries and line features. Therefore, the curvelet transform was proposed as a first-generation extension of the wavelet transform, preserving the advantages of multiresolution and time-frequency localization while introducing the ability to perform multiscale, translation, and multidirectional transformations. This enables the curvelet transform to better describe the singular changes in images, which cannot be effectively expressed by the wavelet transform.

Numbered lists can be added as follows:

- First-generation curvelet. The first-generation curvelet transform primarily involves subband decomposition and a multiscale ridgelet transform, which is a nonadaptive representation method. As shown in Figure 1, the first-generation curvelet includes subband decomposition, smoothing partitioning, normalization, and ridgelet analysis. The decomposition process introduces significant data computation, making it quite complex in digital implementation. Based on this, E.J. Candes and Donoho proposed a second-generation transform three years later, and this transform is an improved curvelet algorithm that is easier to understand and simpler to implement. Furthermore, the two scholars introduced a fast curvelet algorithm that directly divides frequencies, eliminating the need for the ridgelet transform. This construction significantly differs from the first-generation curvelet algorithm, resulting in a reduction in data and computational complexity.

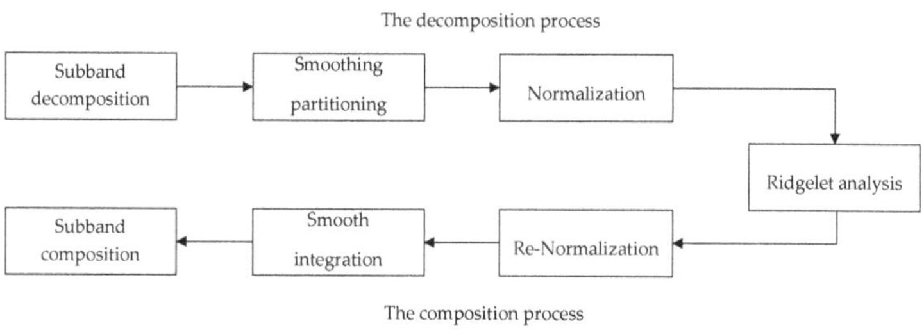

Figure 1. Flowchart of The first-generation curvelet transform algorithm.

- Second-generation curvelet transform. The construction of the second-generation curvelet transform differs from the construction of the first generation transform, but the second-generation transform is an improvement of the first-generation transform. The idea behind the first-generation curvelet transform is to partition the target function space into infinitely small blocks to approximate curves as straight lines; then, local ridgelet analysis is utilized. In contrast, the second-generation curvelet transform only retains the abstract principles of the ridgelet transform, such as the framework and tight frame, while discarding its specific computational methods. The continuous curvelet transform serves as an example.

  The formula for the continuous curvelet transform in the time domain is as follows:

$$C(i,j,k) = \left\langle f, \varphi_{i,j,k} \right\rangle = \int_{R^2} f(x)\overline{\varphi_{i,j,k}(x)}dx \tag{11}$$

where

$$\varphi_{i,j,k}(x) = \varphi_j\left[R_\theta\left(x - x_k^j\right)\right] \tag{12}$$

indicates the position at

$$x_k^{(j)} = R_0^{-1}\left(k_1 \times 2^{-j}, k_2 \times 2^{-j/2}\right) \tag{13}$$

When the angle position of all points at $2(^{-j})$ is considered, the curvelet transform can be obtained by rotating the curvelet basis φ_j by a certain angle θ and shifting it by K.

$$\theta = 2\pi \times 2^{-\lfloor j/2 \rfloor} \times l,\ l = 0, 1, \ldots, 0 \leq \theta_1 \leq 2\pi \tag{14}$$

The parameter $k = (k_1, k_2) \in z^2$ represents the displacement sequence in the 2-dimensional space.

2.1.4. Curvelet Transform Algorithm

The algorithm for the curvelet transform is described as follows:

(1) Perform a J-level decomposition of the original image I to obtain the subband sequences CJ and DJ;
(2) Set the initial block size (i.e., the finest scale block) to BMIN, and let B1 = BMIN;
(3) For j = 1 to J:
  - Divide Dj into subblocks of size BJ with overlapping regions;
  - Apply the discrete ridgelet transform to each subblock;
  - If (j mod 2 = 1), set Bj + 1 = 2Bj; otherwise, set Bj + 1 = Bj.

The curvelet transform itself is a redundant transformation. If the number of wavelet decompositions in the first step is J, then the overall redundancy factor of the entire transform is 16J + 1. Since each step of the above decomposition is reversible, it is possible to construct the corresponding inverse transform algorithm to completely reconstruct the original image.

2.1.5. Analysis of Wavelet Transform Denoising

The wavelet transform method solves the correlation values of the spatial image in a set of established anisotropic bases. The method of wavelet transform denoising is as follows: first, the noisy image is decomposed into subbands of different scales, and the wavelet transform coefficients corresponding to each subband are obtained through the Plancherel operation; then, the obtained wavelet transform coefficients are processed by the hard threshold method (abandoning the smaller transform coefficients and retaining the larger transform coefficients) to filter out noise in the image and retain the edge characteristics of the image. The expression of the hard threshold in the wavelet transform denoising is as follows:

$$C_0(i,h) = \begin{cases} C(i,h), |C(i,h)| \geq T_i \\ 0, |C(i,h)| < T_i \end{cases} \tag{15}$$

where $C(i,h)$ is the wavelet coefficient at scale i and direction h, T_i is the threshold corresponding to different scales, and $C_0(i,h)$ is the wavelet transform coefficient after hard threshold denoising; the expression for selecting the threshold T_i is as follows:

$$T_i = k_i, \sigma \cdot \sigma_i \tag{16}$$

where σ is the noise standard deviation, σ_i is the noise standard deviation corresponding to the subband after transformation, and k_i is the adaptive constant corresponding to each subband. The Monte Carlo algorithm with strong adaptability is used to estimate the noise standard deviation in this method.

2.2. Analysis of Mean Filtering Algorithm

The principle of mean filtering is to calculate the mean of the pixels in the current point neighborhood to replace the gray value of the current point, and this is a linear filtering algorithm. It is equivalent to convolving the image with a filtering template of a certain size, filtering out the objectives with poor correlation, and obtaining a smooth star map background. The formula for mean filtering is as follows:

$$g(x,y) = W(i,j) * f(x,y) \tag{17}$$

where $f(x,y)$ is the original image data, $g(x,y)$ is the filtered image data, and $W(i,j)$ is the filtering template, which is represented as follows:

$$W(i,j) = \frac{1}{M \times N}\begin{bmatrix} 1 & 1 & \cdots & 1 \\ 1 & 1 & \cdots & 1 \\ \vdots & \vdots & \ddots & \vdots \\ 1 & 1 & \cdots & 1 \end{bmatrix}_{M \times N} \tag{18}$$

2.3. Mean Wavelet Transform Denoising Process

In a space target image, the background information is primarily represented by pixels with low grayscale values, while the edges and details of the target are mainly represented by pixels with higher grayscale values, assuming there is no contamination from stray light. The curvelet transform is known for its multi-scale and multi-directional sensitivity, which enables effective capture of detailed features in images. However, in practical scenarios, space target images usually contain significant amounts of non-uniform stray light, which can vary in intensity. When the grayscale value of the background information is similar to that of the target edges, the curvelet transform may struggle to effectively capture these subtle changes. As a result, it becomes challenging to distinguish the target from the background, leading to blurry outcomes.

To address this issue, we introduce the mean filtering algorithm to enhance contrast and aid in distinguishing the target from the background. Subsequently, we apply the curvelet transform for filtering. Nevertheless, mean filtering has its limitations, such as insufficient smoothing of the grayscale values in the background information and potential blurring of the target edges. These limitations restrict the denoising capability of the current transform. Furthermore, the non-uniformity of stray light in the space target image can result in regions with no stray light or severely weak stray light, where pixel values already contain valid target information. However, the mean filtering averages over these regions, potentially leading to blurring of object detail.

Hence, in this paper, we propose an image processing approach based on weighted averaging. We denote the two source images as X and Y, where the X image is denoised directly using the curvelet transform and the Y image is filtered by the mean prior to the current transform. The final composite image, denoted as Z, is obtained by taking the weighted average of the two images. The mean curvelet denoising formula can be expressed as follows:

$$C(Z,p) = KC(X,p) + (1-k)C(Y,p) \tag{19}$$

The weighting coefficient, denoted as k, its expression is as follows:

$$k = \mathrm{a} \times \frac{\sum_{x=M_x}^{N_x} \sum_{y=M_y}^{N_y} \frac{I(x,y)}{2^N}}{(N_x - M_x + 1) \times (N_y - M_y + 1)} \tag{20}$$

M_x, M_y, N_x, N_y represent the coordinates of a pixel in the image, $I(x,y)$ represents the value of the pixel at coordinates (x,y), N represents the number of gray levels in the image, a represents an empirical value within the range of $a \in [0,1]$.

In regions of the background where strong stray light is present, we amplify the k value to enhance the proportion of the denoised image Y, which is obtained after mean filtering, in the composite image Z. Conversely, in regions with weak stray light in the background, we reduce the k value to enhance the proportion of the denoised image X, obtained through direct curvelet filtering, in the composite image Z. The flowchart of the mean wavelet transform denoising algorithm is shown in Figure 2.

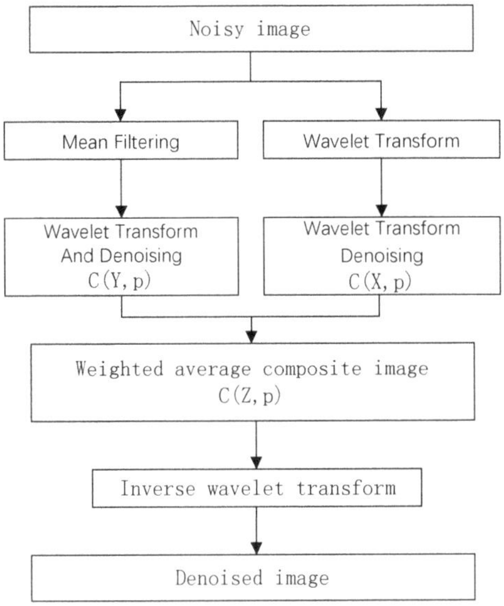

Figure 2. Flowchart of the mean wavelet transform denoising algorithm.

3. Results

This paper verifies the effectiveness of the mean wavelet transform denoising algorithm through two comparison experiments.

3.1. Lena Image of a Hat Part in the Reverse Color Image Experiment

The resolution of the inverse color image of the hat part of the Lena image is a gray image of 512 × 512 with an 8-bit grayscale. We use the mean filtering algorithm, wavelet filtering algorithm and mean wavelet filtering algorithm to remove the white noise in the original image, and the experimental results are shown in Figure 3 and Table 1.

Table 1. Quality of the lena image of a hat part reverse color image and experiment images obtained via different methods.

Method	Peak Signal to Noise Ratio	Note
Original noisy image	21.13	-
Mean filtering	24.45	large amount of distortion
Wavelet filtering	28.95	small amount of distortion
Mean wavelet filtering	29.55	very little distortion

From Figure 3, the mean wavelet transform algorithm restores the slanted lines on the hat more realistically than the other methods.

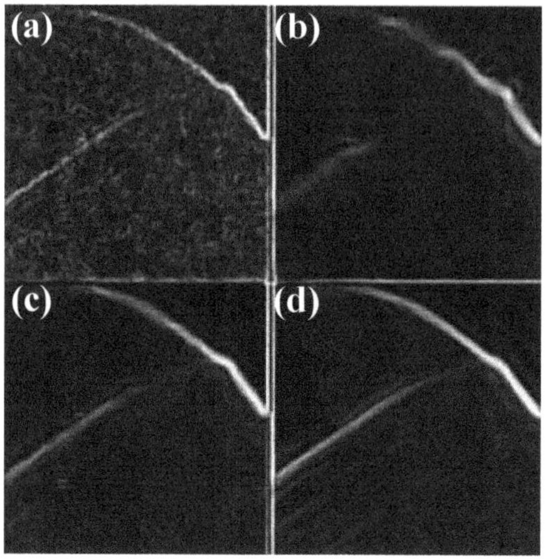

Figure 3. Lena image filtering results: (**a**) original image with noise; (**b**) mean filtered image; (**c**) wavelet transformed filtered image; (**d**) mean wavelet filtered image.

3.2. Real Star Image Experiment

The real star image resolution is a 2048 × 2048 gray image with a gray level of 12 bits. We use the mean filtering algorithm, wavelet filtering algorithm and mean wavelet filtering algorithm to filter out the white noise in the original image. The experimental results are shown in Figure 4 and Table 2.

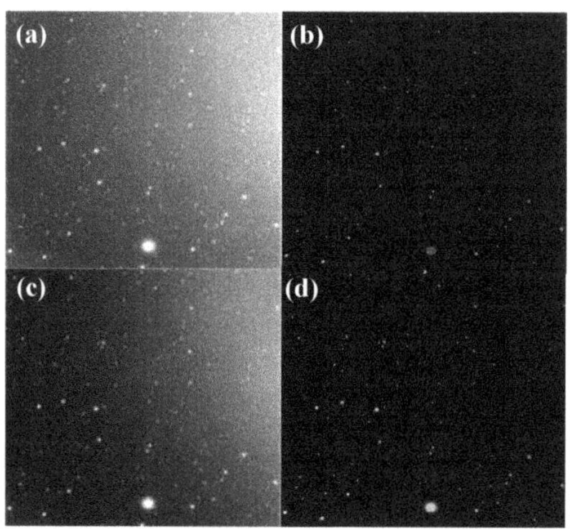

Figure 4. Real star image filtering results: (**a**) original image with noise; (**b**) mean filtering image; (**c**) wavelet filtering image; (**d**) mean wavelet filtering image.

As shown in Figure 4, in the real star image, the mean wavelet transform can not only remove nonuniform noise but also significantly improve the signal-to-noise ratio of the image.

Table 2. Quality of real star image and experimental images obtained via different methods.

Method	Peak Signal to Noise Ratio	Note
Original noisy image	6.20	-
Mean filtering	10.23	small amount of distortion
Wavelet filtering	6.95	large amount of distortion
Mean wavelet filtering	12.94	very little distortion

4. Discussion

This paper proposes the mean-curvelet transform denoising algorithm to address the issues of uneven background intensity, complex noise and interference, and diverse target shapes in images of space targets captured by visible light cameras under the constraints of the imaging system and observation environment. The algorithm applies mean filtering and curvelet transform filtering to the noisy space target images separately and then applies the inverse curvelet transform to the filtered results to generate the final image. The experimental results show that this algorithm effectively improves the signal-to-noise ratio of images.

Furthermore, since the fixed threshold is used in the curvelet transform in this paper, we will study the application of the variable threshold curvelet transform to the algorithm in future experiments.

However, since the curvelet transform is mostly implemented on hardware platforms such as X86 and X64, which cannot tolerate long-term operation in space radiation environments, the algorithm proposed in this paper currently does not have the capability of real-time processing in orbit. In future studies, we will select some hardware platforms with radiation resistance indicators, such as DSP or FPGA, and apply the algorithm on these hardware platforms to achieve real-time processing in orbit.

5. Conclusions

Based on our study, and compared with existing methods, our proposed mean curvelet transform method not only has better suppression performance for images of space targets with uneven background intensity, complex noise and clutter but also maintains good target shape diversity and preserves image details and other information, significantly improving the image quality. Therefore, our proposed algorithm has great application value for denoising space target images. In future research, we will further investigate the performance of the algorithm.

Author Contributions: Project administration, Z.M.; data curation, Q.Z., X.Q., W.L. and S.W.; X.C. wrote the manuscript. All authors have read and agreed to the published version of the manuscript.

Funding: Key Technology R&D Program of Jilin Province, China (No. 20220203053SF); National Natural Science Foundation of China (No. 12103053).

Institutional Review Board Statement: Not applicable.

Informed Consent Statement: Not applicable.

Data Availability Statement: Data are unavailable due to privacy or ethical restrictions.

Acknowledgments: Thanks to all the Project team members.

Conflicts of Interest: The authors declare no conflict of interest.

References

1. Mallat, S.G. A theory for multiresolution signal decomposition: The wavelet representation. *IEEE Trans. Pattern Anal. Mach. Intell.* **1989**, *11*, 674–693. [CrossRef]
2. Donoho, D.L.; Johnstone, I.M. Ideal spatial adaptation by wavelet shrinkage. *Biometrika* **1994**, *81*, 425–455. [CrossRef]
3. Gu, G.Q.; Wang, K.F.; Xu, X. Denoising in digital speckle pattern interferometry using fast discrete curvelet transform. *Imaging Sci. J.* **2014**, *62*, 106–110. [CrossRef]

4. Chen, Z.A.; Hu, Z. Remote sensing image denoising based on improved wavelet threshold algorithm. *Bull. Surv. Mapp.* **2018**, 28–31. [CrossRef]
5. Narendra, P.M. A separable median filter for image noise smoothing. *IEEE Trans. Pattern Anal. Mach. Intell.* **1981**, *3*, 20–29. [CrossRef] [PubMed]
6. Eng, H.L.; Ma, K.K. Noise adaptive soft-switching median filter. *IEEE Trans. Image Process. Publ. IEEE Signal Process. Soc.* **2001**, *10*, 242–251. [CrossRef]
7. Guariglia, E. Harmonic Sierpinski Gasket and Applications. *Entropy* **2018**, *20*, 714. [CrossRef]
8. Zheng, X.; Tang, Y.Y.; Zhou, J. A Framework of Adaptive Multiscale Wavelet Decomposition for Signals on Undirected Graphs. *IEEE Trans. Signal Process.* **2019**, *67*, 1696–1711. [CrossRef]
9. Guido, R.C.; Pedroso, F.; Contreras, R.C.; Rodrigues, L.C.; Guariglia, E.; Neto, J.S. Introducing the Discrete Path Transform (DPT) and its applications in signal analysis, artefact removal, and spoken word recognition. *Digit. Signal Process* **2021**, *117*, 103158. [CrossRef]
10. Guariglia, E.; Silvestrov, S. *Fractional-Wavelet Analysis of Positive Definite Distributions and Wavelets on D′(C) D′(C)[C]//Engineering Mathematics II: Algebraic, Stochastic and Analysis Structures for Networks, Data Classification and Optimization*; Springer International Publishing: Berlin/Heidelberg, Germany, 2016; pp. 337–353.
11. Yang, L.; Su, H.; Zhong, C.; Meng, Z.; Luo, H.; Li, X.; Tang, Y.Y.; Lu, Y. Hyperspectral image classification using wavelet transform-based smooth ordering. *Int. J. Wavelets Multiresolution Inf. Process.* **2019**, *17*, 1950050. [CrossRef]
12. Guariglia, E.; Guido, R.C. Chebyshev Wavelet Analysis. *J. Funct. Spaces* **2022**, *2022*, 5542054. [CrossRef]
13. Berry, M.V.; Lewis, Z.V.; Nye, J.F. On the Weierstrass-Mandelbrot fractal function. *Proc. R. Soc. Lond. Math. Phys. Sci.* **1980**, *370*, 459–484.
14. Beck, A.; Teboulle, M. Fast gradient-based algorithms for constrained total variation image denoising and deblurring problems. *IEEE Trans. Image Process* **2009**, *18*, 2419–2434. [CrossRef]
15. Fu, B.; Li, W.W.; Fu, Y.P.; Song, C.M. An image topic model for image denoising. *Neurocomputing* **2015**, *169*, 119–123. [CrossRef]
16. Yuan, J.; He, J. Blocking sparse method for image denoising. *Pattern Anal. Appl.* **2021**, *24*, 1125–1133. [CrossRef]
17. Liao, X.R. An improved ROF denoising model based on time-fractional derivative. *Front. Inf. Technol. Electron. Eng.* **2020**, *21*, 856–865. [CrossRef]
18. Boyat, A.; Joshi, B.K. Image denoising using wavelet transform and median filtering. In Proceedings of the 2013 Nirma University International Conference on Engineering (NUiCONE), Ahmedabad, India, 28–30 November 2013; pp. 1–6. [CrossRef]
19. Song, Q.; Ma, L.; Cao, J.; Han, X. Image Denoising Based on Mean Filter and Wavelet Transform. In Proceedings of the 2015 4th International Conference on Advanced Information Technology and Sensor Application (AITS), Harbin, China, 21–23 August 2015; pp. 39–42. [CrossRef]
20. Morlet, J. Wave Propagation and Sampling Theory. *Geophysics* **1982**, *47*, 222–236. [CrossRef]
21. Meyer, Y. *Wavelets: Algorithms & Applications*; Society for Industrial and Applied Mathematics: Philadelphia, PA, USA, 1993.
22. Mallat, S. *A Wavelet Tour of Signal Processing: The Sparse Way*; Academic Press: Cambridge, MA, USA, 2009.
23. Candes, E.J. Ridgelets: Theory and Applications. Ph.D. Thesis, Stanford University, Stanford, CA, USA, 1998.
24. Candès, E.J.; Donoho, D.L. Ridgelets: A key to higher-dimensional intermittency? *Philos. Trans. R. Soc. Lond. Ser. Math. Phys. Eng. Sci.* **1999**, *357*, 2495–2509. [CrossRef]

applied sciences

MDPI

Article

Fast and Accurate Visual Tracking with Group Convolution and Pixel-Level Correlation

Liduo Liu [1,2], Yongji Long [1,2], Guoning Li [1], Ting Nie [1], Chengcheng Zhang [1,2] and Bin He [1,*]

1 Changchun Institute of Optics, Fine Mechanics and Physics, Chinese Academy of Sciences, Changchun 130033, China; liuliduo21@mails.ucas.ac.cn (L.L.); liguoning@ciomp.ac.cn (G.L.)
2 University of Chinese Academy of Sciences, Beijing 100049, China
* Correspondence: heb@ciomp.ac.cn

Abstract: Visual object trackers based on Siamese networks perform well in visual object tracking (VOT); however, degradation of the tracking accuracy occurs when the target has fast motion, large-scale changes, and occlusion. In this study, in order to solve this problem and enhance the inference speed of the tracker, fast and accurate visual tracking with a group convolution and pixel-level correlation based on a Siamese network is proposed. The algorithm incorporates multi-layer feature information on the basis of Siamese networks. We designed a multi-scale feature aggregated channel attention block (MCA) and a global-to-local-information-fused spatial attention block (GSA), which enhance the feature extraction capability of the network. The use of a pixel-level mutual correlation operation in the network to match the search region with the template region refines the bounding box and reduces background interference. Comparing our work with the latest algorithms, the precision and success rates on the UAV123, OTB100, LaSOT, and GOT10K datasets were improved, and our tracker was able to run at 40FPS, with a better performance in complex scenes such as those with occlusion, illumination changes, and fast-motion situations.

Keywords: feature fusion; pixel-level correlation; Siamese network; attention mechanism

Citation: Liu, L.; Long, Y.; Li, G.; Nie, T.; Zhang, C.; He, B. Fast and Accurate Visual Tracking with Group Convolution and Pixel-Level Correlation. *Appl. Sci.* **2023**, *13*, 9746. https://doi.org/10.3390/app13179746

Academic Editor: Andrea Prati

Received: 25 July 2023
Revised: 12 August 2023
Accepted: 16 August 2023
Published: 29 August 2023

1. Introduction

As one of the research contents in computer vision, visual object tracking has wide application prospects and value in security surveillance, intelligent transportation, autonomous driving, human–computer interaction, autonomous robotics, marine exploration, military target identification, and tracking. Visual object tracking was first carried out using correlation filtering for tracking, and with the development of deep learning, convolutional neural networks have gradually been widely used due to their powerful feature extraction capabilities. Visual object tracking is usually divided into three parts: using a backbone network to extract the target's features, then correlating the template features with the search, and finally utilizing a classification and regression sub-network to predict the center and bounding box of the target. Siamese networks are widely used in object tracking with this structure.

SiamFC [1] first introduced Siamese networks to object tracking. In SiamFC, the template features are correlated with the search features to find the region with the largest response and complete tracking and evaluation. Since then, many works have been carried out on Siamese networks in object tracking. SiamRPN [2] introduced the RPN (region proposal network) structure of object detection to tracking, constructing two branches—one for the regression of the target bounding box, and the other for the classification of the target—where the multi-scale anchor box improves the performance under object scale changes. SiamRPN++ [3] solved the problem of poor results in deep networks due to the destruction of translation invariance when the network is deepened, successfully using ResNet [4] and MobileNet [5] as the backbone networks. SiamFC++ [6] removes the anchor frame and changes the output prediction to an anchor-free style without presetting the anchor frame.

In recent years, transformer structures have boomed in various fields of computer vision. TransT [7] uses the structure of a transformer as the correlation operation, which improves accuracy. Zhao et al. [8] used a transformer structure as the backbone network and utilized a decoder to reconstruct the target appearance within the search region so that the template is close to the search frame, rather than the search frame being directly related to the template image. In this way, the robustness of the tracker is enhanced, even if the appearance of the target has changed. Gao et al. [9] proposed a one-and-a-half-stream structure that uses an adaptive token division method so that the search and template regions have self-attention and cross-attention, as in a two-stream structure, as well as advanced template interactions with the search region, as in a one-stream structure. This structure outperforms some two-stream and one-stream pipelines.

In object tracking, training datasets usually contain many videos and multiple forms of motion. Some annotations may be less accurate due to occlusion and present similarities; thus, some trackers use data processing methods to improve the performance. Yang et al. [10] analyzed the dataset distribution in a low-level feature space and proposed a sample squeezing method to eliminate redundant samples, making the dataset more abundant and informative and increasing the diversity of the dataset. Qi et al. [11] adaptively obtained a tight enclosing box; when the target is in deformation or rotation, the bounding box cannot tightly enclose the target. They also designed a classifier to determine whether the target is occluded or not, which helps to avoid the collection of occluded samples for tracker updates, and to improve accuracy.

However, there are still some challenges in practical applications. Target appearance changes, illumination variation, and occlusion can affect the effectiveness of tracking.

Generally, different features of the object are extracted in different stages of the network. As shown with HDT [12], combining these features from different layers improves the performance of the tracker. HDT uses an improved hedge algorithm to hedge weak trackers from each layer into a strong tracker. In this work, we consider feature fusion by using a 1×1 convolution to concatenate and fuse features from different stages in the Siamese backbone network, which can improve the algorithm accuracy. Meanwhile, in order to improve the detection speed, we use a group convolution for the dimensionality reduction. A group convolution [13] can exponentially reduce the number of parameters compared with a normal convolution, which can speed up the operation. In the correlation stage, we use a new matching method, namely a pixel-level correlation operation, in the network, which is able to obtain a correlation feature map with a smaller kernel size and a more diverse target representation, reducing the interference of background clutter and preserving the target boundary and scale information, which is beneficial to the subsequent prediction.

The main contributions of this work are as follows:

(1) Feature fusion: we use not only the last layer output feature map for prediction but also the feature map of layers 3, 4, and 5 for feature fusion to output the prediction;
(2) Pixel-level correlation: the template features are decomposed into spatial features and channel features, which are matched with the search features, instead of correlating channel-by-channel;
(3) Speed improvement: we use a group convolution for the dimensionality reduction, which reduces the number of parameters and the use of activation functions and normalization in the backbone to speed up the detection;
(4) New attention module: we designed two new attention modules, namely, a multi-scale feature aggregated channel attention block (MCA) and a global-to-local-information-fused spatial attention block (GSA), enabling the network to focus on certain parts of the features and reduce the attention on useless parts, thus improving the performance and accuracy of the model.

The rest of this paper is organized as follows. In Section 2, we present research on object tracking based on Siamese networks published in recent years. Section 3 outlines the core of our tracker, including four parts to improve accuracy, from the lightness to the robustness of the algorithm. Section 4 is the experimental section, which presents an ablation study and a comparison of the results of different trackers on different datasets to analyze the validity of our work. Finally, we conclude the paper in Section 5.

2. Related Work

This section introduces the development of object tracking and some object trackers that have been reported in recent years. Object tracking algorithms can be divided into two categories: one is based on correlation filtering, and the other is based on deep learning. The methods based on correlation filtering include MOSSE [14], KCF [15], and DSST [16]. Correlation filtering introduces the convolution theorem from the signal domain to object tracking and transforms the template matching problem into a correlation operation in the frequency domain. This method is fast in operation but has average accuracy in complex scenarios. In recent years, with the development of deep learning technology and the establishment of large-scale datasets, object tracking algorithms based on convolutional neural networks have gradually emerged, among which Siamese network-based visual object trackers are particularly remarkable. A Siamese network consists of two sub-networks with the same structure and shared parameters, which are initially used for picture similarity analysis and metric learning. SINT [17] and SiamFC [1] first introduced Siamese networks to the visual object tracking field. SiamFC inputs the template picture and search sample, obtains the template feature map and search feature map, and then slides the template feature map over the search feature map as part of the correlation operation. The point with the largest response on the search feature map is considered the prediction target. SiamFC, as a fully convolutional network, has a simple structure and high tracking speed, and many subsequent works have been based on it. SiamRPN [2] introduced the RPN structure from object detection to the tracking field. One branch judges whether the object is in the foreground or background, and the other branch predicts the bounding box of the target. However, these algorithms only use shallow networks, and the tracking effect worsens for deep networks. Through the use of SiamRPN++ [3], it was found that the accuracy of deep networks is reduced because the strict translational invariance is broken, but allowing the target to be shifted in a certain range near the center point during training can alleviate the impact, enabling the successful application of deep networks in tracking algorithms. SiamFC++ [6] uses an anchor-free prediction head that does not set any anchor parameters, eliminating the effect of preset hyperparameters on the generalization ability of the algorithm. There are also some transformer structures used in visual object tracking that have achieved good results.

Although these works achieved good results, the tracking accuracy decreases and the inference speed becomes slower in the face of occlusion, object scale changes, background clutter, and other situations. In this paper, we adopt feature fusion and some simplified methods for complex scenes to reduce the computational cost and improve accuracy at the same time, using pixel-level correlation to reduce the influence of background clutter and to refine the object bounding box.

3. Proposed Method

In this section, we describe the network framework in detail. As shown in Figure 1, our model mainly consists of a Siamese network backbone and two sub-network detection heads for the bounding box classification and regression. The Siamese backbone network is fine-tuned from ResNet50, inspired by the transformer structure, reducing the use of activation functions and normalization, and instead using channel attention [18] and spatial attention [19] modules in the classification and regression sub-networks to make the network more accurate in extracting features. Moreover, to improve the inference speed, a group convolution and 1×1 convolution are used for the dimensionality reduction in the feature fusion stage; both of them accelerate the computation speed and reduce the inference time. The cross-correlation operation no longer uses depth-wise correlation [3]; template features and search features are correlated in a pixel-level matching model, which can effectively reduce background clutter and allow the model to refine the object boundary ranges and focus more on the target.

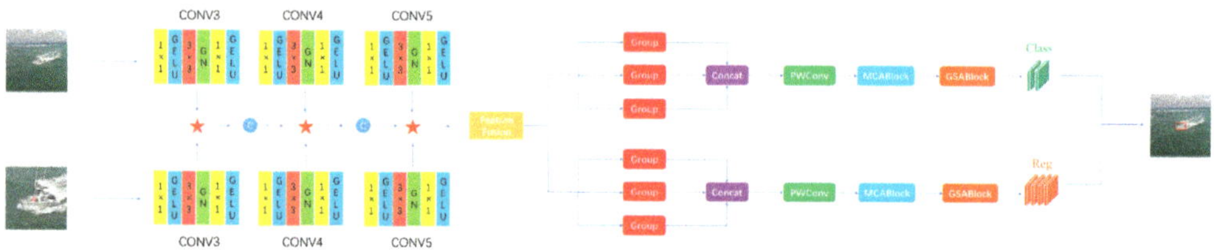

Figure 1. Illustration of our proposed framework. Section 3.1 presents Siamese backbone network, CNN1, CONV3, CONV4, CONV5 represent layer 3, 4, 5 of it. ★ represents the pixel-level correlation method, which is presented in Section 3.2. The feature fusion model is presented in Section 3.3. The classification and regression sub-network using a dual-attention mechanism, CNN2, is presented in Section 3.4.

3.1. Siamese Backbone Framework

Thus far, deep convolutional neural networks have been successfully applied in the field of object tracking. The deepening of these networks has led to improvements in the performance of trackers, such as ResNet [4], ResNeXt [13], and MobileNet [5], which have achieved a good performance. ResNet50, as a classical network, has good robustness and effectiveness and is usually used in trackers as a feature extraction backbone network while modifying the backbone network in order to cater to the accuracy requirements of the tracking task.

Ren et al. [20] proposed Flow Alignment FPN (FAFPN) to align feature maps of different resolutions to solve the semantic misalignment problem when fusing features of different layers. We set the steps of the conv4 and conv5 feature layers to 1 and remove the down-sampling operation so that the output resolution of the last three blocks is the same; meanwhile, to increase the receptive field, the use of a dilated convolution [21] to extract more features has been proven to be effective. Transformers [22], as excellent model architectures, are widely used in various vision tasks. Compared to convolutional neural networks, transformers usually use less activation functions and normalization operations with good results. Inspired by this, a similar method is applied in the backbone.

The original ResNet50 network uses a convolution of 7×7 with a 2-step size in the first layer, following a maximum pooling to complete a 4-fold down-sampling of the input image. The transformer divides the image into patches of the same size and feeds each patch into the network. We change the first layer of the network to a convolution of 4×4 with a 4-step length, with no overlap between convolutions. Compared with the previous one, the convolutional kernel with K = 4 and S = 4 has a smaller kernel size and a larger step size. The computation and parameter numbers are shown in Equations (1) and (2):

$$FLOPs_{old} : \left(\frac{N}{2}\right)^2 \times 7^2 \times 3 \times 64 = 2352N^2 \quad (1)$$

$$FLOPs_{new} : \left(\frac{N}{4}\right)^2 \times 4^2 \times 3 \times 64 = 192N^2, \quad (2)$$

where N denotes the input size, and 3 and 64 are the input and output channels in the first layer of the network, leading to a significant reduction in computation.

Another difference between transformers and CNNs is the use of activation functions and normalization. RELUs are widely used in various CNN networks as simple and efficient activation functions. GELUs, as a variant of RELUs, are used in the latest transformer structures, such as the Swin Transformer and BERT, and can effectively alleviate neuron death and avoid gradient disappearance. Therefore, we use GELUs [23] instead of RELUs.

Traditional convolutional neural networks use an activation function after each layer of convolution. In order to speed up the operation, we remove the activation function after the 3×3 convolution, only using it after the 1×1 convolution.

As for normalization, BN is the most common normalization method, which is widely used in various vision tasks. Meanwhile, the setting of the batch size affects the final result. Models with an insufficient batch size are not suitable for convergence, while there may be a reduction in the generalization ability of models with too large a batch size. Group normalization [24] can be used for the normalization of samples, and it has been used in many application scenarios. We use GN instead of BN and also reduce its use to improve the inference speed. The modified Resnet50 consists of a new bottleneck (see Figure 2), and the inference speed is about 5 FPS faster.

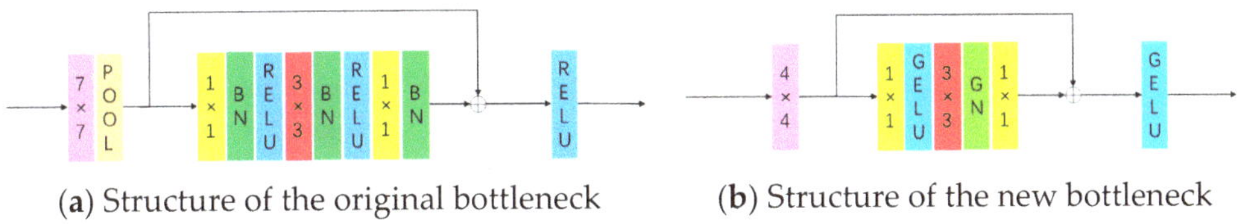

Figure 2. (**a**) original bottleneck using triple activation function and triple normalization. (**b**) new bottleneck using two activation function and one normalization.

3.2. Pixel to Global Correlation

Correlation is the most important part of object tracking, which combines template features with search features and then connects them to the output of the classification and regression sub-networks. Unlike depth-wise correlation [3], which correlates template features with search features channel by channel, in this work, we use pixel to global correlation [25], which decomposes template features and correlates every pixel with the search features to obtain a correlated feature map S. This correlation can effectively suppress background interference, improve the target response on the feature map, and further improve the accuracy of the target bounding box.

The process is shown in Figure 3, where the template features $Z_f \in R^{C \times H_0 \times W_0}$ are first decomposed into spatial feature vectors $Z_s = \{Z_s^1, Z_s^2, \ldots, Z_s^{n_z}\}$, $Z_s^i \in R^{C \times 1 \times 1}$ for each pixel.

$$n_z = H_0 \times W_0 \tag{3}$$

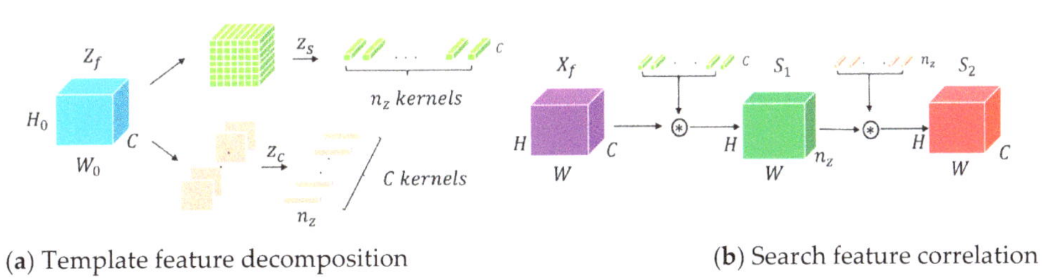

Figure 3. Illustration of pixel to global correlation, where Z_f is the template feature, and X_f is the search feature. (**a**) The template feature is decomposed into feature vectors Z_s and Z_c. Z_s converts the template feature into feature vectors according to each pixel position. Z_c converts the template feature maps of each channel into feature vectors. (**b**) Feature vectors Z_s and Z_c are successively correlated with the search feature X_f to obtain features S_1 and S_2. S_2 is the correlation feature map combining the template and search features.

Similarly, the template features are also converted into channel feature vectors, $Z_c = \{Z_c^1, Z_c^2, \ldots, Z_c^c\}, Z_c^i \in R^{n_z \times 1 \times 1}$, according to the channel dimension. The search features are first correlated with the spatial feature vectors Z_s to obtain feature map S_1 based on Equation (4):

$$S_1 = X_f * Z_s. \quad (4)$$

Then, feature map S_1 is correlated with the channel feature vectors X_f to obtain feature map S_2 based on Equation (5):

$$S_2 = S_1 * Z_c, \quad (5)$$

where $*$ represents the convolution process. Feature map S_2 is obtained after both the channel features and spatial features of the template are correlated. Then, the classification and regression sub-networks complete the target prediction.

Naive correlation [1] and depth-wise correlation [3] use whole template features as kernels to correlate the search features so that the adjacent sliding windows on the feature map produce similar responses, blurring the spatial information. As a refinement method, pixel to global correlation decomposes the template into 1∗1 feature sub-kernels according to the space and channel to correlate the search region, which effectively reduces background interference and further improves the accuracy of the target bounding box, avoiding the blurring of features.

3.3. Feature Fusion

In order to make full use of the features extracted from the backbone network and the advantages of deep networks, features from different layers are used in our feature fusion, and at the same time, in order to speed up the inference, a group convolution [13] is used to first reduce the feature dimensions to simplify the number of parameters and then aggregate the features via a pointwise convolution.

Group convolutions [13] have been widely applied as efficient convolution methods. Their specific process is shown in Figure 4. $C_1 \times H \times W$ is used as the input, and the output is $C_2 \times H \times W$, which represents the channel, height, and width of the convolution. The input is divided into g groups, and each group uses a convolution with a kernel size of $k \times k$ and C_1/g channels. Compared with the number of parameters of an ordinary convolution, i.e., $k \times k \times C_1 \times C_2$, the number of parameters of the group convolution is $k \times k \times C_1 \times C_2/g$, which is $1/g$ of an ordinary convolution, greatly reducing the parameter redundancy. A group convolution is equivalent to decomposing the input and processing the data in parallel, which can speed up the operation. The number of parameters and FLOPs is calculated using Equations (6) and (7):

$$Params_{normal}: k \times k \times C_1 \times C_2, \quad FLOPs_{normal}: k \times k \times C_1 \times C_2 \times H \times W \quad (6)$$

$$Params_{group}: \frac{k \times k \times C_1 \times C_2}{g}, \quad FLOPs_{group}: \frac{k \times k \times C_1 \times C_2 \times H \times W}{g}. \quad (7)$$

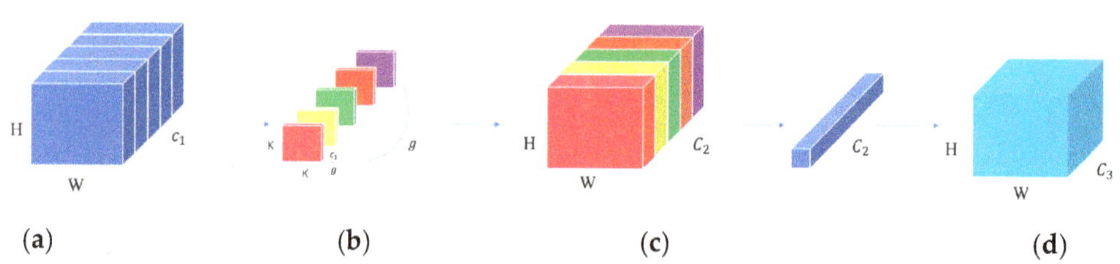

Figure 4. Feature fusion model with a group convolution and pointwise convolution: (**a**) denotes input features, (**b**) denotes group convolution, (**c**) denotes pointwise convolution, and (**d**) denotes output.

Generally speaking, during the tracking process, there may be problems such as illumination changes and scale variation, which require the tracking task to use as much feature information as possible. It is usually considered that in the shallow layer of a network, the network extracts the fine-grained information [26] of the object, such as its color and shape, to help locate the object's position, and as the network deepens, the network extracts the semantic information of the object. Fusing these features from different deep and shallow layers helps to track the target. After correlation, the features of the three stages are concatenated together, and the fusion of the features is implemented using a pointwise convolution [27], which achieves the fusion of cross-channel information quickly and efficiently.

3.4. Classification and Regression Sub-Network

The aim of an attention mechanism is to allow the model to learn how to allocate its own attention and weight the input signal. An attention mechanism scores each dimension of the input and then weights the features according to the score, increasing the weight of interesting parts and decreasing the weight of uninteresting parts, so that the network adaptively highlights the features that are important to the downstream model or task. In this work, two attention modules, namely, channel attention and spatial attention modules, are implemented in the classification and regression sub-network (CNN2), as shown in Figure 5. The features are first reduced in dimensionality via a group convolution [13]; then, a PW convolution [27] is used for feature fusion, and finally the dual channel and spatial attention module is followed.

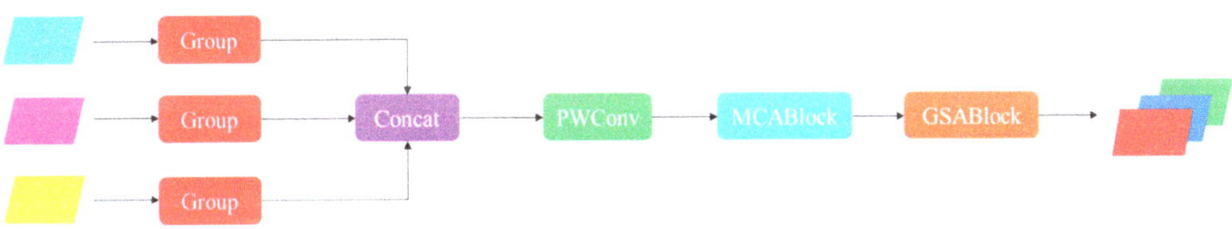

Figure 5. Illustration of the classification and regression sub-network (CNN2).

The multi-scale feature aggregated channel attention block (MCA) is a mechanism for tuning the network at the channel level, as shown in Figure 6. The input features are first divided into four parts, each of which is reduced to half of the original channel via a convolution layer. Two operations are performed independently: one directly uses global average pooling to make the features $1 \times 1 \times C$ in size, with a global perceptual field, aggregating the global features and squeezing information from the channels after the sigmoid activation to obtain the channel weights, which are then multiplied back to the divided features; the other uses an additional convolution layer and then performs the same operation as the former. The four parts adopt the same operation and concatenate together, completing the attention enhancement of the channel dimension, making the network automatically focus on the channels that are important.

The MCA block is based on Equations (8)–(10), where F is the input, S is the spilt operation, Cat is the concatenate operation, δ is the activation function, C_1 and C_2 represent the convolution layers, and GAP stands for global average pooling.

$$F_1 = C_1(S(F)) \tag{8}$$

$$F_{SE1} = Cat(\delta(GAP(F_1)) \times F_1, \delta(GAP(C_2(F_1))) \times C_2(F_1)) \tag{9}$$

$$F_{SE} = Cat(F_{SE1}, F_{SE2}, F_{SE3}, F_{SE4}) \tag{10}$$

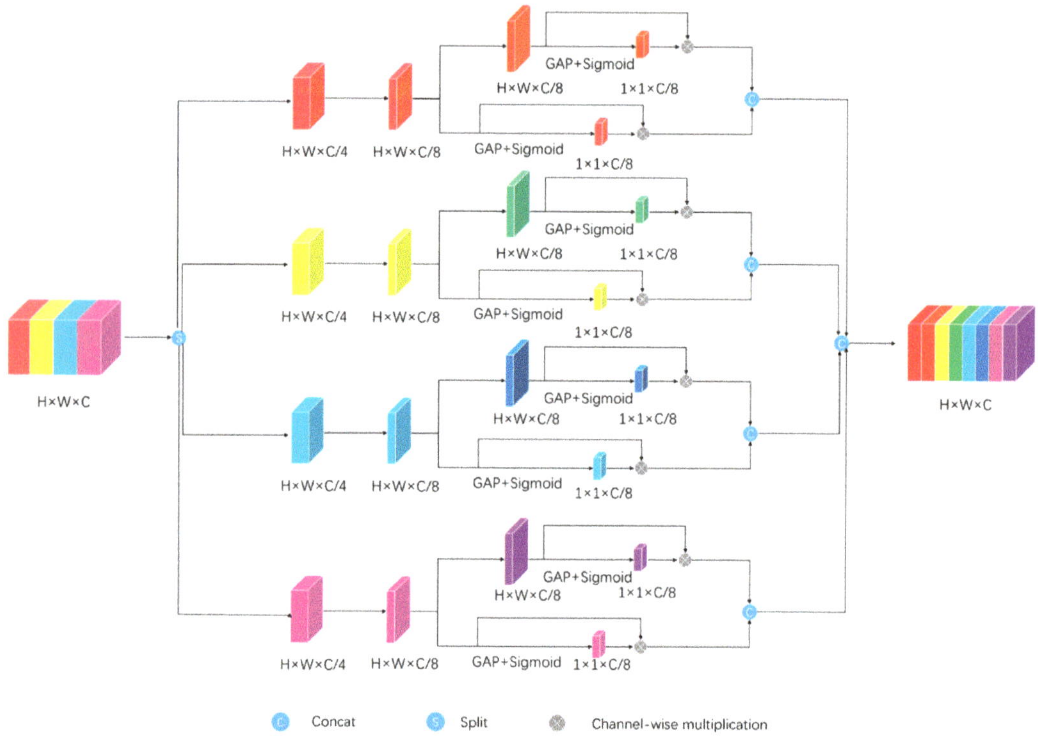

Figure 6. The multi-scale feature aggregated channel attention block (MCA).

The global-to-local-information-fused spatial attention block (GSA) is similar to the channel attention block in that it weights the network from the spatial dimension as shown in Figure 7. The same input features are divided into four parts, using two convolution layers, average pooling, and maximum pooling [28] for each feature point of the network along the channel direction to obtain four 1∗h∗w feature maps. The pooling map and convolution map are concatenated before another convolution layer to obtain weights in the spatial dimension, which are then multiplied back to the input. Two parts are then added to complete the attention enhancement of the spatial dimension, making the network focus on the more important regions. We employ the GSA block in Equations (11)–(13).

$$F_{SPA1} = F \times C_2(Cat(C_1(F), GAP(F))) \tag{11}$$

$$F_{SPA2} = F \times C_4(Cat(C_3(F), GMP(F))) \tag{12}$$

$$F_{SPA} = F_{SPA1} + F_{SPA2}, \tag{13}$$

where GAP and GMP represent average pooling and maximum pooling, F is the input feature, C_1, C_2, C_3, C_4 represent the convolution layers, and Cat is the concatenate operation.

After the template features are correlated with the search features (pixel-level correlation), they are fed into the classification and regression sub-networks (CNN2), which predict whether it is an object or background, along with the bounding box of the target. As shown in Figure 8, the two sub-networks use the same correlation module as the input and do not use separate correlation modules, which also reduces the amount of computation and speeds up the operation of the network. The algorithm finally runs at 40 FPS, which is nearly 9 FPS faster than SiamCAR.

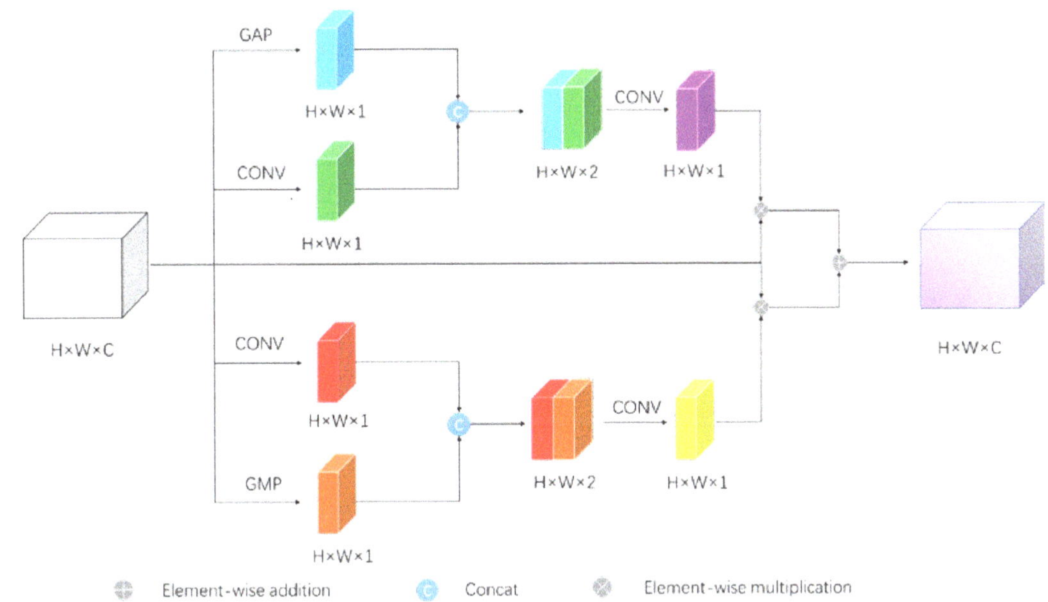

Figure 7. The global to local information fused spatial attention block (GSA).

(**a**) Different correlation model (**b**) Shared correlation model

Figure 8. Different connections between the correlation module and prediction sub-network: (**a**) separate correlation module connected to the classification and regression sub-networks; (**b**) use of a shared correlation module.

4. Experiments

4.1. Implementation Details

The initial model of the backbone was derived from ResNet50 [4] trained on the COCO [29] dataset, a migration learning approach that is commonly used for network training today. We used the Lasot [30], Got10k [31], ImageNet VID [32], and YouTube Bounding Boxes [33] datasets as training sets. The search region was cropped to 255 × 255, and the template region was cropped to 127 × 27 for training. The initial learning rate was 0.001, and 20 training epochs were performed using stochastic gradient descent (SGD). In the first 5 epochs, the learning rate increased from 0.001 to 0.005, and in the last 15 epochs, it gradually decreased from 0.005 to 0.0005. Meanwhile, the parameters of the backbone network were frozen in the first 10 epochs, where only the neck and output parts were trained, and in the last 10 epochs, the parameters of the backbone network were unfrozen, and the network was trained as a whole. Finally, the model was tested and evaluated on the UAV123 [34] and OTB100 [35] datasets.

4.2. Ablation Study

In order to explore the effect of the multi-layer feature fusion, ablation comparison experiments were conducted. Table 1 shows that the use of multi-layer feature fusion is better than just using a single feature, and the effect is better when using the three-layer feature fusion of CONV3, CONV4, and CONV5 than when using the two-layer feature fusion of CONV4 and CONV5, indicating that the features extracted from the different stages of the network are not the same, and fusing multi-layer features is beneficial to improving the tracking accuracy. The correlation method based on pixel matching of the template features also shows an improvement compared to the channel-by-channel correlation method, with an improvement of 0.8% on the UAV123 dataset. The addition of the attention module to the network further improves the effect of the network, and the use of both spatial and channel attention models enables the network to achieve the best effect, with a final accuracy of 65.5% on the UAV123 dataset.

Table 1. Ablation study of the proposed tracker on UAV123. L3, L4, and L5 represent conv3, conv4, and conv5, respectively. DW/Pix stands for depth-wise correlation and pixel to global correlation.

L3	L4	L5	Correlation	MCA Block	GSA Block	AUC
		√	DW			0.616
	√	√	DW			0.620
√	√	√	DW			0.628
√	√	√	Pix			0.636
√	√	√	Pix	√		0.647
√	√	√	Pix	√	√	0.655

In order to analyze the effect of fusing multi-layer features, we tested the model on three datasets. As shown in Table 2, the use of three feature maps from different convolution layers leads to the best results on all three datasets, which shows that the use of multi-layer feature fusion is beneficial to improving the accuracy.

Table 2. Ablation study of the use of feature maps from different layers.

Conv Layers Used	UAV123		OTB100		GOT10K	
	AUC	P	AUC	P	AO	$SR_{0.5}$
Conv5	0.616	0.814	0.690	0.905	0.585	0.680
Conv4, 5	0.620	0.822	0.693	0.907	0.591	0.689
Conv3, 4, 5	0.628	0.827	0.695	0.908	0.594	0.693

Another ablation experiment was conducted to explore the attention mechanism and pixel-level correlation. As shown in Table 3, the baseline uses three convolution layers with pixel-level correlation, while MCA and GSA are the multi-scale feature aggregated channel attention block and the global-to-local-information-fused spatial attention block. Every addition improves the accuracy. In the end, all modules are used, achieving the best performance with an AUC of 65.5% and a precision rate of 85.2%.

Table 3. Ablation study of the attention model and correlation method.

Method	AUC	P_{Norm}	P
Baseline (3layers + pix)	0.636	0.857	0.830
+MCA	0.647	0.869	0.844
+MCA +GSA	0.655	0.876	0.852

4.3. Results on UAV123

UAV123 [34] is a collection of 123 high-definition videos captured using UAVs during aerial photography, containing a variety of targets such as pedestrians, ships, planes, and cars; a variety of scenes including fields, roads, and water, with many activity styles; and occlusions, scale changes, lighting changes, and camera movements in order to increase the tracking challenge. The evaluation metrics include success, precision, and norm precision. Precision is the center position error, using the average center position error of all frames in a sequence to evaluate the performance of the trackers. Success is the proportion of area overlapped between the detection and the real area; generally, the area under the curve is used as its value.

We compared our work with other state-of-the-art trackers, including SiamRPN++ [3], Ocean [36], SiamBAN [37], and SiamGAT [38]. As shown in Figure 9, compared with SiamCAR, our tracker shows a 4.0% improvement in success and a 4.8% improvement in precision.

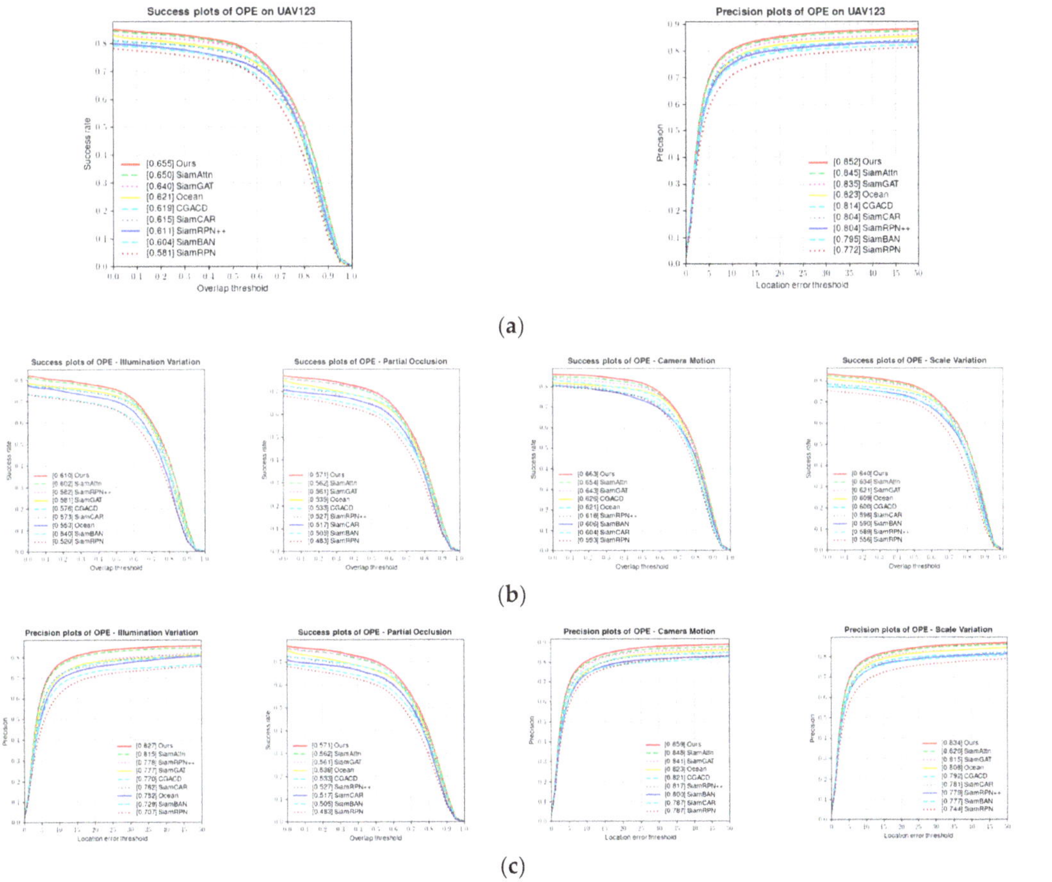

Figure 9. (**a**) Overall success and precision plots of our tracker on UAV123 compared with other trackers. (**b**) Success plot for visual attributes. (**c**) Precision plot for visual attributes.

We also compared the trackers in terms of visual attributes, including illumination changes, occlusion, scale changes, and background clutter, as shown in Figure 10. Our tracker ranks first, which shows that our tracker has the ability to cope with illumination changes, occlusion, and scale changes.

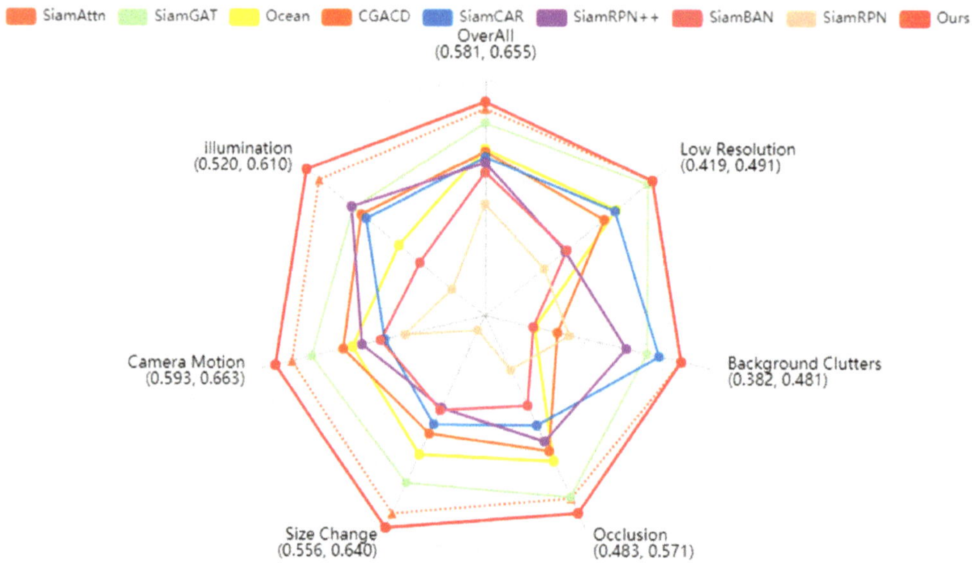

Figure 10. Comparison of success in terms of visual attributes.

4.4. Results on OTB100

OTB100 is a widely used object-tracking dataset. It contains 100 video sequences with attributes such as fast motion, motion blur, and low resolution. We compared our tracker with other state-of-the-art trackers including SiamCAR [39], SiamRPN++ [3], SiamBAN [37], and CFNet [40].

Figure 11 illustrates the success and precision plots of the compared trackers. Our track-er achieves better results than SiamCAR [39] and SiamBAN [37], with a faster speed in terms of scale variation, out-of-plane rotation, low resolution, etc. Our tracker obtains a success rate of 0.701 and a precision rate of 0.914. The integration of the attention and pix-el-level correlation methods enables the tracker to work well in scenarios with low resolution, scale variation, etc.

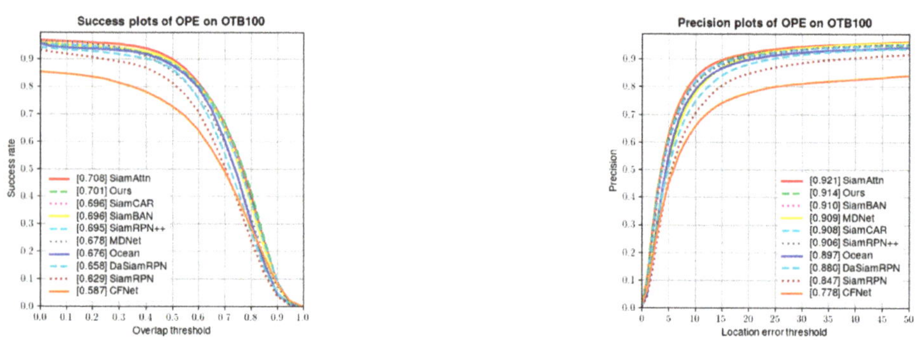

(**a**)

Figure 11. *Cont.*

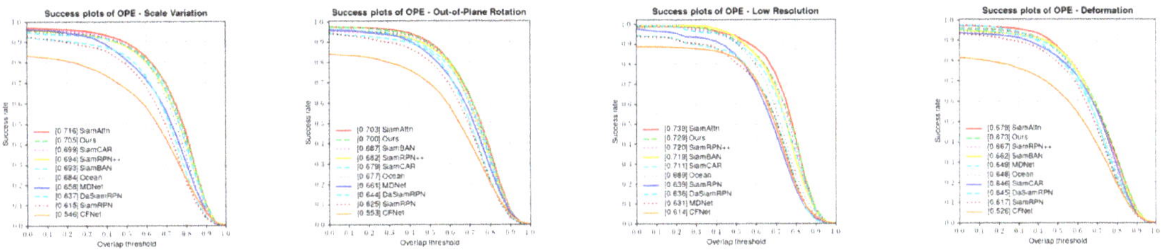

(b)

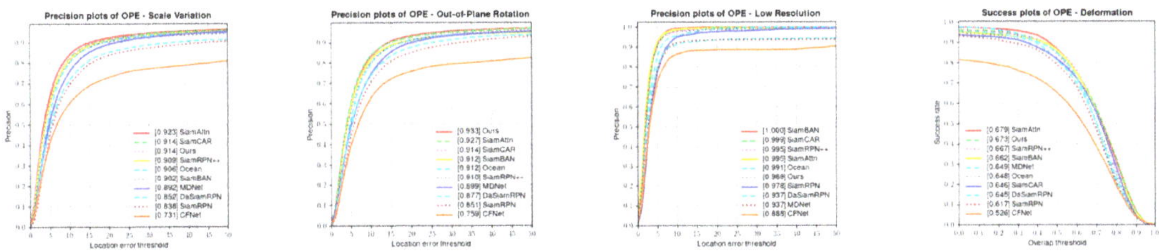

(c)

Figure 11. (**a**) Overall success and precision plots of our tracker on OTB100 compared with other trackers. (**b**) Success plot for visual attributes. (**c**) Precision plot for visual attributes.

4.5. Results on GOT10K and LaSOT

As a large tracking dataset, GOT10K contains more than 10,000 videos, and it is populated with more than 560 categories of moving objects and 87 motion patterns—more than other datasets. We tested our model on the test set. As shown in Table 4, compared with SiamCAR [39], SiamFC++ [6], and Ocean [36], our tracker achieves an AO of 60.7%, which is 1.2% better than that of SiameseFC++ and generally better than that of the other trackers.

Table 4. Comparison with other trackers on the GOT10k test set.

	SiamFC	SiamRPN	SiamRPN++	SiamCAR	SiamFC++	Ocean	Ours
AO	0.374	0.483	0.517	0.569	0.595	0.611	0.607
$SR_{0.5}$	0.404	0.581	0.616	0.670	0.695	0.721	0.713

LaSOT contains 70 object categories and provides an equal number of sequences for each category to mitigate potential category bias, resulting in a collection of 1400 sequences with an average video length of 2512 frames, constituting a high-quality tracking dataset. We tested our tracker on this test set. As shown in Table 5, our tracker outperforms Ocean by 1.2% and has a better performance than the other trackers, which shows its effectiveness and generalizability.

Table 5. Comparison with other trackers on the UAV123, OTB100, and LaSOT datasets in terms of the AUC.

	SiamRPN++	SiamCAR	SiamBAN	CGCAD	PGNet	Ocean	Ours
UAV123	0.611	0.604	0.615	0.623	0.619	0.621	0.655
OTB100	0.695	0.696	0.696	0.691	0.703	0.676	0.701
LaSOT	0.469	0.507	0.514	0.518	0.531	0.560	0.572

Figure 12 shows that our model can track successfully in the face of size variation, occlusion, and low resolution, improving the success and precision rates. The inaccuracy of the boat tracking is due to the fixed viewpoint, and as the boat is traveling from far to near, its size changes rapidly, so the tracker does not work well. Our model aggregates multi-layer features with different receptive fields, which reduces the problem of accuracy degradation due to the change in the size of the object. The person tracking inaccuracy is due to the close distance and high similarity of the two people, resulting in the bounding box containing both. Pixel-level correlation is a more refined correlation method that can refine the bounding box and diminish tracking exceptions caused by background interference. Due to the small size and fast movement of UAVs, tracking errors often occur. The attention module can enhance the feature extraction ability of the network, allowing the network to focus on important features and track successfully. Therefore, our tracker provides a better accuracy than the other algorithms in different situations. Meanwhile, compared to SiamCAR's inference speed of 31FPS, our model runs at 40FPS, representing an improvement of 9FPS, which is an improvement in both speed and success.

(**a**) Car (**b**) Boat (**c**) Person
(**d**) UAV (**e**) Wakeboard (**f**) UAV
(**g**) Person (**h**) Person (**i**) Car

Figure 12. Comparisons of tracking results from different trackers. Targets including person, car, boat, UAV, and images present challenging attributes such as low resolution, occlusion, fast motion, and size variation. Green boxes denote ground truth, yellow boxes are results from SiamCAR, and red boxes are our model results.

5. Conclusions

In this work, we propose a Siamese framework with a group convolution and pixel-level correlation for visual object tracking, with training from end to end, using multi-layer feature fusion and attention mechanisms to improve the feature extraction capability of the network, which works well under fast motion, occlusion, etc. We designed two attention modules: a multi-scale channel attention block (MCA) and a global-to-local spatial attention block (GSA), which enable the network to extract more meaningful features in the classification and regression sub-network. During tracking, pixel-level correlation reduces background interference and provides more refined target boundaries, and it decomposes the template features from the channel and spatial dimensions and uses every pixel feature to correlate the template and search regions. Furthermore, in order to improve the inference speed, our tracker uses a group convolution, which reduces the number of parameters in the network, as well as the use of activation functions and normalization in the backbone. The final inference speed reaches 40FPS, nearly 9FPS faster than that of SiamCAR. Our model achieved a 65.5% success rate and an 85.2% precision rate on the UAV123 dataset, outperforming SianCAR by 4%; a 70.1% success rate and a 91.4% precision rate on the OTB100 dataset; and a 57.2% success rate on LaSOT, outperforming Ocean by 1.2%. Accordingly, our tracker performs better than other trackers and effectively improves the results under lighting changes and occlusion, showing its effectiveness and generalizability.

Author Contributions: Conceptualization, L.L. and B.H.; methodology, L.L.; software, L.L.; validation, L.L., Y.L. and T.N.; formal analysis, G.L.; data curation, L.L.; writing—original draft preparation, L.L.; writing—review and editing, B.H., G.L. and C.Z. All authors have read and agreed to the published version of the manuscript.

Funding: This research was funded by the National Science Foundation of China, grant number 62105328.

Institutional Review Board Statement: Not applicable.

Informed Consent Statement: Not applicable.

Data Availability Statement: Not applicable.

Conflicts of Interest: The authors declare no conflict of interest.

References

1. Bertinetto, L.; Valmadre, J.; Henriques, J.F.; Vedaldi, A.; Torr, P.H. Fully-convolutional siamese networks for object tracking. In Proceedings of the Computer Vision–ECCV 2016 Workshops, Amsterdam, The Netherlands, 8–10 and 15–16 October 2016; pp. 850–865.
2. Li, B.; Yan, J.; Wu, W.; Zhu, Z.; Hu, X. High performance visual tracking with siamese region proposal network. In Proceedings of the IEEE Conference on Computer Vision and Pattern Recognition, Salt Lake City, UT, USA, 18–23 June 2018; pp. 8971–8980.
3. Li, B.; Wu, W.; Wang, Q.; Zhang, F.; Xing, J.; Yan, J. Siamrpn++: Evolution of siamese visual tracking with very deep networks. In Proceedings of the IEEE/CVF Conference on Computer Vision and Pattern Recognition, Long Beach, CA, USA, 15–20 June 2019; pp. 4282–4291.
4. He, K.; Zhang, X.; Ren, S.; Sun, J. Deep residual learning for image recognition. In Proceedings of the IEEE Conference on Computer Vision and Pattern Recognition, Las Vegas, NV, USA, 27–30 June 2016; pp. 770–778.
5. Howard, A.G.; Zhu, M.; Chen, B.; Kalenichenko, D.; Wang, W.; Weyand, T.; Andreetto, M.; Adam, H. Mobilenets: Efficient convolutional neural networks for mobile vision applications. *arXiv* **2017**, arXiv:1704.04861.
6. Xu, Y.; Wang, Z.; Li, Z.; Yuan, Y.; Yu, G. Siamfc++: Towards robust and accurate visual tracking with target estimation guidelines. In Proceedings of the AAAI Conference on Artificial Intelligence, New York, NY, USA, 7–12 February 2020; pp. 12549–12556.
7. Chen, X.; Yan, B.; Zhu, J.; Wang, D.; Yang, X.; Lu, H. Transformer tracking. In Proceedings of the IEEE/CVF Conference on Computer Vision and Pattern Recognition, Nashville, TN, USA, 20–25 June 2021; pp. 8126–8135.
8. Xie, S.; Girshick, R.; Dollár, P.; Tu, Z.; He, K. Aggregated residual transformations for deep neural networks. In Proceedings of the IEEE Conference on Computer Vision and Pattern Recognition, Honolulu, HI, USA, 21–26 July 2017; pp. 1492–1500.
9. Bolme, D.S.; Beveridge, J.R.; Draper, B.A.; Lui, Y.M. Visual object tracking using adaptive correlation filters. In Proceedings of the IEEE Computer Society Conference on Computer Vision and Pattern Recognition, San Francisco, CA, USA, 13–18 June 2010; pp. 2544–2550.
10. Henriques, J.F.; Caseiro, R.; Martins, P.; Batista, J. High-speed tracking with kernelized correlation filters. *IEEE Trans. Pattern Anal. Mach. Intell.* **2014**, *37*, 583–596. [CrossRef] [PubMed]
11. Danelljan, M.; Häger, G.; Khan, F.S.; Felsberg, M. Discriminative scale space tracking. *IEEE Trans. Pattern Anal. Mach. Intell.* **2016**, *39*, 1561–1575. [CrossRef] [PubMed]
12. Tao, R.; Gavves, E.; Smeulders, A.W. Siamese instance search for tracking. In Proceedings of the IEEE Conference on Computer Vision and Pattern Recognition, Las Vegas, NV, USA, 27–30 June 2016; pp. 1420–1429.
13. Hu, J.; Shen, L.; Sun, G. Squeeze-and-excitation networks. In Proceedings of the IEEE Conference on Computer Vision and Pattern Recognition, Salt Lake City, UT, USA, 18–23 June 2018; pp. 7132–7141.
14. Woo, S.; Park, J.; Lee, J.-Y.; Kweon, I.S. Cbam: Convolutional block attention module. In Proceedings of the European Conference on Computer Vision (ECCV), Munich, Germany, 8–14 September 2018; pp. 3–19.
15. Chollet, F. Xception: Deep learning with depthwise separable convolutions. In Proceedings of the IEEE Conference on Computer Vision and Pattern Recognition, Honolulu, HI, USA, 21–26 July 2017; pp. 1251–1258.
16. Zhang, Z.; Peng, H.; Fu, J.; Li, B.; Hu, W. Ocean: Object-aware anchor-free tracking. In Proceedings of the Computer Vision–ECCV 2020: 16th European Conference, Glasgow, UK, 23–28 August 2020; pp. 771–787.

17. Liu, Z.; Mao, H.; Wu, C.-Y.; Feichtenhofer, C.; Darrell, T.; Xie, S. A convnet for the 2020s. In Proceedings of the IEEE/CVF Conference on Computer Vision and Pattern Recognition, New Orleans, LA, USA, 18–24 June 2022; pp. 11976–11986.
18. Wu, Y.; He, K. Group normalization. In Proceedings of the European Conference on Computer Vision (ECCV), Munich, Germany, 8–14 September 2018; pp. 3–19.
19. Devlin, J.; Chang, M.-W.; Lee, K.; Toutanova, K. Bert: Pre-training of deep bidirectional transformers for language understanding. *arXiv* **2018**, arXiv:1810.04805.
20. Lin, M.; Chen, Q.; Yan, S. Network in network. *arXiv* **2013**, arXiv:1312.4400.
21. Liao, B.; Wang, C.; Wang, Y.; Wang, Y.; Yin, J. Pg-net: Pixel to global matching network for visual tracking. In Proceedings of the Computer Vision–ECCV 2020: 16th European Conference, Glasgow, UK, 23–28 August 2020; pp. 429–444.
22. Lin, T.-Y.; Maire, M.; Belongie, S.; Hays, J.; Perona, P.; Ramanan, D.; Dollár, P.; Zitnick, C.L. Microsoft coco: Common objects in context. In Proceedings of the Computer Vision–ECCV 2014: 13th European Conference, Zurich, Switzerland, 6–12 September 2014; pp. 740–755.
23. Fan, H.; Lin, L.; Yang, F.; Chu, P.; Deng, G.; Yu, S.; Bai, H.; Xu, Y.; Liao, C.; Ling, H. Lasot: A high-quality benchmark for large-scale single object tracking. In Proceedings of the IEEE/CVF Conference on Computer Vision and Pattern Recognition, Long Beach, CA, USA, 15–20 June 2019; pp. 5374–5383.
24. Huang, L.; Zhao, X.; Huang, K. Got-10k: A large high-diversity benchmark for generic object tracking in the wild. *IEEE Trans. Pattern Anal. Mach. Intell.* **2019**, *43*, 1562–1577. [CrossRef]
25. Deng, J.; Dong, W.; Socher, R.; Li, L.-J.; Li, K.; Fei-Fei, L. Imagenet: A large-scale hierarchical image database. In Proceedings of the IEEE Conference on Computer Vision and Pattern Recognition, Miami, FL, USA, 20–25 June 2009; pp. 248–255.
26. Ma, C.; Huang, J.-B.; Yang, X.; Yang, M.-H. Hierarchical convolutional features for visual tracking. In Proceedings of the IEEE International Conference on Computer Vision, Santiago, Chile, 7–13 December 2015; pp. 3074–3082.
27. Mueller, M.; Smith, N.; Ghanem, B. A benchmark and simulator for uav tracking. In Proceedings of the Computer Vision–ECCV 2016: 14th European Conference, Amsterdam, The Netherlands, 11–14 October 2016; pp. 445–461.
28. Wu, Y.; Lim, J.; Yang, M.-H. Online object tracking: A benchmark. In Proceedings of the IEEE Conference on Computer Vision and Pattern Recognition, Portland, OR, USA, 23–28 June 2013; pp. 2411–2418.
29. Chen, Z.; Zhong, B.; Li, G.; Zhang, S.; Ji, R. Siamese box adaptive network for visual tracking. In Proceedings of the IEEE/CVF Conference on Computer Vision and Pattern Recognition, Seattle, WA, USA, 13–19 June 2020; pp. 6668–6677.
30. Guo, D.; Shao, Y.; Cui, Y.; Wang, Z.; Zhang, L.; Shen, C. Graph attention tracking. In Proceedings of the IEEE/CVF Conference on Computer Vision and pattern Recognition, Nashville, TN, USA, 20–25 June 2021; pp. 9543–9552.
31. Guo, D.; Wang, J.; Cui, Y.; Wang, Z.; Chen, S. SiamCAR: Siamese fully convolutional classification and regression for visual tracking. In Proceedings of the IEEE/CVF Conference on Computer Vision and Pattern Recognition, Seattle, WA, USA, 13–19 June 2020; pp. 6269–6277.
32. Real, E.; Shlens, J.; Mazzocchi, S.; Pan, X.; Vanhoucke, V. Youtube-boundingboxes: A large high-precision human-annotated data set for object detection in video. In Proceedings of the IEEE Conference on Computer Vision and Pattern Recognition, Honolulu, HI, USA, 21–26 July 2017; pp. 5296–5305.
33. Valmadre, J.; Bertinetto, L.; Henriques, J.; Vedaldi, A.; Torr, P.H. End-to-end representation learning for correlation filter based tracking. In Proceedings of the IEEE Conference on Computer Vision and Pattern Recognition, Honolulu, HI, USA, 21–26 July 2017; pp. 2805–2813.
34. Yu, F.; Koltun, V.; Funkhouser, T. Dilated residual networks. In Proceedings of the IEEE Conference on Computer Vision and Pattern Recognition, Honolulu, HI, USA, 21–26 July 2017; pp. 472–480.
35. Zhao, H.; Wang, D.; Lu, H. Representation Learning for Visual Object Tracking by Masked Appearance Transfer. In Proceedings of the IEEE/CVF Conference on Computer Vision and Pattern Recognition, Vancouver, BC, Canada, 18–22 June 2023; pp. 18696–18705.
36. Gao, S.; Zhou, C.; Zhang, J. Generalized Relation Modeling for Transformer Tracking. In Proceedings of the IEEE/CVF Conference on Computer Vision and Pattern Recognition, Vancouver, BC, Canada, 18–22 June 2023; pp. 18686–18695.
37. Qi, Y.; Zhang, S.; Qin, L.; Yao, H.; Huang, Q.; Lim, J.; Yang, M.-H. Hedged deep tracking. In Proceedings of the IEEE Conference on Computer Vision and Pattern Recognition, Las Vegas, NV, USA, 27–30 June 2016; pp. 4303–4311.
38. Yang, Y.; Li, G.; Qi, Y.; Huang, Q. Release the power of online-training for robust visual tracking. In Proceedings of the AAAI Conference on Artificial Intelligence, New York, NY, USA, 7–12 February 2020; pp. 12645–12652.
39. Qi, Y.; Qin, L.; Zhang, S.; Huang, Q.; Yao, H. Robust visual tracking via scale-and-state-awareness. *Neurocomputing* **2019**, *329*, 75–85. [CrossRef]
40. Ren, H.; Han, S.; Ding, H.; Zhang, Z.; Wang, H.; Wang, F. Focus on Details: Online Multi-object Tracking with Diverse Fine-grained Representation. In Proceedings of the IEEE/CVF Conference on Computer Vision and Pattern Recognition, Vancouver, BC, Canada, 18–22 June 2023; pp. 11289–11298.

applied sciences

MDPI

Article

Development of Smart and Lean Pick-and-Place System Using EfficientDet-Lite for Custom Dataset

Elven Kee *, Jun Jie Chong, Zi Jie Choong and Michael Lau

Faculty of Science, Agriculture and Engineering, SIT Building at Nanyang Polytechnic Singapore, Newcastle University in Singapore, Singapore 567739, Singapore; junjie.chong@newcastle.ac.uk (J.J.C.); zijie.choong@newcastle.ac.uk (Z.J.C.); michael.lau@newcastle.ac.uk (M.L.)

* Correspondence: e.kee@newcastle.ac.uk

Abstract: Object detection for a pick-and-place system has been widely acknowledged as a significant research area in the field of computer vision. The integration of AI and machine vision with pick-and-place operations should be made affordable for Small and Medium Enterprises (SMEs) so they can leverage this technology. Therefore, the aim of this study is to develop a smart and lean pick-and-place solution for custom workpieces, which requires minimal computational resources. In this study, we evaluate the effectiveness of illumination and batch size to improve the Average Precision (AP) and detection score of an EfficientDet-Lite model. The addition of 8% optimized bright Alpha3 images results in an increase of 7.5% in AP and a 6.3% increase in F1-score as compared to the control dataset. Using a training batch size of 4, the AP is significantly improved to 66.8% as compared to a batch size of 16 at 57.4%. The detection scores are improved to 80% with a low variance of 1.65 using a uniform 135-angle lamp and 0 illumination level. The pick-and-place solution is validated using Single-Shot Detector (SSD) MobileNet V2 Feature Pyramid Network (FPN) Lite. Our experimental results clearly show that the proposed method has an increase of 5.19% in AP compared to SSD MobileNet V2 FPNLite.

Keywords: object detection; EfficientDet-Lite; Average Precision

Citation: Kee, E.; Chong, J.J.; Choong, Z.J.; Lau, M. Development of Smart and Lean Pick-and-Place System Using EfficientDet-Lite for Custom Dataset. *Appl. Sci.* **2023**, *13*, 11131. https://doi.org/10.3390/app132011131

Academic Editors: Qi Song and Qinglei Zhao

Received: 9 September 2023
Revised: 2 October 2023
Accepted: 9 October 2023
Published: 10 October 2023

1. Introduction

One of the current trends in advanced manufacturing is to employ Artificial Intelligence (AI) methods to improve the pick-and-place process. The integration of AI and machine vision with pick-and-place operations can significantly improve the manufacturing process. It should be made affordable for Small and Medium Enterprises (SMEs) so that they can leverage the benefits come with this technology without being concerned with allocating a significant financial budget. The fast and smooth integration of machine vision technology with the current pick-and-place operations of SMEs is another crucial aspect that should be taken into consideration.

In this context, any machine vision solution should be developed in a way that the commissioning and installing can be carried out simply and quickly by the field operators of SMEs without special skills. Therefore, one of the current trends in advanced manufacturing is to employ object detection using Deep Learning methods to improve the pick-and-place process. Furthermore, we study the usage of this smart and lean system in low-light environments, such as waferfab manufacturing during the night shift. Die-sorting machines, for example, can be used in low-light environments because they are totally automated and require no user intervention. As such, sufficient ambient lighting is not needed [1], and hence, our aim is to develop a smart and lean pick-and-place solution, which requires minimal computational resources, for custom workpieces.

AI model efficiency has become increasingly important in computer vision, and object detection using machine learning is becoming crucial for pick-and-place operation and warehousing logistics [2,3]. Other than allowing the smart detection of objects without camera recalibration, AI enhances the application's robustness to light changes, shadow, background noise and low-illumination [4].

Continuing from our previous work [5], we aim to improve the pick-and-place operation using an Artificial Intelligence model deployed on an embedded controller. Our target is to develop a smart and lean pick-and-place system that meets the requirements of low power consumption, small memory usage, and fast run time.

Model efficiency has become increasingly important in computer vision to be used on mobile devices and embedded controllers such as Raspberry Pi due to the model's compact size and quick inference. Convolutional Neural Network (CNN) models are crucial for feature extraction and classification, but conventional high-performance algorithms are expensive due to their high-power consumption and memory usage.

TensorFlow has developed several deep learning object identification models that were designed for embedded and mobile devices, hence the name TensorFlow Lite (TFLite). The model's compact size and quick inference for TFLite versions make them appropriate for a variety of applications, including wearable technology and the Internet of Things. A pre-trained TensorFlow model is used to act as a learnt generic model because it has previously been trained on a sizable general dataset. This helps to reduce the quantity of data required to train a model from scratch.

Our project uses the state-of-the-art model EfficientDet-Lite [6], which was developed by Google in 2020. Compared to other state-of-the-art models, EfficientDet-Lite detectors are more precise and need fewer computational resources (working memory, power consumption, and floating-point operations per second or FLOPS) than their predecessors.

TFLite is a model optimization toolkit provided by Google, whose purpose is also to reduce the complexity of deep learning models and speed up the inference time. TfLite's key feature is size reduction, which results in a smaller model and less need for storage space and RAM. TFLite decreases latency by quantization, which helps to simplify calculations during inference with only a little amount of accuracy loss.

TFLite optimizes the model via quantization, clustering, and pruning. The model's accuracy is reduced, as it is quantized from FP64 to lower resolutions (FP32, FP16, INT8). In order to reduce a model's complexity by reducing the number of its unique weights, clustering involves dividing the weights of the trained model's layers into clusters and sharing the centroid of each cluster's weight. Pruning is the process of removing less important model parameters that barely affect the outcomes of predictions. This aids in simplifying the model at the expense of some accuracy loss and offers far higher performance.

EfficientDet uses the same backbone as EfficientNet and adds a bi-directional feature pyramid network (BiFPN) to help in multi-scale feature fusion. It utilizes several optimization and backbone tweaks and a compound scaling method that uniformly scales the resolution, depth and width for all backbones, feature networks and box/class prediction networks at the same time.

Considerable research has already been completed for object detection using AI, laying the foundation for this work. The contribution of every work published previously in all the relevant domains has played a significant role in developing this work. Our pick-and-place approach is similar to a robot collecting a rock from a hopper using a vision-based image processing algorithm [7] and a robot performing pick and place operations on deformable items [8].

Similar to us, some researchers developed novel methods to improve the mean Average Precision. Using Faster-RCNN, Leung [1] developed a vehicle detection approach for insufficient and night-time illumination conditions and improved the mAP values by 0.2; however, it was not meant for a resource-constrained embedded system. Luo [9] achieved an average detection accuracy of 57.51% on EfficientDet-D2 for the road damage detection, which was lower than our 74.1% AP for the Alpha1 dataset. Jain [10] developed "DeepSeaNet" to detect underwater objects with EfficientDet with a high accuracy of 98.63%, but the method was not suitable for lightweight devices due to the complexity of the model.

Similar to our lightweight methodology, Cirjak [11] used EfficientDet-4 on Raspberry Pi to monitor the codling moth population and achieve a high accuracy of 99% with a small dataset of 430 images. Wu [12] adopted EfficientDet to detect textureless objects in an industrial environment, but this method was not suitable for Raspberry Pi due to the heavy computation. Saurabh [13] used a web camera and ABB robot to detect color and perform a color-sorting algorithm; this work has 100% detection but no machine learning implementation. Konaite [14] used another lightweight model, SSD MobileNet V2, on Raspberry Pi to detect barriers for blind people to navigate safely.

EfficientDet is used in non-industrial scenarios such as dental application. Bayaran [15] assessed the diagnostic quality of bitewing radiographs at contact areas between teeth, which can help the oral radiologists provide better radiographic qualities.

Other than EfficientDet, advanced CNN is used in other applications. For example, it is used in implementing traffic signs recognition in a mobile-based application [16], a speech recognition system using TensorFlow [17], a smart surveillance system for night low-illumination objects [18] and a visual feedback algorithm on AlexNet [19].

The effectiveness of illumination to improve the detection scores has not been extensively studied for pick-and-place solutions, especially robots used in warehouses and logistics areas with low illumination, and thus, our research aims to fill this gap.

We implement a smart and lightweight object detector requiring minimal computational resources which controls a pick-and-place system. The illumination effect on the EfficientDet-Lite model deployed in Raspberry Pi are investigated and compared. Our contributions include the following:

- The addition of 8% optimized bright Alpha3 images resulted in a 7.5% increase in Average Precision and a 6.3% increase in F1-score.
- Obtain high detection scores over 80% and low variance of 1.65 by using 135-degree angle and level 0 illumination in accordance with Japanese Industrial Standard (JIS).
- In-depth analysis of EfficientDet-Lite models with training batch sizes 4, 8, and 16. Batch size 4 had the best performance with an overall mean of 66.8% and low standard deviation of 6.23%

The remainder of this paper is organized as follows: In Section 2, we introduce the project setup for a smart and lean pick-and-place solution as well as the data collection and training process. In Section 3, we test our proposed method on the custom dataset and compare it with some other state-of-the-art methods to show the effectiveness and advantages of our method. Conclusions are drawn in Section 4.

2. Materials and Methods

2.1. Materials and Measurements Setup

Figure 1 shows the project setup for our smart and lean pick-and-place solution. The framework of the pick-and-place solution is mainly divided into two parts; object detection and location using an embedded system, and workpiece placement using a robot arm. A Universal Robot 3 (UR3) collaborative robot is used to perform the pick-and-place solution. A Logitech 2D camera of 5 Megapixels (MP)is mounted on the arm of UR3 and connected to the USB port of a Raspberry Pi. Once the object is detected, the General-Purpose Input–Output (GPIO) sends a signal to control the hardwired signal of the UR, as proposed by our previous work [5]. Using 2D cameras is beneficial to the pick-and-place system, as they are more affordable and have good accuracy [20]. Compared to other vision solutions, a 2D camera detection engine is easy to use, thus increasing their adaptability and flexibility to different custom-made objects.

Figure 2 shows the flowchart for our pick-and-place solution using the Raspberry Pi. Using OpenCV on Raspberry Pi, the data images are captured, and the images are annotated using the online Roboflow tool. The dataset is then preprocessed, augmented, and run on Google Colab for model building and training. Using TensorFlow's Model Maker [21], the model is built, trained, and evaluated before being converted to TensorFlow Lite. The lite version consumes less memory compared to the original version.

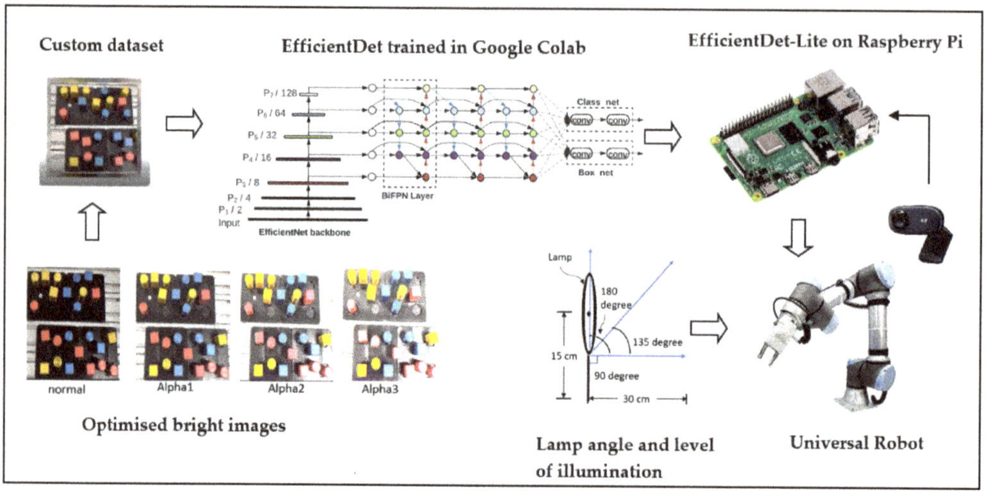

Figure 1. Project setup of our smart and lean pick-and-place solution.

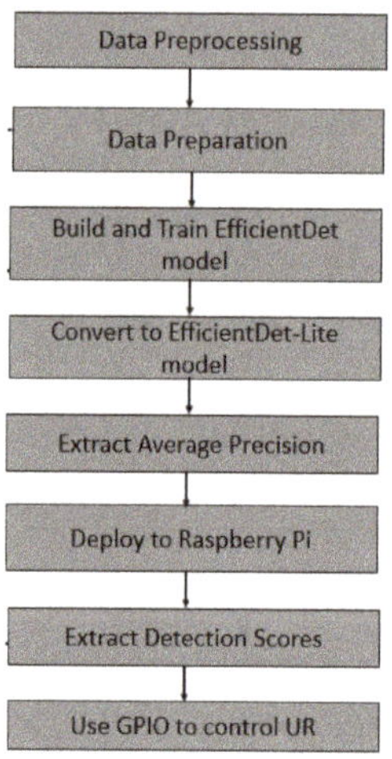

Figure 2. Flowchart with dataset, batch size and lamp control.

This project used 2 datasets for training: Dataset 1 for optimized bright and illumination level and Dataset 2 for batch test. The mean and variance AP are then extracted and compared. For the batch test, we compare the AP for batch sizes 4, 8 and 16. For the illumination test, we use 3 lamp angles and compare the best detection scores.

The project uses custom datasets for the initial training process. Three different colors (yellow, blue, red) are used, and each color has 2 different shapes (cylinder, cube), as shown in Figure 3. The project uses Roboflow [22] as an online annotation and data splitting tool. Using the Label Assist tool, the highest mAP is used to annotate images, and the confidence is lowered to 20%. The confidence level is lowered to make the annotation visible for all objects. The overlap is reduced to 50% to enable detection of workpiece with poor confidence. These values are chosen after trial and error in order for the AI tool to add most of the annotations automatically. For faster annotation, the zoom and lock view functions are used to ensure more accurate annotation as the workpiece looks bigger. A Roboflow's Healthcheck is performed to ensure the distribution of workpieces in the dataset.

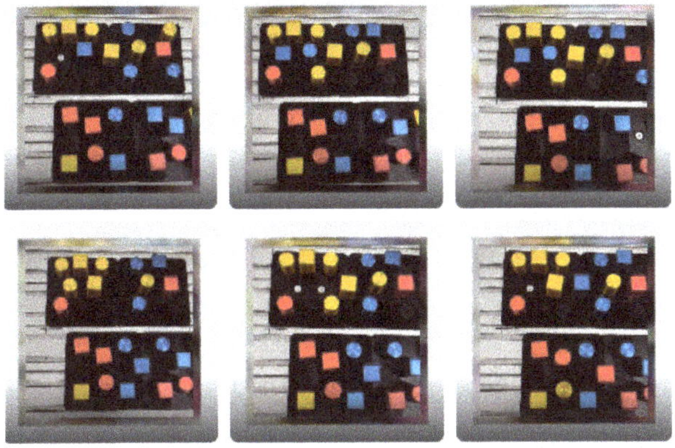

Figure 3. Custom dataset.

According to our previously published research [5], EfficientDet-Lite4 has a large file size and may not be suitable for Raspberry Pi due to its memory limitations. Hence, in this project, we specifically use EfficientDet-Lite 0 to 3 as our models, and these four architectures are trained using the TensorFlow Lite Model Maker [23] library.

All Tensorflow models are developed and evaluated using Google Colab [23] with a Graphic Processing Unit (GPU) Hardware Accelerator. The proposed framework aims to be a complete object detector that only needs minimal data processing before and after model detection and classification. After training, the EfficientDet Lite models are quantized (weights of 8-bit integer) using Post-Training Quantization.

For our object detection, the evaluation criteria are the mean Average Precision (mAP) and F1-score. According to the validation dataset COCO2017, its mAP is the same as Average Precision in Tensorflow Model Maker [23]. For our pick-and-place application, ARmax10 is chosen as the Recall value as we expect to have a maximum of 10 detections per pick-and-place application.

The formula of the mean Average Precision is given as below:

$$mAP = \frac{1}{n}\sum\nolimits_{k=1}^{k=n} AP_k \tag{1}$$

where AP_k is the Average Precision of class k, and n is the number of classes.

The F1-score is used to evaluate the models' accuracy, since it allows for the simultaneous maximization of two metrics that are well known in this field: Precision, which measures the detections of objects, and Recall, which measures the objects that are detected. The F1-score is calculated based in the mAP and Recall value in the formula below:

$$\text{F1} = \frac{2 * mAP * ARmax10}{mAP + ARmax10} \tag{2}$$

To speed up the training, the number of epochs is fixed at 50 for the 4 models. The threshold for detection is set at 30%. The standard deviation of the AP is calculated as follows:

$$\sigma = \sqrt{\frac{\sum(x_i - \mu)^2}{N}} \tag{3}$$

where σ is the population standard deviation, N is the size of the population, x_i is the value of an AP and μ is the population mean of all APs.

The variance of the AP is calculated as follows:

$$S^2 = \frac{\sum(x_i - \mu)^2}{N} \quad (4)$$

where S^2 is the population variance, N is the size of the population, x_i is the value of an AP and μ is the population mean of all APs.

Data preprocessing decreases the training time and increases the performance of the AI model by applying image transformations to all images in this dataset, as shown in Table 1. To prevent overfitting of the model, the data augmentation process is shown in Table 2. Both data preprocessing and augmentation are completed using the online Roboflow tool.

Table 1. Preprocessing operations applied to the original images.

Model	Setting	Description	Batch Size
Auto-Orient	Activated	Rotate image 15° counter-clockwise	Discard EXIF rotations and standardize
Resize	416 × 416	Resize all the images to square size	416 is divisible by 16

Table 2. Augmentation operations applied to the original images.

Model	Setting	Description	Comments
Rotation	−15°	Rotate image 15° counter-clockwise	Add variability to perspective to be more
Rotation	15°	Rotate image 15° clockwise	resilient to camera's angle
Shear	Horizontal 15°	Shear image horizontally by 15°	Add variability to perspective to be more
Shear	Vertical 15°	Shear image vertically 15°	resilient to camera's pitch and yaw

This project uses an augmentation technique to increase the number of images from a small number of images. Two datasets were built: Dataset 1 with 124 images to study the illumination effect and Dataset 2 with 82 images to study effect of batch size. Dataset 2 is a small dataset consisting of 82 images, and it is used for batch testing. Our objective is to determine whether there is a visible increase in mAP due to batch size despite the small datasets. Table 3 presents the two datasets utilized for the three test trials.

Table 3. Augmented dataset for illumination, optimized brightness and batch size test.

Test Experiment	Dataset	Test Application	TensorFlow Model	Batch Size	Number of Original Images	Number of Augmented Images	Augmented Ratio
1	Dataset 1	Optimized bright	EfficientDet-Lite 2	8	124	1006	8.12
2	Dataset 1	Illumination level	EfficientDet-Lite 2	8	124	1006	8.12
3	Dataset 2	Batch size	EfficientDet-Lite 0 EfficientDet-Lite 1 EfficientDet-Lite 2 EfficientDet-Lite 3	4, 8, 16	82	333	4.05

For test experiments 1 and 2, the model chosen is EfficientDet-Lite2 with a batch size of 8, as it is lightweight and has a good trade-off between training speed and accuracy. For test experiment 3, we vary the batch size to 4, 8 and 16, and we also vary the 3 Tensorflow models from EfficientDet-Lite 0 to 3.

2.2. Image Optimization Process to Improve Mean Average Process

Table 4 shows the distribution of the workpiece in Dataset 1, where there are a total of 963 objects. The blue cylinder has highest distribution at 35.61%, which is followed by the blue cube at 20.87%. The red cylinder has the lowest distribution at 13.33%.

Table 4. Distribution of workpiece in Dataset 1.

Total Objects	Number of Blue Cube	Number of Blue Cylinder	Number of Yellow Cylinder	Number of Yellow Cube	Number of Red Cube	Number of Red Cylinder
963 (100%)	201 (20.87%)	343 (35.61%)	128 (13.29%)	162 (16.82%)	166 (17.23%)	129 (13.33%)

The proposed pick-and-place method aims to be a complete object detector that only needs minimal data processing before and after model detection and classification. Therefore, we choose a small number of images to evaluate the effectiveness of this approach. This study emphasizes the necessity of rapid and accurate label annotations, as well as fast data processing with small datasets, using a smart and "lean" approach.

Figure 4 shows the augmentation process, where 124 original images are augmented to 327 images using the Flip-90 degree, Flip-180 degree and Rotate-15 degrees manipulation. Similarly, the 10 bright images (8% of the total images) are augmented to 24 images. Using the Roboflow online tool, the images are merged to produce 1006 augmented image after undergoing image shearing manipulation.

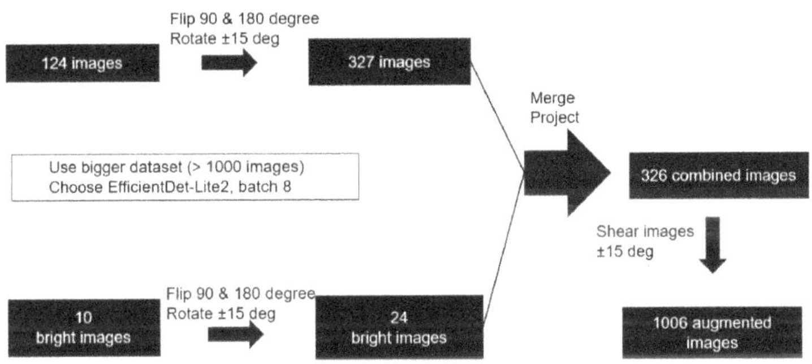

Figure 4. Process of augmentation for Dataset 1.

For the bright images, 10 images of different alpha values are added to the dataset, as shown in Table 5. Using the addWeighted function from OpenCV, a brightness of alpha(α) parameter is overlayed onto the 10 images. The addWeighted function is a function that helps by adding two images (f0 and f1) by passing varying α values of 1, 2 and 3 into the formula below.

$$g(x) = (1 - \alpha)f0(x) + \alpha f1(x) \quad (5)$$

Table 5. Optimization process with Alpha datasets.

Optimization Process	Control Group	Alpha1 Dataset	Alpha2 Dataset	Alpha3 Dataset
Base image + 10 normal images	124 + 10 normal			
Base image + 10 bright Alpha1 images		124 + 10 bright level 1		
Base image + 10 bright Alpha2 images			124 + 10 bright level 2	
Base image + 10 bright Alpha3 images				124 + 10 bright level 3

The datasets are named as follows. Alpha1 has 124 base images with 10 bright images. Alpha2 has 124 base images with 10 bright images, and Alpha3 has 10 bright images. The optimization is shown in Figure 5, where the results of brightening are clearly visible.

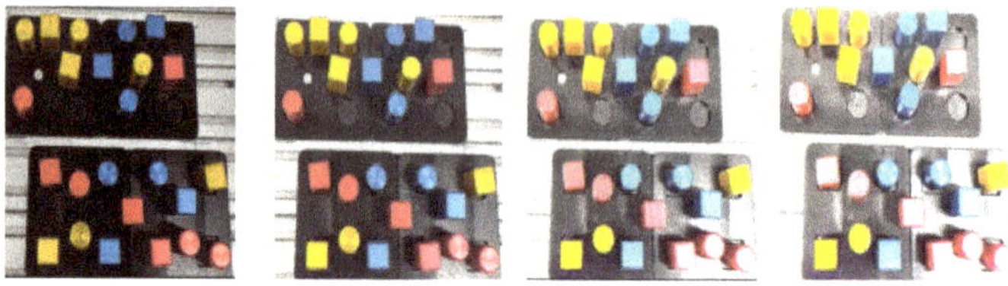

Figure 5. Optimized bright image using OpenCV: (**a**) normal; (**b**) Alpha1; (**c**) Alpha2; (**d**) Alpha3.

2.3. Illumination Level Setup to Improve Detection Scores

According to the Japanese Industrial Standard (JIS) Z 9110-1979 [24], the recommended level of illumination shall be as shown in Table 6 below. Using an illuminance (lux) meter and a commercially available lamp, the appropriate level of illumination is determined for a specific work location in a typical factory environment, such as the packing, assembly and inspection area.

Table 6. JIS recommend level of illumination.

Illumination Level	Lux Range	Work Areas
Level 0	Less than 5	Darkroom and indoor emergency stairways
Level 1	150 to 300	Wrapping and packing
Level 2	300 to 750	Assembly, test and ordinary visual work
Level 3	750 to 1500	Inspection, selection and precise visual work
Level 4	1500 to 3000	Inspection, selection and extremely precise visual work

It is important to note that the detection scores are affected by the surface reflection since the workpieces are made of metallic materials. Rather than mounting the lamp directly above the workpieces, the lamp is positioned on the side so that the light does not reflect directly from the metallic surface. As shown in Figure 6, the base of the lamp is positioned 30 cm horizontally and 25 cm vertically from the workpieces, with 3 angles of 180 degrees, 90 degrees and 135 degrees from the lowest vertical point lamp. This project focuses on pick-and-place applications within darkrooms, corridors, and poorly illuminated areas with less than 300 Lux levels, as it is less extensively studied.

Table 7 shows the on-site lux measurement as well as classification of areas for the proposed robot implementation in an assembly line.

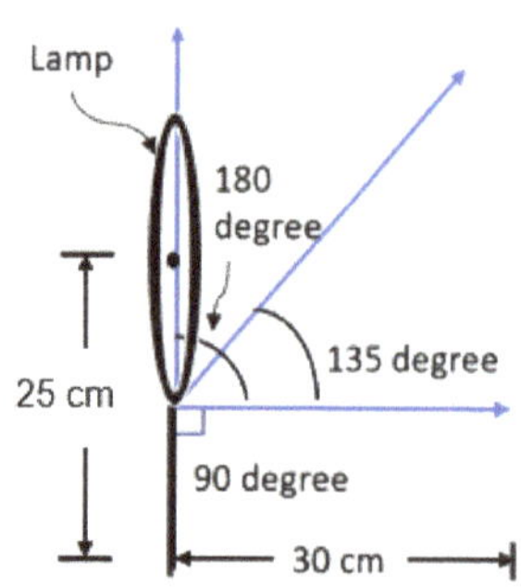

Figure 6. Setup of angled lamp.

Table 7. Classification of work areas.

Illumination Level	On-Site Lux Measurement	Work Areas	Application
0	6	Darkroom, indoor emergency stairways	Robot in INDOOR in darkroom
1	242	Wrapping and packing	Robot in INDOOR doing packing
2	663	Assembling, testing and ordinary visual work	Robot in INDOOR doing assembly
3	950	Inspection, selection and precise visual work	Robot in OUTDOOR doing inspection
4	1212	Inspection, selection and extremely precise visual work	Robot in OUTDOOR/DIRECT SUNLIGHT doing detailed inspection

To ensure reliability and determine the average detection score, a minimum of three readings are taken. Figure 7 illustrates the setup of three different lamp angles (180, 135 and 90). A lamp angle has two adjustable levels of illumination—lower than 150 lux and between 150 and 300 lux.

(a) (b) (c)

Figure 7. Illumination setup for varying levels: (**a**) 180 degree; (**b**) 135 degree; (**c**) 90 degree.

To prevent the direct reflection of the light by the metallic surface, the lamp is mounted at the side instead of directly above the workpieces. The reason for this setup is to reduce the direct glare from the lamp while providing enough illumination, as shown in Figure 8. The number of frames per second (FPSs) can be seen on the top-left corner of the image. The speed is typically in the range of 0.8 frames per second as the Python code used to display the results using OpenCV uses a portion of the processing time.

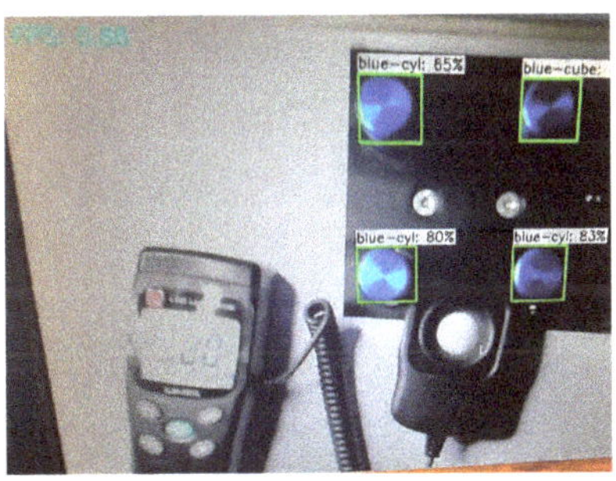

Figure 8. Best detection scores at different illumination levels.

2.4. Training Batch Size Configuration to Improve Mean Average Precision

Table 8 below defines their training settings. The batch size is set based on the GPU Random Access Memory (RAM) with the default learning rate and input resolution. The Tensorflow model is trained with training data with batch sizes of 4, 8 and 16, respectively. In this project, a batch size of 4 is abbreviated to "Batch4", a batch size of 8 is abbreviated to "Batch8", and a batch size of 16 is abbreviated to "Batch16".

Table 8. Training setting of EfficientDet Lite0 to Lite3.

Model	Input Resolution	Learning Rate	Batch Size	Epochs
EfficientDet-Lite0	320 × 320	0.08	4, 8, 16	50
EfficientDet-Lite1	384 × 384	0.08	4, 8, 16	50
EfficientDet-Lite2	448 × 448	0.08	4, 8, 16	50
EfficientDet-Lite3	512 × 512	0.08	4, 8, 16	50

The model training and validation are completed in Google Colab (Python 3.8) using Tensorflow Model Maker on free GPU. The Tensorflow model is trained with training data with batch sizes of 4, 8 and 16, respectively. To speed up the training, the number of epochs is fixed at 50 for the 4 models.

Dataset 2 is used for the batch size training; it is deliberately chosen to be a small dataset consisting of 82 images and 1022 objects. As shown in Table 9, it has a healthy distribution percentage of the workpieces. The small dataset allows fast training times and practical data preparation. Its purpose is to determine an optimal number of batch sizes with a good trade-off between the accuracy of the detected objects and training speed.

Table 9. Equal distribution of workpiece in Dataset 2.

Total Number of Objects	Number of Blue Cube	Number of Blue Cylinder	Number of Yellow Cylinder	Number of Yellow Cube	Number of Red Cube	Number of Red Cylinder
1022 (100%)	136 (13.3%)	135 (13.2%)	216 (21.1%)	208 (20.3%)	150 (14.6%)	177 (17.3%)

3. Results

3.1. Results of Optimized Bright Images on Average Precision

Table 10 shows the results of the evaluation metrics using EfficientDet-Lite2, and training was completed for 50 epochs. The results demonstrate that Dataset 3 has the highest F1-score, which is to be expected given that it has the highest AP and AR Max10. The best performance comes from Alpha3, as it results in a 10.2% increase in Average Precision and a 6.3% increase in F1-score as compared to the Control Dataset.

Table 10. The Average Precision and calculation of F1-score.

Average Precision	Control Dataset (%)	Alpha1 (%)	Alpha2 (%)	Alpha3 (%)
AP (mAP)	73.5	75.7	70.9	81.0 (+7.5%)
AP Tflite	73.3	74.1	69.5	79.0
AR Max10	76.7	78.6	78.5	81.9
F1-score	75.1	77.1	74.5	81.4 (+6.3%)

The color of the workpiece has a direct effect on the accuracy, as shown in Table 11. The blue cube performs the best under Alpha3 with an Average Precision increase of 24.6% while the red cube performs the worst with the AP dropping by 28.6% in the Alpha2 dataset and further to less than the threshold value of 30% in the Alpha3 dataset. One explanation could be that red colors reflect more light, whilst blue colors absorb more. The blue cube has the highest AP (90.4%) in the Alpha3 dataset, while the red cube has the lowest AP in the Alpha2 (45%) and in Alpha3 datasets (29%). As our goal here is to have consistent high detection scores with low variance, we choose Alpha1 instead of Alpha3 for our subsequent projects. The average AP for Alpha1 is 0.8% higher than the control dataset. Hence, Alpha1 is chosen to increase the AP while maintaining low variance.

Table 11. Average Precision with Alpha datasets.

Average Precision	Control Dataset (%)	Alpha1 (%)	Alpha2 (%)	Alpha3 (%)
Yellow cube	78.9	78.1	78.9	76.2
Yellow cylinder	76.3	72.2	82.7	83.3
Red cube	73.6	85.1	45.0 (−28.6%)	29.0
Red cylinder	73.4	65.9	55.0	66.7
Blue cube	65.8	66.1	77.6	90.40 (+24.6%)
Blue cylinder	71.9	77.3	73.8	78.4
Overall APs	73.3	74.1 (+0.8%)	69.5	65.8
Variance of APs (%)	16.4	0.47	1.924	3.99

Figure 9 shows that in general, all the APs for Alpha1 and Alpha2 are higher than that of the COCO2017 dataset, which is 33.97%. For Alpha3, only the red cube has a lower AP than the validation mAP. The overall mean of Alpha1 is 6.2% higher than that of Alpha2. Therefore, for a high and consistent object detection approach, we use the Alpha1 dataset in our pick-and-place solution.

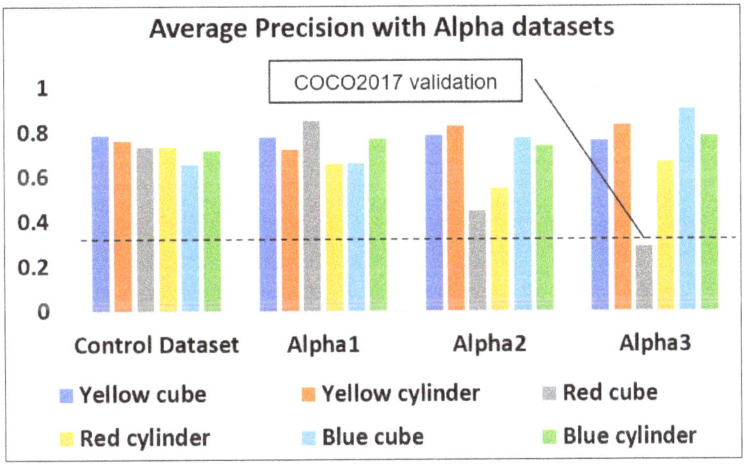

Figure 9. Average Precision with Alpha datasets.

3.2. Results of Illumination Level on Detection Scores

Table 12 shows the results of various illumination angles and levels. The average is taken from the three best detection scores for all the three different lamp angles (180, 135 and 90). Each angle has two levels of illumination—a lux level less than 150 and a lux level in between 150 and 300. The lux measurements are taken to ensure that the detection occurs during the specified illumination level, as per the JIS specification. The threshold for detection is set at 30% and, hence, if there is any detection less than 30%, it will be shown as 'Nil' below.

Table 13 shows the average detection score of workpieces and the variance. The 135-degree level 1 has the highest average detection score (81.17%), which is followed by the 135-degree level 0 at 80.50%. The 180-degree level 1 has the lowest detection scores of 77.94%. The lowest variance is 1.65% at the 135-degree level, and the highest variance is 25.08% for the 90-degree level 1.

Table 12. Detection scores and lux level for the different angles and intensities of the lamps.

Angle and Lux Level of Lamp	Class	Reading 1 (%)	Reading 2 (%)	Reading 3 (%)	Average Reading (%)	Measured Lux Value (lm/m^2)
180° Level 0	Red cylinder	85	77	75	79.00	25
	Red cube	83	80	69	77.33	23.5
	Yellow cylinder	85	77	78	80.00	25.7
	Yellow cube	91	77	85	84.33	25.7
	Blue cylinder	77	78	75	76.67	24.1
	Blue cube	77	73	75	75.00	24.1
180° Level 1	Red cylinder	80	85	75	80.00	187.9
	Red cube	80	62	65	69.00	187.8
	Yellow cylinder	85	80	85	83.33	188.6
	Yellow cube	92	57	86	78.33	188.9
	Blue cylinder	75	73	83	77.00	188.1
	Blue cube	83	70	83	80.00	187.5
135° Level 0	Red cylinder	78	83	86	82.33	50
	Red cube	80	83	83	82.00	48
	Yellow cylinder	78	80	83	80.33	50
	Yellow cube	77	78	85	80.00	49
	Blue cylinder	77	82	77	78.67	50
	Blue cube	85	71	83	79.67	49
135° Level 1	Red cylinder	75	87	83	81.67	175
	Red cube	86	83	86	85.00	172
	Yellow cylinder	76	75	83	78.00	177
	Yellow cube	80	89	83	84.00	174
	Blue cylinder	73	80	78	77.00	175
	Blue cube	75	83	86	81.33	176
90° Level 0	Red cylinder	89	51	89	76.33	28
	Red cube	91	80	88	86.33	25
	Yellow cylinder	91	39	83	71.00	28
	Yellow cube	94	48	85	75.67	26
	Blue cylinder	80	57	78	71.67	28
	Blue cube	89	70	69	76.00	26
90° Level 1	Red cylinder	90	65	85	80	206
	Red cube	93	76	83	84	189
	Yellow cylinder	91	49	82	73.67	209
	Yellow cube	92	62	89	81	192
	Blue cylinder	80	53	86	73	207
	Blue cube	88	80	65	77.67	201

Table 13. Detection scores and Lux level for 180-degree level 1 lamp.

Class	180° Level 0 (%)	180° Level 1 (%)	135° Level 0 (%)	135° Level 1 (%)	90° Level 0 (%)	90° Level 1 (%)
Yellow cube	79.00	80.00	82.33	81.67	76.33	80
Yellow cylinder	77.33	69.00	82.00	85.00	86.33	84
Red cube	80.00	83.33	80.33	78.00	71.00	73.67
Red cylinder	84.33	78.33	80.00	84.00	75.67	81
Blue cube	76.67	77.00	78.67	77.00	71.67	73
Blue cylinder	75.00	80.00	79.67	81.33	76.00	77.67
Average	78.72	77.94	80.50	81.17	76.17	78.22
Variance	8.86	19.75	1.65	8.40	25.08	15.43

Hence, to increase the average detection score to over 80%, as shown in Figure 10 below, we propose to use a 135-degree lamp. We can observe that the illumination distribution is more even and that there is less surface reflection with a 135-degree lamp of level 0 illumination because of its low variance.

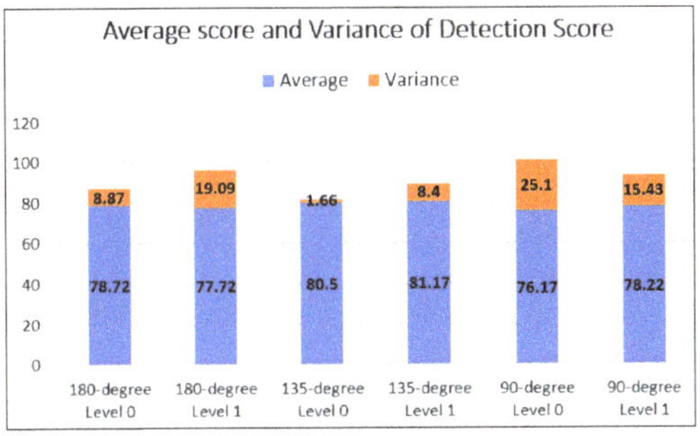

Figure 10. Bar graph of detection scores with average and variance.

3.3. Results of Variation of Batch Size on Average Precision

A comparison of Average Precision across Efficient-Det Lite models is built in order to investigate which model performs the best with a small dataset. As shown in Table 14 below, Batch4 has the highest overall AP at 66.8%, and it has the lowest standard deviation at 6.23%. In contrast, Batch16 has the lowest overall AP at 57.4% and highest standard deviation at 7.87%. According to Keskar [25], this is because the large-batch methods tend to converge to sharp minimizers of the training and testing function and as is well known, sharp minima lead to poorer generalization. In contrast, small-batch methods consistently converge to flat minimizers due to the inherent noise in the gradient estimation. As compared to Batch16, both Batch4 and Batch8 have higher average APs of 66.1 ± 0.7% and standard deviations of 6.245 ± 0.015%.

Table 14. Variation of batch size.

TFLite Model	Average Precision	Batch4 (%)	Batch8 (%)	Batch16 (%)
EfficientDet-Lite0	Yellow cube	72.1	64.4	66.3
	Yellow cylinder	47.9	53.2	40.2
	Red cube	66.2	71.4	73.2
	Red cylinder	62.7	64.8	56.6
	Blue cube	63.4	62.2	55.6
	Blue cylinder	59.1	59.9	54.4
EfficientDet-Lite1	Yellow cube	56.9	73.7	55.5
	Yellow cylinder	72.5	50.3	40.7
	Red cube	68.6	70.4	58.2
	Red cylinder	68.1	64.0	53.9
	Blue cube	63.9	63.8	49.6
	Blue cylinder	71.4	59.3	50.7
EfficientDet-Lite2	Yellow cube	73.5	70.8	67.7
	Yellow cylinder	62.4	59.0	49.2
	Red cube	72.3	75.8	70.7
	Red cylinder	68.4	63.0	55.5
	Blue cube	63.3	63.8	58.3
	Blue cylinder	70.8	70.3	56.8
EfficientDet-Lite3	Yellow cube	72.5	69.9	59.3
	Yellow cylinder	57.6	57.6	54.7
	Red cube	73.8	71.8	66.0
	Red cylinder	66.0	68.9	57.4
	Blue cube	63.1	62.5	55.5
	Blue cylinder	68.4	67.2	53.3
	Average of APs	66.8	65.4	57.4
	Standard deviation of APs	6.23	6.26	7.87

From Table 15 and Figure 11 below, the overall results show that the red cube has the highest AP at 69.9% and the second lowest standard deviation at 4.48%. The red cylinder has the second highest AP at 62.4% with a standard deviation at 5.09%. In comparison to other classes, the yellow cylinder has the lowest AP at 53.8% and the highest standard deviation at 8.69%. This is consistent with our prior findings [26], which show that neutral-colored things such as yellow cubes have lower APs than strong-colored objects such as red cubes. As the objects for this pick-and-place action have a metallic surface, yellow objects suffer from surface reflection from external lighting.

Table 15. Average Precision according to workpieces.

Batch Size	TFLite Model	Yellow Cube (%)	Yellow Cylinder (%)	Red Cube (%)	Red Cylinder (%)	Blue Cube (%)	Blue Cylinder (%)
4	EfficientDet-Lite0	72.1	47.9	66.2	62.7	63.4	59.1
8		64.4	53.2	71.4	64.8	62.2	59.9
16		66.3	40.2	73.2	56.6	55.6	54.4
4	EfficientDet-Lite1	56.9	72.5	68.6	68.1	63.9	71.4
8		73.7	50.3	70.4	64.0	63.8	59.3
16		55.5	40.7	58.2	53.9	49.6	50.7
4	EfficientDet-Lite2	73.5	62.4	72.3	68.4	63.3	70.8
8		70.8	59.0	75.8	63.0	63.8	70.3
16		67.7	49.2	70.7	55.5	58.3	56.8
4	EfficientDet-Lite3	72.5	57.6	73.8	66.0	63.1	68.4
8		69.9	57.6	71.8	68.9	62.5	67.2
16		59.3	54.7	66.0	57.4	55.5	53.3
	Average of APs	66.9	53.8	69.9	62.4	60.4	61.8
	Standard Deviation of APs	6.24	8.69	4.48	5.09	4.434	7.13

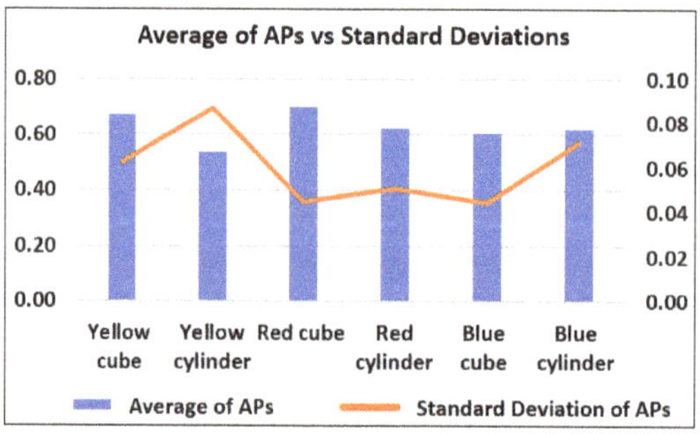

Figure 11. Comparison of Average Precision across Efficientdet-Lite models.

3.4. Statistical Analysis on Variation of Batch Size

As we observed in Table 14 above, the Average APs of Batch4 and Batch8 are quite similar. Therefore, a statistical analysis is conducted to see whether there is a significant difference between the two groups of values. We utilize the Mann–Whitney U method as it is one of the most commonly used non-parametric statistical tests [27]. Developed by Mann and Whitney in 1947, this non-parametric test is frequently used for small samples of data that are not normally distributed [28].

In Mann–Whitney U test, the null hypothesis states that the medians of the two respective groups are not different. As for the alternative hypothesis, it states that one median is larger than the other or that the two medians differ. If the null hypothesis is not rejected, it means that the median of each group of observations is similar. If the null hypothesis is rejected, it means the two medians differ.

We apply the Mann–Whitney U test to our Batch4 and Batch8 as the number of samples are small [29], less than 30, and the AP results are not normally distributed. Our null hypothesis (H0) and alternative hypothesis (H1) follow:

H0. *The median of APs is equal between Batch4 and Batch8 APs.*

H1. *The median of APs is not equal between Batch4 and Batch8 APs.*

Using SciPy which is a Python library used for scientific computing and technical computing, we obtain a p-value of 0.448. Since the p-value (0.448) is above the 0.05 significance level, we fail to reject the null hypothesis.

We conclude there is not enough evidence to suggest a significant difference in medians between the two datasets. As the standard deviation only differs by 0.008, we recommend using Batch8 instead of Batch4 for faster training and evaluation.

3.5. Performance Validation

We use the pre-trained model SSD MobileNet V2 FPNLite and compared the performance regarding the Average Precision and detection score. Table 16 shows the performance results of our selected EfficientDet-Lite2 model and SSD MobileNet V2 FPNLite, using the same training data for each of them. We compare the performance of the SSD MobileNet to that of our EfficientDet-Lite2 model with the Alpha1 dataset and found that our results were significantly better. Compared to SSD MobileNet V2 FPNLite, our model shows a significant improvement of accuracy for the yellow cube at 40.4% and red cube at 40.1%.

Table 16. Comparison of APs for custom object detection model.

Class	EfficientDet-Lite2 with Alpha1 Dataset (%)	SSD MobileNet V2 FPNLite (%)	Improvement of Accuracy (%)
Yellow cube	78.1	37.7	40.4
Yellow cylinder	72.2	35.8	36.4
Red cube	85.1	45.0	40.1
Red cylinder	65.9	43.4	22.5
Blue cube	66.1	48.8	17.3
Blue cylinder	77.3	37.7	39.6
Overall mean	74.1	41.4	32.7

Table 17 shows the comparison of detection scores for the custom dataset running on our EfficientDet-Lite2 model and SSD MobileNet V2 FPNLite. The detection scores are taken with the project setup of the 135-degree lamp and level 0 illumination. When compared to SSD MobileNet V2, our model significantly improves the overall detection scores by 5.19%. Comparing EfficientDet-Lite2 to SSD MobileNet V2, significant improvements are observed for the red cube (+19.33%) and red cylinder (+20.66%).

Table 17. Comparison of detection scores with SSD MobileNet V2.

Class	135-Degree Lamp with Level 0 Illumination (%)	SSD MobileNet V2 FPNLite (%)	Comparison of Detection Scores (%)
Yellow cube	80.00	91.50	−11.5
Yellow cylinder	80.33	94.67	−14.34
Red cube	82.00	62.67	+19.33
Red cylinder	82.33	61.67	+20.66
Blue cube	79.67	83.00	−3.33
Blue cylinder	78.67	58.33	+20.34
Overall mean	80.50	75.31	+5.19

3.6. Comparison mAP with COCO2017 Validation Dataset

Tensorflow Lite Model Maker provides the performance of each EfficientDet-Lite model with the mean Average Precision evaluated on the COCO2017 validation dataset. Hence, for additional validation, we compare our AP with the COCO2017 dataset, as shown in Table 18. As indicated in Section 3.1, we choose the Alpha1 dataset since it performs the best in AP rather than Alpha2 or Alpha3 datasets. The results show that all the Alpha1 datasets of the EfficientDet-Lite models outperform the COCO2017 dataset, with EfficientDet-Lite2 outperforming it by 51.13%. As a result, we chose EfficientDet-Lite2 for our future projects.

Table 18. Comparison of Average Precision with COCO2017 dataset.

Model Architecture	COCO2017 Dataset (%)	Alpha1 Dataset (%)	Improvement of Accuracy (%)
EfficientDet-Lite0	25.69	78.1	52.41
EfficientDet-Lite1	30.55	72.2	41.65
EfficientDet-Lite2	33.97	85.1	51.13
EfficientDet-Lite3	37.7	65.9	28.20
Overall AP	31.98	75.33	43.35

4. Discussions

The aim of the our project is to develop a smart and lean pick-and-place system for a lightweight embedded controller such as Raspberry Pi. The advantage of our system is that the robot could perform well in low-illumination area such as the wafer-cutting system. In this study, we evaluate the effectiveness of illumination and batch size to improve the Average Precision and detection scores of the EfficientDet model. This study is important because our control algorithm utilizes high and consistent detection scores to establish the location of the workpiece and regulate the arm movement. The improvement of the Average Precision and detection scores depends on many factors and features; this study focused on the illumination angle and level as an important feature of the control of lighting. The results of the detection score are subject to ambience lighting and noise, which may vary significantly if the workplace is located in an open area. Therefore, the application of Deep Learning for custom object detection will aid in reducing this variation significantly.

This study has several limitations; for instance, the quality of the dataset plays a part in determining the Average Precision. We observed that certain workpieces have surface reflection from external illumination, which has an impact on the Average Precision. In this study, we used the Japanese industrial standards for the system's lux levels for illumination. We acknowledge that different regions and industries may have varying standards and requirements for illumination levels, such as the Illuminating Engineering Society of North America (IESNA) and the European Standard EN12464-1 [25].

Currently, our method works well for application for illumination levels 0 and 1. For future projects, we would like to expand the application to other use cases of higher levels of illuminations. In order to attain the necessary high Average Precision and detection scores, we will take into account additional dataset preparation, optimization, and reinforcement learning.

5. Conclusions

In this work, we have successfully developed a novel method to develop a smart and lean pick-and-place system for custom workpieces. We evaluated the effectiveness of illumination and batch size to improve the Average Precision and detection score of the EfficientDet-Lite model, and we used it in our novel approach to develop a smart algorithm for the Raspberry Pi to control the Universal Robot.

Using a lightweight embedded system, we developed advanced pick-and-place robotic systems, enhancing automation in manufacturing processes by accurately detecting custom objects and controlling robot arm movement. This project has improved the detection of objects with deep learning and implemented the state-of-the-art (SOTA) EfficientDet-Lite model on Raspberry Pi. The validation on another SOTA model, SSD MobileNet V2 FPN-Lite, has shown that our object detection significantly improves the Average Precision and overall detection scores. Our experimental results clearly show that the proposed method has an increase of 5.19% in AP compared to the SSD model. In fact, all the APs obtained during the evaluation test exceed the AP from the COCO2017 validation test.

In the EfficientDet-Lite2 model, the addition of 8% optimized bright Alpha3 images resulted in an increase of 7.5% in Average Precision and an increase of 6.3% in F1-score. We observed that the blue cube has the highest AP in the Alpha3 dataset, while the red cube has the lowest AP in the Alpha2 and Alpha3 datasets. As our goal here is to have a consistent high detection score with low variance, we choose Alpha1 instead of Alpha3 for our subsequent projects.

As a result of using Batch4, the overall AP across all EfficientDet-Lite models increased significantly to 66.8% as compared to Batch8 at 65.4% and Batch16 at 57.4%. In order to speed up the training for our subsequent projects, we use Batch8 for training rather than Batch4. This decision is validated by the Mann–Whitney U statistical analysis, which shows that Batch4 and Batch8 do not have any significant differences in AP.

In order to increase the overall detection score to over 80%, we utilized a 135-degree lamp and level 0 illumination. As such, this study presents a very good starting point for the development of a better object detection for pick-and-place robots by using the effect of illumination. This is important because the control algorithm [4] that Raspberry Pi uses to move the arm of the universal robot depends on high and consistent detection scores to establish the location of the workpiece. In the future, we will extend the system to identify angular rotations and improve real-time detection as well as address the issue of surface reflection from external lights.

Author Contributions: Conceptualization, E.K. and J.J.C.; methodology, E.K., Z.J.C. and M.L.; software, E.K.; validation, E.K., Z.J.C. and M.L.; formal analysis, E.K., Z.J.C. and M.L.; investigation, E.K.; writing—original draft preparation, E.K.; writing—review and editing, E.K. and J.J.C.; supervision, Z.J.C.; funding acquisition, J.J.C. All authors have read and agreed to the published version of the manuscript.

Funding: This research received no external funding.

Institutional Review Board Statement: Not applicable.

Informed Consent Statement: Not applicable.

Data Availability Statement: All data that support the findings of this study are included within the article.

Conflicts of Interest: The authors declare no conflict of interest.

References

1. Leung, H.K.; Chen, X.-Z.; Yu, C.-W.; Liang, H.-Y.; Wu, J.-Y.; Chen, Y.-L. A deep-learning-based vehicle detection approach for insufficient and nighttime illumination conditions. *Appl. Sci.* **2019**, *9*, 4769. [CrossRef]
2. Bencak, P.; Vincetič, U.; Lerher, T. Product Assembly Assistance System Based on Pick-to-Light and Computer Vision Technology. *Sensors* **2022**, *22*, 9769. [CrossRef]
3. Yin, X.; Fan, X.; Zhu, W.; Liu, R. Synchronous AR Assembly Assistance and Monitoring System Based on Ego-Centric Vision. *Assem. Autom.* **2019**, *39*, 1–16. [CrossRef]
4. Zhao, W.; Jiang, C.; An, Y.; Yan, X.; Dai, C. Study on a Low-Illumination Enhancement Method for Online Monitoring Images Considering Multiple-Exposure Image Sequence Fusion. *Electronics* **2023**, *12*, 2654. [CrossRef]
5. Kee, E.; Jie, C.J.; Jie, C.Z.; Lau, M. Low-cost and sustainable Pick and Place solution by machine vision assistance. In Proceedings of the 25th International Conference on Mechatronics Technology (ICMT), Kaohsiung, Taiwan, 18–21 November 2022.
6. Tan, M.; Pang, R.; Le, Q.V. Efficientdet: Scalable and efficient object detection. In Proceedings of the IEEE/CVF Conference on Computer Vision and Pattern Recognition, Seattle, WA, USA, 13–19 June 2020; pp. 10781–10790.
7. Kim, H.; Choi, Y. Lab Scale Model Experiment of Smart Hopper System to Remove Blockages Using Machine Vision and Collaborative Robot. *Appl. Sci.* **2022**, *12*, 579. [CrossRef]

8. Jørgensen, T.B.; Jensen, S.H.N.; Aanæs, H.; Hansen, N.W.; Krüger, N. An adaptive robotic system for doing pick and place operations with deformable objects. *J. Intell. Robot. Syst.* **2019**, *94*, 81–100. [CrossRef]
9. Luo, H.; Li, C.; Wu, M.; Cai, L. An Enhanced Lightweight Network for Road Damage Detection Based on Deep Learning. *Electronics* **2023**, *12*, 2583. [CrossRef]
10. Jain, S. DeepSeaNet: Improving Underwater Object Detection using EfficientDet. *arXiv* **2023**, arXiv:2306.06075.
11. Čirjak, D.; Aleksi, I.; Lemic, D.; Pajač Živković, I. EfficientDet-4 Deep Neural Network-Based Remote Monitoring of Codling Moth Population for Early Damage Detection in Apple Orchard. *Agriculture* **2023**, *13*, 961. [CrossRef]
12. Wu, C.; Chen, L.; Wu, S. A Novel Metric-Learning-Based Method for Multi-Instance Textureless Objects' 6D Pose Estimation. *Appl. Sci.* **2021**, *11*, 10531. [CrossRef]
13. Chakole, S.; Ukani, N. Low-Cost Vision System for Pick and Place application using camera and ABB Industrial Robot. In Proceedings of the 2020 11th International Conference on Computing, Communication and Networking Technologies (ICCCNT), Kharagpur, India, 1–3 July 2020.
14. Konaite, M.; Owolawi, P.A.; Mapayi, T.; Malele, V.; Odeyemi, K.; Aiyetoro, G.; Ojo, J.S. Smart Hat for the blind with Real-Time Object Detection using Raspberry Pi and TensorFlow Lite. In Proceedings of the International Conference on Artificial Intelligence and Its Applications, Virtual, 9–10 December 2021.
15. Barayan, M.A.; Qawas, A.A.; Alghamdi, A.S.; Alkhallagi, T.S.; Al-Dabbagh, R.A.; Aldabbagh, G.A.; Linjawi, A.I. Effectiveness of Machine Learning in Assessing the Diagnostic Quality of Bitewing Radiographs. *Appl. Sci.* **2022**, *12*, 9588. [CrossRef]
16. Benhamida, A.; Várkonyi-Kóczy, A.R.; Kozlovszky, M. Traffic Signs Recognition in a mobile-based application using TensorFlow and Transfer Learning technics. In Proceedings of the IEEE 15th Conference of Systems of Systems of Engineering, Budapest, Hungary, 2–4 June 2020.
17. Dua, S.; Kumar, S.S.; Albagory, Y.; Ramalingam, R.; Dumka, A.; Singh, R.; Rashid, M.; Gehlot, A.; Alshamrani, S.S.; AlGhamdi, A.S. Developing a Speech Recognition System for Recognizing Tonal Speech Signals Using a Convolutional Neural Network. *Appl. Sci.* **2022**, *12*, 6223. [CrossRef]
18. Kim, I.S.; Jeong, Y.; Kim, S.H.; Jang, J.S.; Jung, S.K. Deep Learning based Effective Surveillance System for Low-Illumination Environments. In Proceedings of the 2019 Eleventh International Conference on Ubiquitous and Future Networks (ICUFN), Zagreb, Croatia, 2–5 July 2019.
19. Nagata, F.; Miki, K.; Watanabe, K.; Habib, M.K. Visual Feedback Control and Transfer Learning-Based CNN for a Pick and Place Robot on a Sliding Rail. In Proceedings of the 2021 IEEE International Conference on Mechatronics and Automation (ICMA), Takamatsu, Japan, 8–11 August 2021; pp. 697–702.
20. Malik, A.A.; Andersen, M.V.; Bilberg, A. Advances in machine vision for flexible feeding of assembly parts. *Procedia Manuf.* **2019**, *38*, 1228–1235. [CrossRef]
21. TensorFlow Lite Model Maker. Available online: https://www.tensorflow.org/lite/models/modify/model_maker (accessed on 5 September 2023).
22. Roboflow. Available online: https://roboflow.com (accessed on 6 September 2023).
23. Google Colab Notebook. Available online: https://colab.research.google.com (accessed on 5 September 2023).
24. *JIS Z 9110:1979*. Recommended Levels of Illumination. Japanese Standards Association: Tokyo, Japan, 2008.
25. Keskar, N.S.; Mudigere, D.; Nocedal, J.; Smelyanskiy, M.; Tang, P.T. On large-batch training for deep learning: Generalization gap and sharp minima. *arXiv* **2016**, arXiv:1609.04836.
26. Kee, E.; Chong, J.J.; Choong, Z.J.; Lau, M. A Comparative Analysis of Cross-Validation Techniques for a Smart and Lean Pick-and-Place Solution with Deep Learning. *Electronics* **2023**, *12*, 2371. [CrossRef]
27. Kasuya, E. Mann-Whitney U test when variances are unequal. *Anim. Behav.* **2001**, *61*, 1247–1249. [CrossRef]
28. Nachar, N. The Mann-Whitney U: A test for assessing whether two independent samples come from the same distribution. *Tutor. Quant. Methods Psychol.* **2008**, *4*, 13–20. [CrossRef]
29. Geweke, J.F.; Singleton, K.J. Interpreting the likelihood ratio statistic in factor models when sample size is small. *J. Am. Stat. Assoc.* **1980**, *75*, 133–137. [CrossRef]

applied sciences

MDPI

Article

Developing a Static Kinematic Model for Continuum Robots Using Dual Quaternions for Efficient Attitude and Trajectory Planning

Yunfei Li, Qiuhao Wang and Qian Liu *

School of Biomedical Engineering, Hainan University, Haikou 570228, China; yunfeili@hainanu.edu.cn (Y.L.); wangqiuhao95@163.com (Q.W.)
* Correspondence: qliu@hainanu.edu.cn

Abstract: Kinematic modeling is essential for planning and controlling continuum robot motion. The traditional Denavit Hartenberg (DH) model involves complex matrix multiplication operations, resulting in computationally intensive inverse solutions and trajectory planning. Solving position and orientation changes in continuum robots using the double quaternion rule can reduce computational complexity. However, existing dual quaternion methods are direct equational transformations of DH rules and do not give a complete modeling process. They usually require more interpretability when applying continuum robot kinematic modeling. This paper uses the dual quaternion method to establish a kinematic model of a continuum robot. It uses a two-section continuum robot model to compare the advantages of dual quaternion and traditional modeling methods. In addition, this paper proposes a five-polynomial interpolation algorithm based on the dual quaternion method for trajectory planning of continuum robots. This method accurately models spatial bending and torsional motions of singularity-free continuum robots.

Keywords: Denavit–Hartenberg; continuum robot; dual quaternion Jacobian matrix; trajectory planning

check for updates

Citation: Li, Y.; Wang, Q.; Liu, Q. Developing a Static Kinematic Model for Continuum Robots Using Dual Quaternions for Efficient Attitude and Trajectory Planning. *Appl. Sci.* **2023**, *13*, 11289. https://doi.org/10.3390/app132011289

Academic Editors: Qi Song and Qinglei Zhao

Received: 11 September 2023
Revised: 11 October 2023
Accepted: 11 October 2023
Published: 14 October 2023

1. Introduction

Researchers have become increasingly enthusiastic about continuum robots in recent years because of their excellent mechanical properties when operating in unique environments. The continuum robot is a flexible, continuous, multi-segmented robotic system inspired by the skeletal structure of biological organisms. In contrast with conventional rigid multi-joint robots, continuum robots employ a soft, deformable structure composed of numerous interconnected and continuous flexible segments. These segments can be actuated internally or externally using stimuli such as gas, liquid, or motors, facilitating smooth, seamless, and flexible motion and deformation. For example, continuum robots can perform surgical operations under minimally invasive and non-invasive conditions of the human body [1–4], target detection and fault diagnosis in narrow intervals [5–7], and grasp targets in high-pressure underwater environments, such as the deep sea [8]. They were developed from studying structures in nature that can be freely bent, twisted, and elongated, such as the arms of octopuses, the tongues of mammals and reptiles, and the trunks of elephants [9–11]. The diversity of potential applications of the continuum robot leads to various designs [12], which are reflected in the structure and the matching drive. From the physical form, the continuum robot is divided into the following formats: a single flexible pipe or rod with uniform stiffness [13], a series of flexible concentric tubes [14], a series of parallel truss platforms [15], flexible continuum pipes with multiple open slots [16,17], and a plurality of elastic material disks stacked. Drive models include pneumatic, traction line, electrochemical, and other drive modes. However, they all exhibit continuum curvature in a continuum robot, i.e., a continuum changes curvature along the main chain's length. Furthermore, unlike conventional manipulators, which consist mainly of rigid elements resulting in only changing themselves at discrete points in their structure, continuum robots can theoretically change any position in their system [18,19], which leads to challenging kinematic and dynamic modeling of continuum robots and further leads to difficulties in real-time dynamic control.

Continuum robots have widespread use in quasi-static environments where dynamic models may not be applicable [20]. Researchers have employed various approaches to solving the kinematics of continuum robots, such as utilizing motion combinations of fake rigid manipulators to simulate their motion and applying the Denavit–Hartenberg (DH) model, which was initially developed for rigid manipulators' accuracy in emulating continuum robot behavior; kinematic and shape correspondence between super-redundant manipulators and desired spatial profiles have been introduced [21,22]. Recently, the incorporation of continuum curvature into a modified DH modeling procedure using differential geometry has provided a comprehensive approach to modeling continuum robots [23]. Building upon this work, the researcher has proposed the variable reality of the central axis, associating the driving variable with the central axis curve to modify and enhance existing ideas [11]. The Jacobian matrix of the model and the corresponding kinematic control method have also been discussed. However, special numerical treatment is required when approaching these models' straight (zero-curvature) cross-section configuration. The researcher expanded the driver variables, employing the Taylor series to address this issue and, thus, preventing model invalidity at zero curvature [24–26]. Nevertheless, modeling a multi-system or multi-joint manipulator arm using the above modeling approach becomes difficult, as the method of obtaining the end pose by multiplying the pose matrix places a significant workload on the system, and the relationship between each part of the system and the global coordinate system must be constantly considered. To address the challenge of dealing with sections with nearly straight deformation, the researcher has proposed using dual quaternions to solve this problem. Although the dual quaternion method offers increased efficiency in representing changes in the position and spatial elements of the robot, the existing approach is directly converted from the DH rule based on mathematical rules without considering the perspective of manipulator motion. This leads to limited interpretability of the dual quaternion method when applied to the kinematic modeling of manipulators in scientific journal articles.

Building upon previous research, this paper explores the dual quaternion method from the standpoint of kinematics in order to tackle continuum manipulator problems. The solution is established based on the definition, and the merits of the dual quaternion method are emphasized by comparing its computational efficiency with traditional DH model-based algorithms. The paper is structured as follows. Section 2 introduces the operational rules of dual quaternions and derives the principles for representing spatial rotation and displacement using dual quaternions. Section 3 illustrates the modeling of forward and inverse kinematics for single and multiple joints employing the dual quaternion method using the standard continuum manipulator model. Section 4 corroborates the results through simulation and experimental testing. Finally, conclusions are drawn in Section 5.

2. Materials and Methods

2.1. Dual Quaternion Rule

Quaternions are fourth-order hypercomplex numbers often used to describe changes in four-dimensional hyperplanes and vectors in graphics. Quaternions are generally represented in the form $a + b\boldsymbol{i} + c\boldsymbol{j} + d\boldsymbol{k}$, where a, b, c, and d are real numbers, and $\boldsymbol{i}$, $\boldsymbol{j}$, and $\boldsymbol{k}$ are basic quaternions. Quaternions can be composed of a scalar part and a vector part. $\boldsymbol{q}$ is a quaternion represented as $\boldsymbol{q} = (r, \boldsymbol{v})$, where r is a scalar defined in the real number field, and $\boldsymbol{v}$ is a three-dimensional vector. $\boldsymbol{q}^*$ is the conjugate of $\boldsymbol{q}$, represented by $\boldsymbol{q}^* = (r, -\boldsymbol{v})$. $\boldsymbol{q}_1$ and $\boldsymbol{q}_2$ are two quaternions. The result and outer product of those are shown in Equations (1) and (2). The product of two quaternions is called the Grassmann product and is denoted by the symbol $\otimes$.

$$(r_1, \boldsymbol{v}_1) + (r_2, \boldsymbol{v}_2) = (r_1 + r_2, \boldsymbol{v}_1 + \boldsymbol{v}_2) \tag{1}$$

$$(r_1, \boldsymbol{v}_1)(r_2, \boldsymbol{v}_2) = (r_1 r_2 - \boldsymbol{v}_1 \cdot \boldsymbol{v}_2, r_1 \boldsymbol{v}_2 + r_2 \boldsymbol{v}_1 + \boldsymbol{v}_1 \times \boldsymbol{v}_2) \tag{2}$$

The dual numbers are a system of hypercomplex numbers, which are expressions of the form $c + d\varepsilon$, where c and d are real numbers, and ε is a symbol taken to satisfy. When c and d are replaced by quaternions using real numbers, the dual numbers are called dual quaternions. A dual quaternion can be represented in the form of $\hat{q}$, which can be written as $\hat{q} = \boldsymbol{q}_r + \varepsilon \boldsymbol{q}_d$. Among them, $\boldsymbol{q}_r$ and $\boldsymbol{q}_d$ are two quaternions, respectively, referred to as the imaginary and real parts of dual quaternions. $\hat{q}^*$ represents the dual quaternion conjugate, as shown in Equation (3).

$$\hat{q}^* = \boldsymbol{q}_r^* + \varepsilon \boldsymbol{q}_d^* \tag{3}$$

2.2. Dual Quaternion Representation of Rigid Body Motion

Rigid body motions describing elements of solid geometry, such as points, lines, and surfaces in space, can be represented by dual quaternions. As shown in Equation (4), this means that the dual quaternion is used to represent a straight line A that changes into a straight line B after rotation and translation in space, where $\hat{\boldsymbol{A}}$ and $\hat{\boldsymbol{B}}$ are the Plücker forms of straight lines A and B, respectively. $\hat{q}$ is the dual quaternions representing the angle of rotation θ around axis l. It can also be written in the form of Equation (5), where the derivation process is given in Appendix A.

$$\hat{\boldsymbol{B}} = \hat{q}^* \otimes \hat{\boldsymbol{A}} \otimes \hat{q} \tag{4}$$

$$\hat{q} = \left(\cos\left(\frac{\theta}{2}\right), \sin\left(\frac{\theta}{2}\right) l\right) \& + \varepsilon\left(-\frac{d}{2}\sin\left(\frac{\theta}{2}\right), \sin\left(\frac{\theta}{2}\right)\boldsymbol{m} + \frac{d}{2}\cos\left(\frac{\theta}{2}\right) l\right) \tag{5}$$

2.3. Physical Model of Continuum Robot

The general kinematic equations of a tendon-driven continuum robot arm are established. A specific example is presented to demonstrate the application of the derived kinematic equations, in which a tendon-driven continuum robot is considered. As illustrated in Figure 1, this continuum robot comprises two independent single-section manipulators, namely Sections 1 and 2. Each manipulator section is constructed using a flexible disc as its primary structure, with the discs connected by springs, referred to as tendons. These tendons are secured at predetermined positions along the arc length of the robot. The end of the arm is equipped with a multi-traction line attached to the discs. By pulling these traction lines, a load is applied to the spring through the disc, resulting in the corresponding bending of the robot arm.

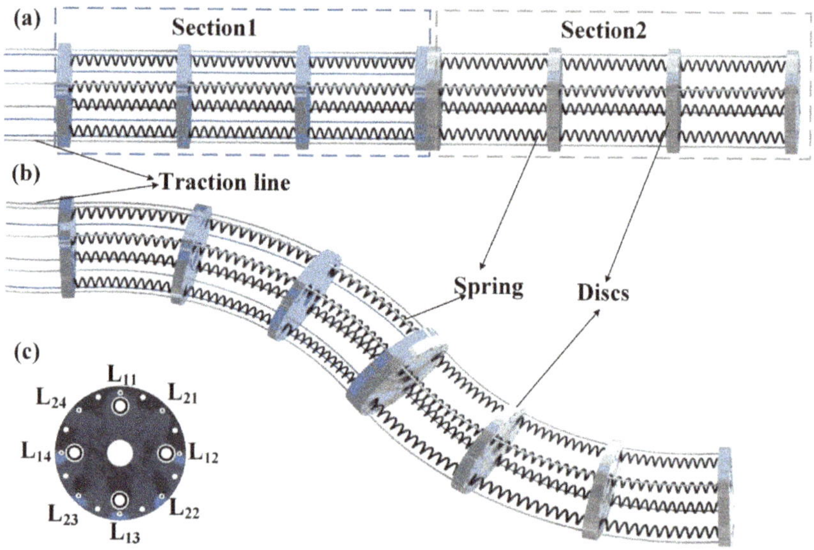

Figure 1. Structure of a continuum tendon-driven robot. (**a**) Schematic diagram of the robotic arm model in the natural state; (**b**) Schematic diagram of the robotic arm model in the driven state; (**c**) Cross-sectional schematic diagram of the robotic arm model. Among them, L_{ij} represents the number of the driving lines, i represents the ith robotic arm, and j represents the jth driving lines.

3. Kinematic Model of Continuum Robotics

3.1. Center Axis Curve Parameters

Due to the arrangement of the tendons (discs and springs), the robotic arm is driven in line, and these continua exhibit a telescopic movement or bend into a circular shape. Therefore, the continuum arm's central axis can be described in space precisely as a circular arc with a variable radius of curvature and length. As shown in Figure 2a, the diagram on the left shows the state of the continuum arm of the section when it is not driven, i.e., $t = 0$. The central axis of the disc is a straight line with four drive lines of length $L_{ij}^{0}(j = 1,\ 2,\ 3,\ 4)$. After the continuous manipulator is driven for time t, the state is shown in Figure 2b. The line length becomes $L_{ij}^{t}(j = 1,\ 2,\ 3,\ 4)$. Let the change in rope length between the driven state and the undriven state be $l_{ij}\ (t)$, as shown in Equation (6).

$$l_{ij}\ (t) = L_{ij}^{0} - L_{ij}^{t} \tag{6}$$

When the continuum arm is driven, the overall curve is assumed to be circular based on continuum curvature [24]. The radius of the curvature is described by $\rho_i \in (0, \infty)$, and the bending angle is described by $\varphi_i \in \left(0, \frac{\pi}{2}\right)$, which is on a plane that forms an angle $\delta_i \in (-\pi, \pi)$ with the x-axis as a whole in space. The curve parameters in joint space variables are given by Equations (7)–(9). A comprehensive derivation of these variables is provided in Appendix B.

$$\varphi_i = \frac{1}{2R_i}\sqrt{(l_{i4} - l_{i2})^2 + (l_{i3} - l_{i1})^2} \tag{7}$$

$$\delta_i = \arctan\left(\frac{2(l_{i1} - l_{i2})}{(l_{i1} - l_{i3})} - 1\right) \tag{8}$$

$$\rho_i = \frac{2\sqrt{(l_{i4} - l_{i2})^2 + (l_{i3} - l_{i1})^2}}{R_i \sum l_{ij}} \tag{9}$$

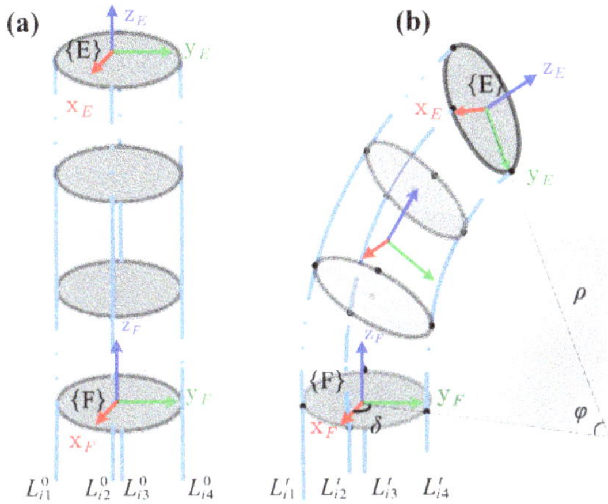

Figure 2. The Structure diagram of the single-section mechanical arm is driven and not driven. (**a**) Schematic diagram of the driving line of the single-section robotic arm when it is not driven, where L_{ij}^0 represents the rope length. (**b**) Schematic diagram of the driving line of the robotic arm after driving time t, where L_{ij}^t represents the rope length.

3.2. Coordinate Systems and Dual Quaternion Transformations of Points and Lines

A single-segment continuous robot is used to model using the dual quaternion method. The forward kinematics of the robot are to solve its end pose after driving. The physical model of the single-section robotic arm when driven is shown in Figure 3a.

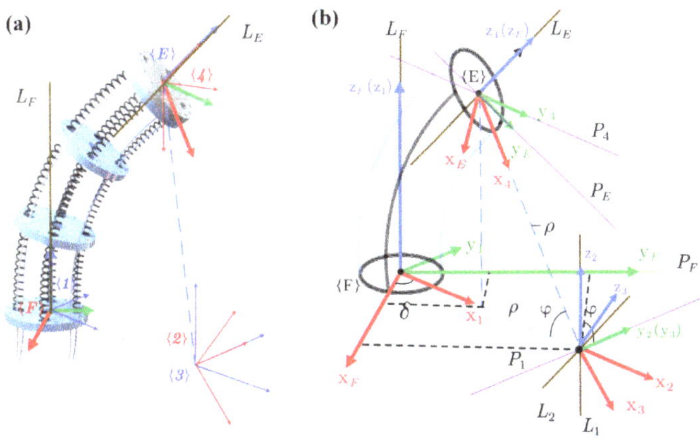

Figure 3. (**a**) Physical model of a single-section robotic arm in the driving state. Among them, {F}{1}{2}{3}{4}{E} are the coordinate systems, respectively, and L_E and L_F represent the straight lines where the z-axis of the {F} and {E} coordinate systems are located, respectively. (**b**) Mathematical model of a single-section robotic arm in the driving state. It describes the z-axis and y-axis in the coordinate system {F}, that is, the straight lines L_F and P_F, which after coordinate transformation become the straight lines L_E and P_E in the coordinate system {E}.

As shown in Figure 3, if {F} and {E} are two reference frames, while $\hat{q}_E, \hat{q}_F$ are the dual quaternion of those reference frames relative to a fixed coordinate system in space, then the relative position relationship between these two coordinate systems is called $\hat{q}_{EF}$, which is represented by (10) and can be obtained from (4); a more detailed derivation process can be found in Appendix C.

$$\hat{q}_{EF} = \hat{q}_E \otimes \hat{q}_F = \left[cos\frac{\varphi}{2}, -sin\delta sin\frac{\varphi}{2}, cos\delta sin\frac{\varphi}{2}, 0\&, 0, 0, 0, \rho \cdot sin\frac{\varphi}{2}\right] \tag{10}$$

Let the lines in the front-end coordinate system {F} where the y-axis and z-axis lie be L_F, P_F and the direction vectors be $\boldsymbol{l}_F$ and $\boldsymbol{p}_F$, respectively. By Euler's theorem, the line L_F in the coordinate system {F} becomes L_1 after a rotation around the z-axis and a translation ρ along the axis x_1, and then L_2 around the axis y_2, before a translation ρ along the new x_3 axis becomes L_E in the coordinate system {E}. Similarly, P_F can become P_E. The moment vectors are $\boldsymbol{m}_L, \boldsymbol{m}_P$, respectively, which are expressed in the Plücker coordinate system as $\boldsymbol{L}_F = (\boldsymbol{l}_F, \boldsymbol{m}_L), \boldsymbol{P}_F = (\boldsymbol{p}_F, \boldsymbol{m}_F)$. They can be expressed as $\hat{L_F} = \boldsymbol{l}_F + \varepsilon \boldsymbol{m}_L, \hat{P_F} = \boldsymbol{p}_F + \varepsilon \boldsymbol{m}_F$ by a dual quaternion. Substituting (10) into (3), we can obtain the relationship between the straight line L_F and L_E on the coordinate system {F} and {E} as (11).

$$\hat{\boldsymbol{L}}_F = \hat{q}^*_{EF} \otimes \hat{\boldsymbol{L}}_E \otimes \hat{q}_{EF} \tag{11}$$

Similarly, the relationship between P_F and P_E is (12).

$$\hat{\boldsymbol{P}}_F = \hat{q}^*_{EF} \otimes \hat{\boldsymbol{P}}_E \otimes \hat{q}_{EF} \tag{12}$$

According to Plück's law, the intersection points of the two are the position of the end coordinate system, and the pose can be expressed as $[\boldsymbol{l}_F, \boldsymbol{p}_F, \boldsymbol{l}_F \times \boldsymbol{p}_F]$.

3.3. Kinematic Equations of Continuum Manipulator

As shown in Figure 4, three identical single-section robotic arms are connected in series to form an overall number: $i-1$, i, $i+1$. Then, the central axis is the z-axis direction on the front-end disk of this multi-section robotic arm. Next, establish a coordinate system and record it as {F}, and set up a coordinate system on the end disc with the central axis as the direction of the z-axis and record it as {E}. Assume that the center point at the top of the segment $(i+1)$ is O, which is expressed as $\boldsymbol{O}_{E(i+1)} = (0,0)$ in the coordinate system $\left\{E_{(i+1)}\right\}$.

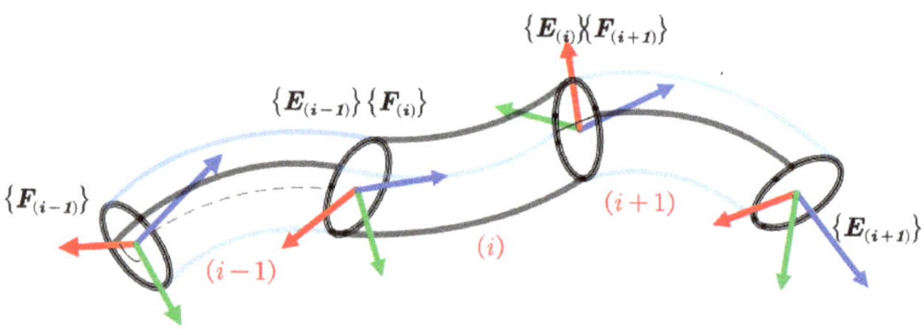

Figure 4. The coordinate system of a multi-section continuous robot in a driven state. $\left\{F_{(i-1)}\right\}$ represents the coordinate system at the front end of the continuous robot in the first section, while $F_{(i)}$ represents it in the second section. Due to physical model limitations, the i-th manipulator's front-end coordinate system differs from the 1-section robot arm, but $F_{(i)}$ and $E_{(i-1)}$ remain the same. Thus, a total of $i+1$ coordinate systems are needed for the i-section robotic arm.

This point is denoted $\boldsymbol{O}_{F(i+1)} = \left(\boldsymbol{O}_{F(i+1)}, \boldsymbol{m}_{F(i+1)}\right)$ in the coordinate system $\left\{F_{(i+1)}\right\}$, because $\left\{F_{(i+1)}\right\}$ and $\left\{E_{(i)}\right\}$ are the same in space. Therefore, the end position of the robot arm in the section $i+1$ can be expressed as $\hat{\boldsymbol{O}}_{E(i)}$ in the end coordinate system of the section i.

$$\hat{\boldsymbol{O}}_{E(i+1)} = \hat{q}^*_{(i+1)i} \otimes \hat{\boldsymbol{O}}_{E(i)} \otimes \hat{q}_{(i+1)i} \tag{13}$$

Similarly, the position of the central axis point of the front end of the multi-segment continuous arm in the coordinate system of the end of the first continuous arm is:

$$\hat{O}_{E(i-1)} = \hat{q}^*_{(i+1)i} \otimes \hat{q}^*_{(i-1)i} \otimes \hat{O}_{E(i+1)} \otimes \hat{q}_{(i-1)i} \otimes \hat{q}_{(i+1)i} \tag{14}$$

Then the z and y axis directions of the frontmost position point are:

$$\hat{L}_{E(i-1)} = \hat{q}^*_{(i+1)i} \otimes \hat{q}^*_{(i-1)i} \otimes \hat{L}_{E(i+1)} \otimes \hat{q}_{(i-1)i} \otimes \hat{q}_{(i+1)i} \tag{15}$$

$$\hat{P}_{E(i-1)} = \hat{q}^*_{(i+1)i} \otimes \hat{q}^*_{(i-1)i} \otimes \hat{P}_{E(i+1)} \otimes \hat{q}_{(i-1)i} \otimes \hat{q}_{(i+1)i} \tag{16}$$

Through (15) and (16), the expression of the end position of the overall mechanical arm can be obtained in the first section of the robotic arm, and the complete forward kinematic equation can be obtained.

3.4. Control the Motion of the Robotic Arm through the End Position

In the previous section, the forward kinematic equations of the continuum arms were derived using the dual quaternion method. The Jacobian matrix of each part of the manipulator is first solved to solve the inverse kinematics numerically.

Let the expression of the Plücker form of the coordinates of a point $\boldsymbol{P}_0$ in the coordinate system {0} be $\hat{\boldsymbol{P}}_0$, and the dual quaternion relationship between the coordinate system {0} and the coordinate system {1} is $\hat{q}_i$.

Let the expression of the Plück form of the coordinates of a point $\boldsymbol{P}_0$ in the coordinate system {0} be $\hat{\boldsymbol{P}}_0$, and the dual quaternion relationship between the coordinate system {0} and the coordinate system {1} is $\hat{q}_i$, then the point can be expressed in the coordinate system {1} for (17).

$$\hat{P}_1 = Q_1 = \hat{q}_i^* \otimes \hat{P}_0 \otimes \hat{q}_i \tag{17}$$

Writing (18) as a vector pattern gives (18).

$$Q^1 = Q_1 = [1, \boldsymbol{\vartheta}_1] + [0, \varepsilon \boldsymbol{x}_1] \tag{18}$$

The complete forward kinematics of ith section relative to {0}, denoted by Q_i, is given by (19).

$$\hat{P}_i = Q^i = \hat{q}_i^* \otimes \cdots \hat{q}_1^* \otimes \hat{P}_0 \otimes \hat{q}_1 \otimes \cdots \hat{q}_i \tag{19}$$

Write (19) as a vector, as shown in (20). Here $\hat{\boldsymbol{\vartheta}} \in \mathrm{R}^4$ is the rotation and $\hat{x} \in \mathrm{R}^4$ is the displacement.

$$Q^i = Q_1 \cdots \otimes Q_i = [1, \boldsymbol{\vartheta}_{1,2\ldots i}] + [0, \varepsilon \boldsymbol{x}_{1,2\ldots i}] \tag{20}$$

$$Q^i = Q_1 \otimes Q_2 \otimes \cdots \otimes Q_i = Q^{i-1} \otimes Q_i \tag{21}$$

Putting $\hat{q} = [q_r, \varepsilon q_d]$ into (21), we can obtain (22) and (23), which are the dual quaternion representations of the position and pose matrix of the ith robotic arm.

$$\hat{\boldsymbol{\vartheta}}_{1,2\ldots i} = \hat{\boldsymbol{\vartheta}}_{1,2\ldots i-1} \otimes \hat{q}_{i,r} \tag{22}$$

$$\hat{x}_{1,2\ldots i} = \hat{x}_{1,2\ldots i-1} \otimes \hat{q}_{i,r} + \hat{\boldsymbol{\vartheta}}_{1,2\ldots i-1} \otimes \hat{q}_{i,d} \tag{23}$$

We need to use the obtained position and pose to obtain partial derivatives of the joint variables, so as to obtain the velocity Jacobian matrix of the manipulator vector. Let J_i^{ϑ} and J_i^{x}, respectively, be the position and pose quaternion Jacobians from (24) and (25).

$$J_i^{\vartheta} = \frac{\partial(\boldsymbol{\vartheta}_{1,2,\ldots i})}{\partial(\delta, \theta)} \in \mathrm{R}^{4\times 2i} \tag{24}$$

$$J_i^x = \frac{\partial(x_{1,2,\ldots i})}{\partial(\delta,\theta)} \in \mathrm{R}^{4\times 2i} \tag{25}$$

Solve the derivatives of joint variables for (22) and (23) to obtain (26) and (24), where the formula for derivation is (24) and (25).

$$J_i^{\vartheta} = J_{i-1}^{\vartheta} \otimes q_{i,r} \tag{26}$$

$$J_i^x = J_{i-1}^x \otimes q_{i,r} + J_{i-1}^{\vartheta} \otimes q_{i,d} \tag{27}$$

The Jacobians derived in (26) and (27) are only valid in the Plück coordinate system, so we need to transform the Plück coordinate system into the inertial coordinate system.

The Cartesian angular velocity, $\omega_x \in \mathbf{R}^3$ relative to {0}, can be recovered from the quaternion velocities as (28).

$$\omega_x = 2\overline{q} \otimes \dot{q} \tag{28}$$

The partial derivative of the angular momentum can be used to obtain the Jacobian matrix of the angular velocity using (29).

$$J_i^{\omega} = \frac{\partial(\omega)}{\partial(\delta,\theta)} \in \mathbf{R}^{3\times 2i} \tag{29}$$

Putting (28) into (29) can obtain the Jacobian matrix of the angular velocity of the manipulator to the joint variable represented by the dual quaternion in the inertial space, that is (30).

$$J_i^{\omega} = 2\left[\widetilde{\vartheta}_i\right] J_i^{\vartheta} \tag{30}$$

$$\left[\widetilde{\vartheta}_i\right] = \begin{bmatrix} a_0 & -a_3 & a_2 & -a_1 \\ a_3 & a_0 & -a_1 & -a_2 \\ -a_2 & a_1 & a_0 & -a_3 \end{bmatrix} \tag{31}$$

Similarly, the linear velocity is recovered from the component as in (31). The Jacobian matrix of the manipulator speed is expressed as (33) in the Cartesian coordinate system. Putting (32) into (33) can obtain the Jacobian matrix of the velocity of the manipulator to the joint variable represented by the dual quaternion in the inertial space, that is (34).

$$v_i = 2\left[\widetilde{\vartheta}_i\right]\dot{x}_i - 2[\widetilde{x}_i]\dot{\vartheta}_i \tag{32}$$

$$J_i^{v} = \frac{\partial(v_i)}{\partial(\delta,\theta)} \in \mathbf{R}^{3\times 2i} \tag{33}$$

$$J_i^{V} = 2\left[\widetilde{\vartheta}_i\right] J_i^{v} - 2[\widetilde{x}_i] J_i^{\vartheta} \tag{34}$$

When solving the pose and position of the manipulator at the same time, the overall Jacobian matrix should be (35).

$$J = \begin{bmatrix} J_i^V \\ J_i^{\omega} \end{bmatrix} \tag{35}$$

Since the kinematics of the continuous manipulator are generally high-order polynomials, it is impossible to solve the closed solution of the complete task space position or orientation of the multi-section continuous manipulator. Therefore, numerical solutions or metaheuristic algorithms are mainly used to solve the inverse kinematics of the manipulator. This paper uses the pseudo-inverse iterative numerical solution method to solve the inverse kinematic Equation (36) used for the inverse position solution.

$$J^{\dagger} = J^{\mathrm{T}}\left(JJ^{\mathrm{T}}\right)^{-1} \ (\mathbf{J} \in \mathbf{R}^{\mathrm{m}\times\mathrm{n}}) \tag{36}$$

4. Simulation Results and Discussion

To evaluate the accuracy of the dual quaternion model compared to traditional kinematic models, specifically the DH and DH Taylor expansion models, we conducted a comparative analysis using the same driving variable. As demonstrated in Figure 5, our findings reveal that the error computed by the dual quaternion model aligns closely with those of the DH and DH Taylor expansion models. This corroborates the precision and reliability of our proposed methodology. These results underscore the potential of dual quaternions for enhancing the accuracy of kinematic models for continuum robots, laying the groundwork for future research in robotics and related fields. To prove the improvement of calculation speed using dual quaternion modeling, we use the same solution algorithm to solve the same target and compare the calculation time of dual quaternions: DH and DH Taylor. We anticipate the endpoint of the robotic arm to traverse from its initial position, P1 (0, 0, 960 mm), ultimately arriving at the desired position, P2 (369.8146 mm, 345.8315 mm, 702.9017 mm), as shown in Table 1. We used the optimization toolbox in MATLAB, and the CPU was an Intel(R) Xeon(R) W-2245 CPU @ 3.90 GHz 3.91 GHz processor for calculation. From Table 1, we can see that, in the numerical method, the dual quaternion model solves the target position with high precision and a short calculation time of 0.45 s; compared with the traditional DH method of 1.15 s, the calculation time is doubled. In standard meta-heuristic algorithms, the dual quaternion models are shortened by more than half.

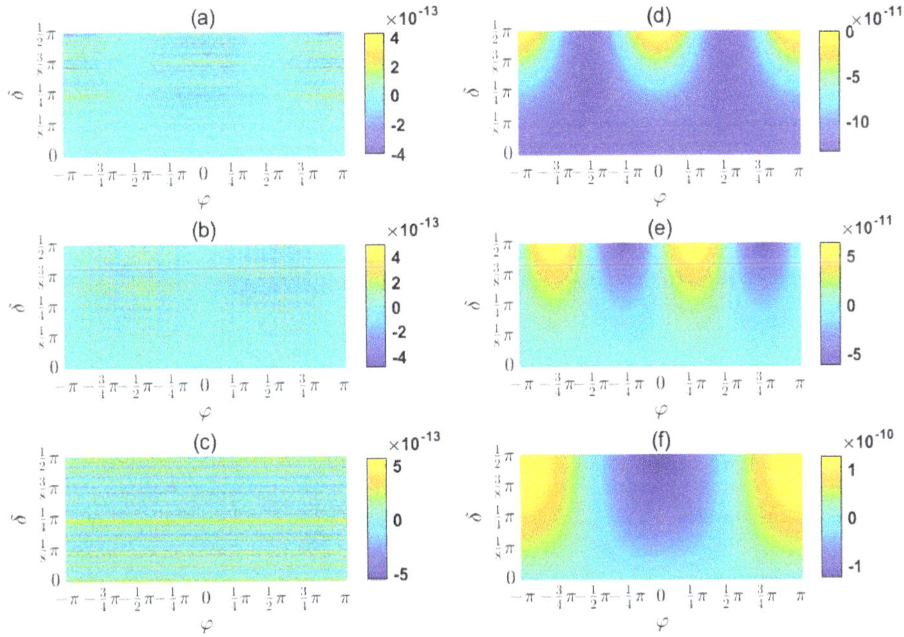

Figure 5. The two-section driven manipulator contains four driving variables: $\varphi_1 \in [-\pi, \pi]$, $\delta_1 \in [0, \pi/2]$, $\varphi_2 = 0$, and $\delta_2 \neq 0$. The spatial coordinates solved by the DH Taylor expansion series model are used as standard results to compare the errors in the results of the dual quaternion model and the DH model. (**a**) The dual quaternion model and the standard result solve the error in the x-coordinate direction between the coordinates. (**b**,**c**) are the errors in the y-axis and z-axis with the standard result, respectively. (**d**) The errors between the DH model solution coordinates and the standard result in the x-coordinate direction, (**e**,**f**), are the errors on the y-axis and z-axis, respectively.

To achieve smooth angular velocity and acceleration changes at the end of the robotic arm during operation, a quintic polynomial interpolation algorithm based on dual quaternions is proposed for motion planning of the robotic arm. The simulated movement of the robotic arm end, as illustrated in Figure 6, demonstrates that the velocity and acceleration of the variables are continuum and smooth during the robotic arm's movement from point P1 to point P2, without any abrupt changes. This indicates that the robotic arm's motion is not subject to speed distortion and can operate seamlessly.

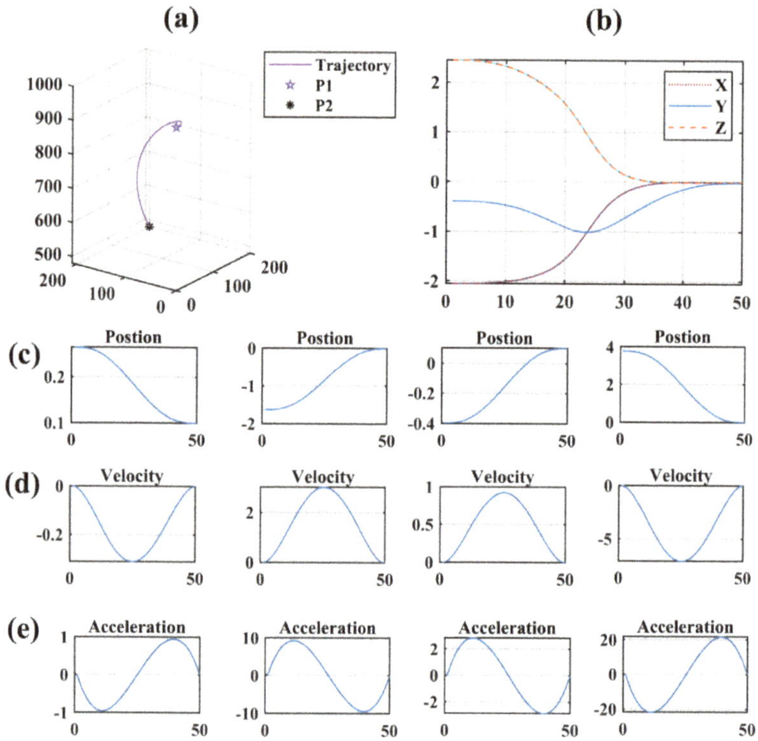

Figure 6. Use the fifth-order polynomial interpolation algorithm to plan the trajectory of the robotic arm and solve the problem of the end of the two-section continuous robot moving from point P_1 to point P_2 in space. (**a**) Schematic diagram of the motion trajectory of the two-section continuous robot. The path is planned through the fifth-order polynomial interpolation algorithm; that is, the continuous robot needs to move 50 steps from point P_1 to point P_2 according to the interpolation sequence. (**b**) The error between the continuous robot's actual path and the algorithm's path (**c**–**e**), respectively, represents the changes in the angle, angular velocity, and angular acceleration of the driving amount when the continuous robot moves from point P_1 to point P_2 and sequentially moves 50 interpolation trajectory points. The *x*-axis represents the 50 trajectory points.

Table 1. Comparison of calculation time and quantity under different models.

Model	Optimization Algorithm	Actual Position	Iterations	Time(s)
DH	Genetic algorithm	[369.8147, 345.8322, 702.9025]	25	3.43
	Simulated annealing algorithm	[369.8150, 345.8317, 702.9021]	5500	62.15
	Numerical solution Algorithm	[369.8143, 345.8315, 702.9017]	32	1.15
	Particle Swarm Optimization	[369.8125, 345.8306, 702.9003]	225	3.36
DH Taylor expansion	Genetic algorithm	[372.2827, 347.7579, 706.3873]	400	4.50
	Simulated annealing algorithm	[369.8165, 345.8329, 702.9044]	7700	52.30
	Numerical solution Algorithm	[369.8146, 345.8319, 702.9018]	31	0.41
	Particle Swarm Optimization	[369.8101, 345.8311, 702.8985]	380	4.60
dual quaternions	Genetic algorithm	[369.8161, 345.8352, 702.9070]	350	2.39
	Simulated annealing algorithm	[369.8118, 345.8293, 702.8986]	3900	28.50
	Numerical solution Algorithm	[369.8146, 345.8319, 702.9018]	32	0.45
	Particle Swarm Optimization	[366.5362, 345.7464, 701.7661]	85	1.18

5. Conclusions

This paper establishes a kinematic model of a continuous robot based on dual quaternions. It derives it from the perspective of the transformation process of geometric elements in space: linear rotation and translation of space. First, the kinematic equations of the line-pulled continuum robot are derived by applying the dual quaternion method. Secondly, the kinematic model of the multi-section complete robotic arm was further established, and the inverse kinematic solution was performed based on the numerical solution method. Finally, this paper proposes a trajectory planning process for a continuum robot using the five-polynomial dual quaternion method.

Author Contributions: Conceptualization, methodology, software, validation, Y.L.; formal analysis, data curation, Q.W.; funding acquisition, Q.L. All authors have read and agreed to the published version of the manuscript.

Funding: This research was funded by Major Science and Technology Project of Hainan Province grant number ZDKJ202006.

Institutional Review Board Statement: Not applicable.

Informed Consent Statement: Not applicable.

Data Availability Statement: Provide data upon reasonable request.

Conflicts of Interest: The authors declare no conflict of interest.

Appendix A

Let there be any two vectors $\boldsymbol{a}$,$\boldsymbol{b}$ in the space, where the vector $\boldsymbol{a}$ translates the distance l along the p axis, and then rotates θ to become the vector $\boldsymbol{b}$. Define the dual angle notation, $\hat{\theta} = \theta + \varepsilon l$, which relates an arbitrary 3D spatial line $\boldsymbol{a}$ to a given 3D spatial line $\boldsymbol{b}$ by a rotation θ about a unique axis and with a translation l along the same axis.

Here, r_a,r_b represent the vectors from any arbitrary point in space, denoted as O, to points $\boldsymbol{a}$ and $\boldsymbol{b}$, respectively. The symbols $\boldsymbol{a}'$, $\boldsymbol{b}'$ signify the cross products. The cross products $\boldsymbol{a}'$ and $\boldsymbol{b}'$ can be represented by $\boldsymbol{a}$ and $\boldsymbol{b}$ and their corresponding r_a and r_b, as shown in Equation (A1).

$$\boldsymbol{a}' = r_a \times \boldsymbol{a}, \boldsymbol{b}' = r_b \times \boldsymbol{b} \tag{A1}$$

We can write the vectors a and b representing line segments $0A$ and $0B$ in Plücker form as $\hat{\boldsymbol{a}}$ and $\hat{\boldsymbol{b}}$, respectively.

$$\hat{\boldsymbol{a}} = \boldsymbol{a} + \varepsilon(r_a \times \boldsymbol{a}), \hat{\boldsymbol{b}} = \boldsymbol{b} + \varepsilon(r_b \times \boldsymbol{b}) \tag{A2}$$

Decomposing the vector $\boldsymbol{b}$ along the orthogonal directions of $\boldsymbol{a}$ and $\boldsymbol{p} \times \boldsymbol{a}$ gives Equation (A3).

$$\boldsymbol{b} = \cos\theta \boldsymbol{a} + \sin\theta(\boldsymbol{p} \times \boldsymbol{a}) \tag{A3}$$

Considering the relationship depicted in the figure, where $\boldsymbol{r}_b = \boldsymbol{r}_a + \boldsymbol{l}$, and combining it with the above equation, we can simplify it to Equation (A4).

$$\hat{\boldsymbol{b}} = (\cos\theta - \varepsilon l \sin\theta)(\boldsymbol{a} + \varepsilon \boldsymbol{r}_a \times \boldsymbol{a}) + (\sin\theta + \varepsilon l \cos\theta)\{[\boldsymbol{p} + \varepsilon(\boldsymbol{r}_a \times \boldsymbol{p})] \times \boldsymbol{a}\} \tag{A4}$$

The inverse Taylor series transformation is utilized for the change process, as shown in Equation (A5), which ultimately simplifies Equation (A6).

$$\cos\,\theta - \varepsilon \sin\,\theta l = \cos\,\hat{\boldsymbol{\theta}}, \sin\,\theta + \varepsilon \cos\theta l = \sin\,\hat{\boldsymbol{\theta}} \tag{A5}$$

$$\hat{\boldsymbol{b}} = \cos\hat{\boldsymbol{\theta}}\hat{\boldsymbol{a}} + \sin\hat{\boldsymbol{\theta}}(\hat{\boldsymbol{p}} \times \hat{\boldsymbol{a}}) \tag{A6}$$

Let $\cos\hat{\boldsymbol{\theta}}$ be a dual number, and $\sin\hat{\boldsymbol{\theta}}(\hat{\boldsymbol{p}} \times \hat{\boldsymbol{a}})$ be a dual vector, then applying the concept of dual quaternion, we can see that the elements in coordinate system $\{A\}$ are transformed into coordinate system $\{B\}$ after a rotation angle θ after a translation distance l around an axis is expressed as $\hat{\Lambda} = \cos\hat{\boldsymbol{\theta}} + \sin\hat{\boldsymbol{\theta}}\hat{p}$, where: $\hat{\boldsymbol{b}} = \Lambda(\hat{\boldsymbol{a}})$. Further, if $\hat{\boldsymbol{q}}_{ab}$ is equal to the Equation (A7), the $\hat{\boldsymbol{b}}$ can be expressed as $\hat{\boldsymbol{q}}^* \otimes \hat{\boldsymbol{a}} \otimes \hat{\boldsymbol{q}}$.

$$\hat{q} = \left(\cos\left(\frac{\theta}{2}\right), \sin\left(\frac{\theta}{2}\right)l\right) + \varepsilon\left(-\frac{d}{2}\sin\left(\frac{\theta}{2}\right), \sin\left(\frac{\theta}{2}\right)m + \frac{d}{2}\cos\left(\frac{\theta}{2}\right)l\right) \tag{A7}$$

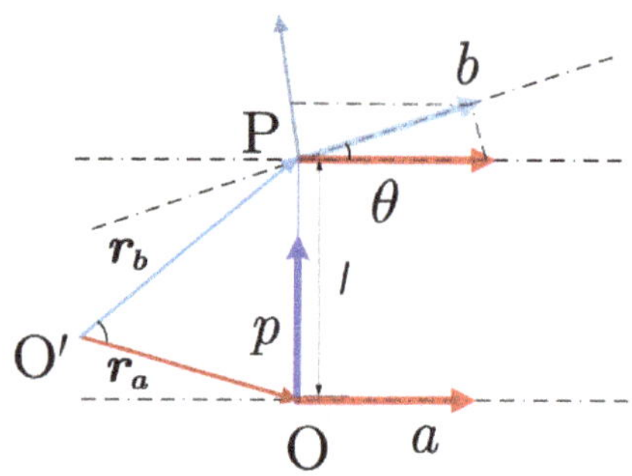

Figure A1. Schematic diagram of rotation and translation transformation of a geometric straight line in space.

Appendix B

The curvature radius ρ_{ij} and the center bracket's curvature radius, which are formed by the changes in each pull line during the driving process, are denoted as R. This is the radius of the circle where the drive line is located. Given that $\rho_i \varphi_i = l_i, \rho_{ij}\varphi_i = l_{ij}$, we can calculate ρ_{ij} and l_{ij} according to Equations (A8) and (A9), respectively.

$$\rho_{ij} = \rho_i + R_i \cos\left(\delta_i\right) \tag{A8}$$

$$l_{ij} = l_i + R_i \varphi_i \cos\,\delta_{ij} \tag{A9}$$

The cables are evenly distributed in the cross-section of the base, with intervals of 90°. Specifically, we have $\delta_{i1} = \delta_i; \delta_{i2} = \delta_i + \frac{\pi}{2}; \delta_{i3} = \delta_i + \pi; \delta_{i4} = \delta_i + \frac{3\pi}{2}$. Consequently, the sum of cosine values of $\sum_i \cos\delta_{ij} = 0$. Considering the relationship between the main arc length and the four chord lengths, we find that $l_i = \frac{1}{4}\sum_i l_{ij}$. When the robot joint only undergoes a bending angle δ while the rotation angle is 0, the transformation of the length of the first drive line can be expressed as Equation (A10).

$$\nabla l_{i1} = l_i - l_{i1} = (\rho_i - \rho_{i1})\delta_i = \nabla \rho_{i_1} \delta_i = R_i \delta_i \tag{A10}$$

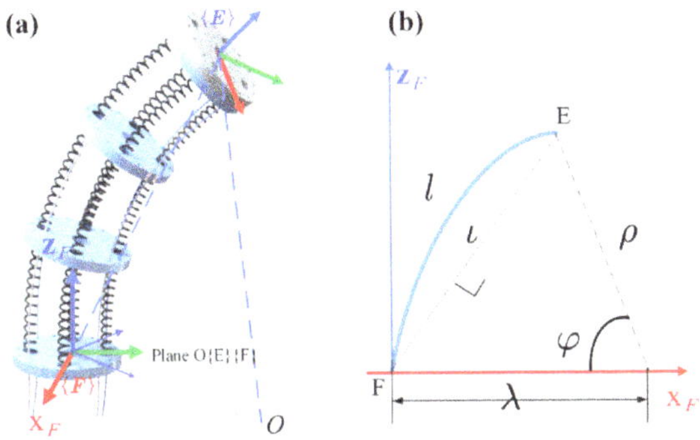

Figure A2. (**a**) Physical model of a single-section robotic arm in the driving state. (**b**) Project Plane O{E}{F} that drives the rear robotic arm onto plane FZ_FX_F.

φ_i, θ_i and the variation of the four drive ∇l_{ij} when the traction arm bends and twists at the same time. Substitute the rope drives 1 and 2 into $l_{ij} = l_i + R_i\varphi_i \cos \delta_{ij}$ to obtain $l_{i1} = l_i + R_i\varphi_i \cos \delta_i; l_{i2} = l_i - R_i\varphi_i \sin(\delta_i)$. In the same way, substitute 1 and 3 to obtain $l_{i1} = l_i + R_i\varphi_i \cos \delta_i; l_{i3} = l_i - R_i\varphi_i \cos(\delta_i)$. By solving for δ_i as shown in Equation (A11), we can determine its value.

$$\delta_i = arc \tan\left(\frac{2(l_{i1} - l_{i2})}{(l_{i1} - l_{i3})} - 1\right), \delta_i \in \left(-\frac{\pi}{2}, \frac{\pi}{2}\right), l_i \in R \tag{A11}$$

We can substitute the rope drives 2 and 4 into $l_{ij} = l_i + R_i\varphi_i\cos \delta_{ij}$ to obtain: $l_{i2} = l_i - R_i\varphi_i\sin \delta_i; l_{i4} = l_i + R_i\varphi_i\sin \delta_i$. In the same way, substitute 1, 3 to obtain: $l_{i1} = l_i + R_i\varphi_i \cos \delta_i; l_{i3} = l_i - R_i\varphi_i \cos(\delta_i)$. By solving for φ_i as shown in Equation (A12), we can determine its value. And because $\rho_i\varphi_i = l_i, l_i = \frac{1}{4}\sum_i l_{ij}$, by solving for ρ_i as shown in Equation (A13), we can determine its value.

$$\varphi_i = \frac{1}{2R_i}\sqrt{(l_{i4} - l_{i2})^2 + (l_{i3} - l_{i1})^2} \tag{A12}$$

$$\rho_i = \frac{R_i \sum_i l_{ij}}{2\sqrt{(l_{i4} - l_{i2})^2 + (l_{i3} - l_{i1})^2}} \tag{A13}$$

Appendix C

The coordinate system {1} relative to {F} is obtained by rotating the coordinate system along the axis l_{z1} by an angle δ.

Then, $q_{F1,r} = \left[\cos \frac{\delta}{2}, \boldsymbol{l_{z1}}\sin \frac{\delta}{2}\right]$ and $q_{F1,d} = \frac{1}{2}(p_i - q_i \otimes p_i \otimes q_i^*) \otimes q_i$, as shown in Equation (A14), the dual quaternion representation.

$$\hat{q}_{F1} = \left[\cos \frac{\delta}{2}\, 0\, 0 \sin \frac{\delta}{2}\, 0\, 0\, 0\, 0\right] \tag{A14}$$

The coordinate system {2} is translated relative to coordinate system {1} along the axis $\boldsymbol{l_{x1}}$, given by $[\rho\cos \delta\; \rho \sin\; \delta\; 0]$.

We have $q_{12,r} = [1,0,0,0]^{\mathrm{T}}$ and $q_{12,d} = [0, \rho\cos\delta, \rho\sin\delta, 0]$. Thus, the dual quaternion representation $\hat{q}_{12}$ is given by Equation (A15).

$$\hat{q}_{F1} = \left[\cos\frac{\delta}{2}\ 0\ 0\ \sin\frac{\delta}{2}\ 0\ 0\ 0\ 0\right] \tag{A15}$$

The coordinate system {3} relative to {2} is obtained by rotating the coordinate system along the axis l_{y1} by an angle φ.

Then, $q_{23,r} = \left[\cos\frac{\varphi}{2}, l_{y1}\sin\frac{\varphi}{2}\right]$ and $q_{23,d} = \left[0,0,0,\rho\sin\frac{\varphi}{2}\right]$, as shown in Equation (A16), the dual quaternion representation.

$$\hat{q}_{23} = \left[\cos\frac{\varphi}{2}, -\sin\delta\sin\frac{\varphi}{2}, \cos\delta\sin\frac{\varphi}{2}, 0,0,0,0, \frac{\rho}{2}\sin\frac{\delta}{2}\sin\frac{\varphi}{2}\right] \tag{A16}$$

The coordinate system {4} is translated relative to coordinate system {3} along the axis l_{x3}, given by $[\rho\cos\delta, \rho\sin\delta, 0]$.

We have $q_{34,r} = [1,0,0,0]^{\mathrm{T}}$ and $q_{34,d} = [0, -\rho\cos\delta\cos\varphi, -\rho\sin\delta\sin\varphi, \rho\sin\varphi]$. Thus, the dual quaternion representation $\hat{q}_{34}$ is given by Equation (A17).

$$\hat{q}_{34} = \left[1,0,0,0,0,-\frac{\rho}{2}\cos\delta\cos\varphi, -\frac{\rho}{2}\sin\delta\cos\varphi, \frac{\rho}{2}\sin\varphi\right] \tag{A17}$$

The coordinate system {E} relative to {4} is obtained by rotating the coordinate system along the axis l_{z4} by an angle $-\delta$.

Then, $q_{4E,r} = \left[\cos\left(\frac{-\delta}{2}\right), l_{z4}\sin\left(\frac{-\delta}{2}\right)\right]$ and $q_{4E,d} = [0, -\rho\sin\delta(\cos\varphi - 1)\sin\frac{\delta}{2}, \rho\cos\delta(\cos\varphi-1)\sin\frac{\delta}{2}, 0]$, as shown in Equation (A18), the dual quaternion representation.

$$\begin{aligned}\hat{q}_{4E} = [&\cos\tfrac{\delta}{2}, -\cos\delta\sin\varphi\sin\tfrac{\delta}{2}, -\sin\delta\sin\varphi\sin\tfrac{\delta}{2}, -\cos\varphi\sin\tfrac{\delta}{2}, 0, -\rho\sin\delta\sin\tfrac{\delta}{2}(\cos\varphi-1),\\ &\rho\cos\delta\sin\tfrac{\delta}{2}(\cos\varphi-1), 0]\end{aligned} \tag{A18}$$

By concatenating the coordinate transformations described in the above equations, we can obtain the transformation dual quaternion between the two coordinate systems, represented as Equation (A19).

$$\hat{q}_{EF} = \left[\cos\frac{\varphi}{2}, -\sin\delta\sin\frac{\varphi}{2}, \cos\delta\sin\frac{\varphi}{2}, 0,0,0,0, \rho\cdot\sin\frac{\varphi}{2}\right] \tag{A19}$$

References

1. Dou, W.; Zhong, G.; Cao, J.; Shi, Z.; Peng, B.; Jiang, L. Soft Robotic Manipulators: Designs, Actuation, Stiffness Tuning, and Sensing. *Adv. Mater. Technol.* **2021**, *6*, 2100018. [CrossRef]
2. Xu, K.; Zhao, J.; Fu, M. Development of the SJTU Unfoldable Robotic System (SURS) for Single Port Laparoscopy. *IEEE/ASME Trans. Mechatron.* **2015**, *20*, 2133–2145. [CrossRef]
3. Su, H.; Qi, W.; Schmirander, Y.; Ovur, S.E.; Cai, S.; Xiong, X. A Human Activity-Aware Shared Control Solution for Medical Human–Robot Interaction. *Assem. Autom.* **2022**, *42*, 388–394. [CrossRef]
4. Jin, S.; Lee, S.K.; Lee, J.; Han, S. Kinematic Model and Real-Time Path Generator for a Wire-Driven Surgical Robot Arm with Articulated Joint Structure. *Appl. Sci.* **2019**, *9*, 4114. [CrossRef]
5. Liu, T.; Mu, Z.; Xu, W.; Yang, T.; You, K.; Fu, H.; Li, Y. Improved Mechanical Design and Simplified Motion Planning of Hybrid Active and Passive Cable-Driven Segmented Manipulator with Coupled Motion. In Proceedings of the 2019 IEEE/RSJ International Conference on Intelligent Robots and Systems (IROS), Macau, China, 3–8 November 2019; pp. 5978–5983.
6. Wang, M.; Dong, X.; Ba, W.; Mohammad, A.; Axinte, D.; Norton, A. Design, Modelling and Validation of a Novel Extra Slender Continuum Robot for in-Situ Inspection and Repair in Aeroengine. *Robot. Comput.-Integr. Manuf.* **2021**, *67*, 102054. [CrossRef]
7. Yeshmukhametov, A.; Koganezawa, K.; Yamamoto, Y.; Buribayev, Z.; Mukhtar, Z.; Amirgaliyev, Y. Development of Continuum Robot Arm and Gripper for Harvesting Cherry Tomatoes. *Appl. Sci.* **2022**, *12*, 6922. [CrossRef]
8. Xu, F.; Wang, H.; Wang, J.; Au, K.W.S.; Chen, W. Underwater Dynamic Visual Servoing for a Soft Robot Arm with Online Distortion Correction. *IEEE/ASME Trans. Mechatron.* **2019**, *24*, 979–989. [CrossRef]
9. Liu, Y.; Ge, Z.; Yang, S.; Walker, I.D.; Ju, Z. Elephant's Trunk Robot: An Extremely Versatile Under-Actuated Continuum Robot Driven by a Single Motor. *J. Mech. Robot.* **2019**, *11*, 051008. [CrossRef]

10. Qin, G.; Ji, A.; Cheng, Y.; Zhao, W.; Pan, H.; Shi, S.; Song, Y. A Snake-Inspired Layer-Driven Continuum Robot. *Soft Robot.* **2022**, *9*, 788–797. [CrossRef]
11. Walker, I.D.; Dawson, D.M.; Flash, T.; Grasso, F.W.; Hanlon, R.T.; Hochner, B.; Kier, W.M.; Pagano, C.C.; Rahn, C.D.; Zhang, Q.M. Continuum Robot Arms Inspired by Cephalopods. In *Unmanned Ground Vehicle Technology VII*; Gerhart, G.R., Shoemaker, C.M., Gage, D.W., Eds.; SPIE: Orlando, FL, USA, 2005; p. 303.
12. Webster, R.J.; Jones, B.A. Design and Kinematic Modeling of Constant Curvature Continuum Robots: A Review. *Int. J. Robot. Res.* **2010**, *29*, 1661–1683. [CrossRef]
13. Rucker, D.C.; Webster, R.J., III. Statics and Dynamics of Continuum Robots with General Tendon Routing and External Loading. *IEEE Trans. Robot.* **2011**, *27*, 1033–1044. [CrossRef]
14. Till, J.; Aloi, V.; Riojas, K.E.; Anderson, P.L.; Webster Iii, R.J.; Rucker, C. A Dynamic Model for Concentric Tube Robots. *IEEE Trans. Robot.* **2020**, *36*, 1704–1718. [CrossRef]
15. Lafmejani, A.S.; Doroudchi, A.; Farivarnejad, H.; He, X.; Aukes, D.; Peet, M.M.; Marvi, H.; Fisher, R.E.; Berman, S. Kinematic Modeling and Trajectory Tracking Control of an Octopus-Inspired Hyper-Redundant Robot. *IEEE Robot. Autom. Lett.* **2020**, *5*, 3460–3467. [CrossRef]
16. Amanov, E.; Nguyen, T.D.; Burgner-Kahrs, J. Tendon-Driven Continuum Robots with Extensible Sections—A Model-Based Evaluation of Path-Following Motions. *Int. J. Robot. Res.* **2019**, *40*, 027836491988604. [CrossRef]
17. Chitalia, Y.; Deaton, N.J.; Jeong, S.; Rahman, N.; Desai, J.P. Towards FBG-Based Shape Sensing for Micro-Scale and Meso-Scale Continuum Robots with Large Deflection. *IEEE Robot. Autom. Lett.* **2020**, *5*, 1712–1719. [CrossRef]
18. Walker, I.D. Continuous Backbone "Continuum" Robot Manipulators. *ISRN Robot.* **2013**, *2013*, 726506. [CrossRef]
19. Samadikhoshkho, Z.; Ghorbani, S.; Janabi-Sharifi, F. Coupled Dynamic Modeling and Control of Aerial Continuum Manipulation Systems. *Appl. Sci.* **2021**, *11*, 9108. [CrossRef]
20. Sedal, A.; Bruder, D.; Bishop-Moser, J.; Vasudevan, R.; Kota, S. A Continuum Model for Fiber-Reinforced Soft Robot Actuators. *J. Mech. Robot.* **2018**, *10*, 024501. [CrossRef]
21. Singh, I.; Amara, Y.; Melingui, A.; Mani Pathak, P.; Merzouki, R. Modeling of Continuum Manipulators Using Pythagorean Hodograph Curves. *Soft Robot.* **2018**, *5*, 425–442. [CrossRef]
22. Ma, X.; Chiu, P.W.-Y.; Li, Z. Shape Sensing of Flexible Manipulators with Visual Occlusion Based on Bezier Curve. *IEEE Sens. J.* **2018**, *18*, 8133–8142. [CrossRef]
23. Kołota, J.; Kargin, T.C. Comparison of Various Reinforcement Learning Environments in the Context of Continuum Robot Control. *Appl. Sci.* **2023**, *13*, 9153. [CrossRef]
24. Godage, I.S.; Walker, I.D. Dual Quaternion Based Modal Kinematics for Multisection Continuum Arms. In Proceedings of the 2015 IEEE International Conference on Robotics and Automation (ICRA), Seattle, WA, USA, 26–30 May 2015; pp. 1416–1422.
25. Araújo, A.L. *SMART 2015 7th ECCOMAS Thematic Conference on Smart Structures and Materials, Azores, Portugal, 3–6 June 2015*; IDMEC: Lisboa, Portugal, 2015; ISBN 978-989-96276-8-0.
26. Seleem, I.A.; El-Hussieny, H.; Assal, S.F.M.; Ishii, H. Development and Stability Analysis of an Imitation Learning-Based Pose Planning Approach for Multi-Section Continuum Robot. *IEEE Access* **2020**, *8*, 99366–99379. [CrossRef]

applied sciences

MDPI

Article

Analysis of the Slanted-Edge Measurement Method for the Modulation Transfer Function of Remote Sensing Cameras

Jian Yu [1,2], Yu Zhang [1,*], Biao Qi [1], Xiaotian Bai [1,2], Wei Wu [1,2] and Hongxing Liu [1]

1 Changchun Institute of Optics, Fine Mechanics and Physics, Chinese Academy of Sciences, Changchun 130033, China
2 University of Chinese Academy of Sciences, Beijing 100049, China
* Correspondence: zhangyu@ciomp.ac.cn

Abstract: The modulation transfer function (MTF) serves as a crucial technical index for assessing the imaging quality of remote sensing cameras, which is integral throughout their entire operational cycle. Currently, the MTF evaluation of remote sensing cameras primarily relies on the slanted-edge method. The factors influencing the slanted-edge method's effectiveness are broadly classified into two categories: algorithmic factors and image factors. This paper innovatively comprehensively analyzes the influencing factors of the slanted-edge method and proposes an improved slanted-edge method to calculate the MTF testing method of remote sensing cameras, which is applied to the MTF testing of remote sensing cameras. Since the traditional algorithm can only be applied in the small angle situation, this paper proposes a new method of slanted-edge method test calculation based on the optimal oversampling rate (OSR) adaptive model of the slanted edge and uses simulation experiments to verify the reliability of the algorithm model through the deviation of the slanted-edge angle calculation and MTF measurement, and the results show that the algorithm improves the accuracy of the MTF measurement compared with the ISO-cos and OMINI-sine methods. Then, the effects of the slanted-edge angle, image region of interest (ROI), as well as image contrast and signal-to-noise ratio (SNR) on the accuracy of the MTF calculation by the slanted-edge method were quantitatively analyzed as the constraints of the slanted-edge method test. Based on the laboratory target experiment, the algorithm flow and various influencing factors obtained in the simulation stage are verified, and the experimental results are more consistent with the various test results obtained in the simulation stage. Consequently, the slanted-edge method introduced in this paper is applicable for future remote sensing camera MTF testing. This approach offers a valuable reference for on-orbit focusing, satellite operational condition monitoring, lifespan estimation, and image restoration.

Keywords: remote sensing camera; MTF; slanted-edge method; accuracy analysis; influencing factors

Citation: Yu, J.; Zhang, Y.; Qi, B.; Bai, X.; Wu, W.; Liu, H. Analysis of the Slanted-Edge Measurement Method for the Modulation Transfer Function of Remote Sensing Cameras. *Appl. Sci.* **2023**, *13*, 13191. https://doi.org/10.3390/app132413191

Academic Editor: Mihailo Ristic

Received: 19 November 2023
Revised: 4 December 2023
Accepted: 8 December 2023
Published: 12 December 2023

1. Introduction

In the lifecycle of remote sensing cameras, from system installation and long-distance transportation to launch impact and in-orbit operation, various factors such as the space environment and component aging lead to a gradual decline in imaging system performance [1]. This deterioration results in diminished image quality and information content. High-quality space images are crucial for delivering accurate and comprehensive data, necessitating regular in-orbit quality assessments of remote sensing cameras to monitor performance shifts. The MTF is a critical parameter for evaluating the imaging quality of optical systems [2]. It quantifies the extent of information attenuation during object imaging, offering higher accuracy and a more objective and comprehensive assessment than traditional methods. As an ideal performance index for orbiting remote sensing cameras, MTF measurements are also vital for predicting the lifespan of satellite imaging systems [3], assisting in satellite in-orbit adjustments, and facilitating image recovery and reconstruction.

MTF detection methods vary depending on the target used, including the three-bar method [4], slanted-edge method [5], slit method [6], and Siemens Star method [7,8]. Currently, laboratory testing of camera MTF predominantly employs the three-bar and slanted-edge methods, as other methods face limitations due to challenges in target processing, extended testing durations, and poor repeatability, leading to less frequent application. The slanted-edge method, a primary technique for remote sensing camera MTF detection [9], involves rotating a slanted-edge target at a specific angle relative to the detector's rows or columns. This rotation projects and rearranges the pixels of the slanted-edge image based on a predefined rule [10], thereby enhancing the image's sampling rate and mitigating under-sampling issues. Recognized as an indirect approach for measuring the MTF, the slanted-edge method has been successfully implemented in the in-orbit MTF measurement of satellites such as IKONOS [11,12], Orbview [13], GF-4 [14], and Quickbird [15]. The International Organization of Standards endorses this method as well (ISO12233:2023) [16], which involves analyzing a camera-captured slanted-edge target image to determine the edge spread function (ESF) [17], differentiating the ESF to acquire the line spread function (LSF), and then applying Fourier transform on the LSF and taking the modes to compute the MTF [18,19]. Offering a comprehensive evaluation at various spatial frequencies in a single instance and characterized by rapid detection, the slanted-edge method has garnered significant scholarly interest [20].

The accuracy of the slanted-edge method hinges on precise estimation of the image edge angle, as this directly impacts the MTF calculation results. Masaoka et al.'s [21] approach utilizes the Sobel operator and the Hough transform for edge angle estimation, examining how estimation errors affect the method's accuracy. Roland [22] focused on the impact of inclination estimation errors on the stability and repeatability of MTFs obtained via this method. A critical step in the slanted-edge method's algorithm involves differentiating the ESF to derive the LSF, a process that can amplify noise. Consequently, noise impact analysis and suppression techniques have become a focal area of research. To mitigate noise's impact on accuracy, scholars have suggested fitting the ESF with a function model. Tzannes and Mooney [23] and J. M. Mooney, for instance, identified the midpoint of each data row as half the mean gray value between the dark and light regions of the slanted-edge image, constructing an oversampled ESF fitted with a Fermi function to counter noise effects. Hwang et al. [24] enhanced MTF detection accuracy by refining the ISO 12233 slanted-edge method with a similar Fermi function fitting of the ESF, while Masaoka [25] achieved higher accuracy through an oversampling ESF, employing image rotation, cumulative distribution function fitting, and projection.

The analysis above reveals that the MTF testing of remote sensing cameras using the slanted-edge method is significantly influenced by the chosen calculation method, with varying algorithmic models leading to considerable deviations in results [26]. In this paper, according to the working state of the space camera on orbit and the imaging characteristics, combined with the relevant theories of optoelectronic imaging systems, the optimal algorithm based on the slanted-edge optimal OSR adaptive model's slanted-edge method calculation is proposed through the study of the sub-pixel edge detection fitting algorithm and edge diffusion function extraction. The GLCM method is used to detect the sub-pixel edges and accurately locate the edge position, and then the edge points are projected, and the sub-pixel arrays are sampled, fitted, and smoothed with a variable OSR for different angles to obtain the ESF, and the differentiation of the ESF curves is used to obtain the LSF, and the smoothing of the LSF curves is performed by using the Tukey window function. Finally, the Fourier transform mode normalization is applied to the LSF curve to obtain the MTF curve. In order to verify the reasonableness of the algorithm, simulation experiments are used to verify the response of the imaging system to the spatial frequency through the mathematical function to simulate the response of the imaging system to the spatial frequency and obtain the slanted-edge image and the theoretical MTF curve, respectively, using the ISO-cos, OMNI-sine, and the algorithm proposed in this paper to calculate the MTF curve of the slanted-edge image and compare it with the theoretical MTF curve, which shows that the MTF curve of the proposed algorithm at each frequency has a higher similarity to the theoretical MTF curve, verifying the reasonableness of the algorithm in this paper. This study shows that the MTF curves at each frequency of the proposed algorithm have high similarity with the theoretical MTF curves, which verifies the reasonableness of the algorithm in this paper.

Subsequently, the image factors affecting the calculation of the slanted-edge method are analyzed. It establishes the relationship between the slanted-edge angle and ROI in the MTF calculation results through simulation. The simulation results are then compared in terms of accuracy and stability to ascertain the optimal slanted-edge angle range and the ideal ROI range for the slanted-edge method. Under optimal conditions for the slanted-edge angle and image ROI, simulations are conducted to obtain images with varying contrasts and SNRs. These simulations aim to investigate the impact of these varying contrasts and SNRs on the calculation results. In the laboratory, the MTF test validation system was constructed. This system utilized a customized slanted-edge target to capture test images via the optical system imaging process. Subsequently, the MTF test was conducted to validate the efficacy of the slanted-edge test method proposed in this paper.

The rest of this paper is structured as follows: In Section 2, the foundational concept of the MTF and the procedural aspects of MTF calculation through the slanted-edge method are introduced. This section also delves into the analysis of algorithmic factors affecting the results of the slanted-edge method, presenting processing steps of the algorithm based on a slanted-edge optimal OSR adaptive model. Section 3 scrutinizes the impact of image factors on the results of the slanted-edge method calculation and proposes constraints for this calculation method. Section 4 presents and analyzes the results of the experiments conducted. Finally, Section 5 offers a comprehensive summary of the entire paper and outlines directions for future research.

2. Measurement Principles and Methods

Per the principles of Fourier optical imaging, in a specific imaging system, when an ideal point light source serves as the input function, the resulting output is not a concentrated point but rather a distinct spot [27]. This phenomenon is known as the point spread function (PSF) [28], exemplified in Figure 1.

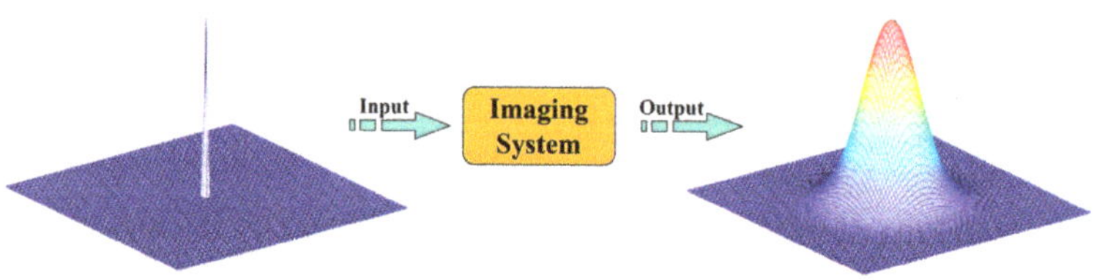

Figure 1. Formation of point spread function.

The PSF characterizes the response of an imaging system to an ideal point source, illustrating the diffusion of this source on the imaging plane and thereby unveiling the system's spatial resolution. The optical transfer function (OTF) [29,30], as delineated in Equation (1), emerges from the two-dimensional Fourier transform of the PSF [10]. It embodies the imaging system's response to various spatial frequencies and is composed of two components: the amplitude and the phase parts.

$$OTF(u,v) = \mathcal{F}\{PSF(x,y)\} \tag{1}$$

Here, $\mathcal{F}$ is the Fourier transform, x, y are the spatial coordinates, and u, v are the frequency coordinates.

The OTF provides a comprehensive description of an imaging system's capacity to transfer spatial details, encompassing both contrast and phase information [31]. The MTF, constituting the magnitude component of the OTF, specifically reflects the system's proficiency in transferring contrast across various spatial frequencies. Accordingly, the MTF is formulized as shown in Equation (2):

$$MTF(u,v) = |OTF(u,v)| \quad (2)$$

Similarly, the LSF is the light intensity distribution captured on the image plane after imaging of a line source, either $\delta(x)$ or $\delta(y)$. The LSF is essentially the integral of the PSF along a specified direction, and its one-dimensional Fourier transform provides the cross-section profile of the MTF in that direction, as demonstrated in Equation (3):

$$\begin{aligned} LSF_x(x) &= \int PSF(x,y)dy \\ LSF_y(y) &= \int PSF(x,y)dx \end{aligned} \quad (3)$$

When the input function is a step function, typically exemplified by the slanted-edge image region of a remote sensing image, its resultant output function is the ESF. This output represents the two-dimensional light intensity distribution post-optical system processing. The ESF can be viewed as the integral of the LSF, establishing a relationship between differentiation and integration, as illustrated in Equation (4). Furthermore, the one-dimensional Fourier transform of the LSF corresponds to the MTF for that specific direction [32], forming the theoretical foundation of the slanted-edge method used in MTF calculation.

$$\begin{aligned} ESF(x) &= \int_{-\infty}^{x} LSF(x')dx' \\ LSF(x) &= \frac{d}{dx}ESF(x) \end{aligned} \quad (4)$$

As delineated by the aforementioned theory, the interrelationship among PSF, LSF, ESF, and MTF is depicted in Figure 2. Consequently, by acquiring any one of LSF, PSF, or ESF, the MTF can be derived based on this established relationship [33,34].

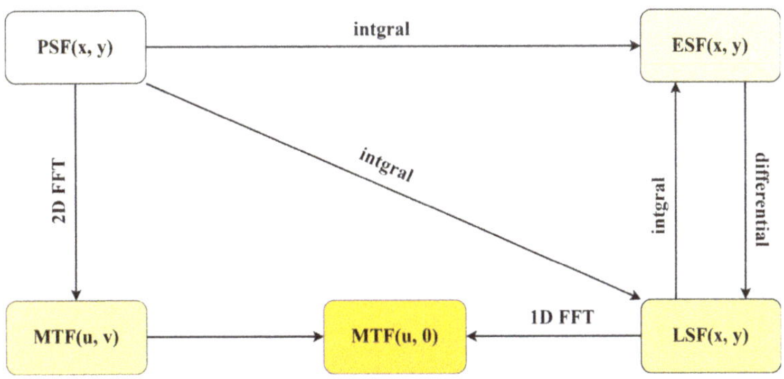

Figure 2. Relationship between PSF, ESF, LSF, and MTF.

The principal procedures for computing the MTF using the slanted-edge method [14], as applied in remote sensing imagery, are illustrated in Figure 3:

(1) Identify and capture an image of the slanted-edge calculation region in the remote sensing image, ensuring it conforms to specific criteria; the edge should be straight and form a slight angle with the detector's image element direction.
(2) Conduct image preprocessing on the slanted-edge region to minimize random noise, followed by sub-pixel slanted-edge detection to ascertain the sub-pixel location of the slanted edge.
(3) Gather data points adjacent to the slanted edge and perform multi-line alignment using the precise sub-pixel slanted-edge positions from step 2, resulting in the ESF of densely sampled points.

(4) Compute the derivative of the ESF to generate the LSF curve.
(5) Apply the Fourier transform to the LSF, yielding the MTF curve.

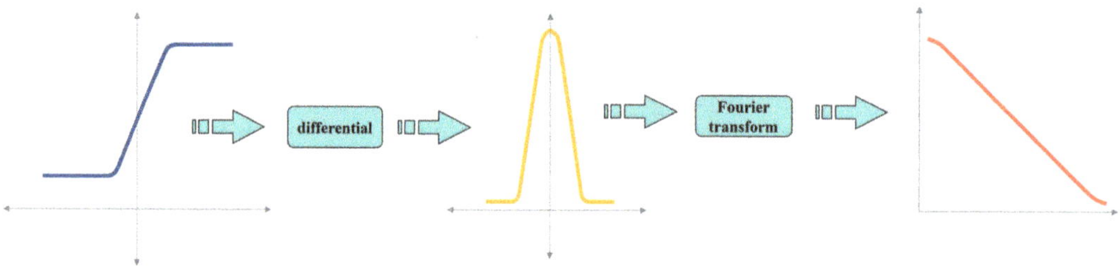

Figure 3. Schematic calculation of slanted-edge method.

2.1. Sub-Pixel Edge Detection and Fitting

The sub-pixel edge detection step is pivotal in the slanted-edge method, as accurate determination of the sub-pixel edge position is essential for proper alignment and fitting in subsequent stages. Presently, sub-pixel edge detection techniques can be broadly classified into three categories: moment-based, interpolation-based, and edge model function fitting-based approaches. In this paper, we employ one method from each category for sub-pixel edge detection using the slanted-edge method: gray-level co-occurrence matrix (GLCM) [35], bilinear interpolation (BLI) [36], and hyperbolic tangent fitting (HTF) [37]. Since sub-pixel edge detection is executed for each row of data points, the methods used are inherently one-dimensional. We simulate, compare, and analyze the accuracy and influencing factors of these methods to identify the most effective sub-pixel edge detection technique for the slanted-edge method.

2.1.1. Gray-Level Co-Occurrence Matrix (GLCM) Method

The GLCM sub-pixel edge detection method operates on the principle of invariant GLCM, comparing the actual edge with an ideal edge model to accurately localize the image edge. This method presupposes that the one-dimensional ideal edge model resembles a step function, $u(x)$, with gray values g_1 and g_2 on either side of the edge and a step height of h. The ideal edge configuration is therefore represented as follows:

$$E(x) = (g_1 - g_2) \times u(x-h) + g_1 \tag{5}$$

where ω_1, ω_2 are the proportions of pixel points with gray values g_1, g_2, respectively, relative to the total number of pixel points in the entire edge region, adhering to the equation: $\omega_1 + \omega_2 = 1$. Assuming x_j represents the GLCM value at an edge point, we consider the first three moments of this value.

$$\overline{m} = \sum_{j=1}^{2} \omega_j g_j^i = \frac{1}{n}\sum_{j=1}^{n} x_j^i \tag{6}$$

Here, n is the total number of pixel points, with the actual edge density denoted as $\omega = k/n$. This value of ω can be deduced through calculation as follows:

$$\begin{gathered} \omega_1 = 0.5\left[1+\bar{s}\sqrt{1/(4+\bar{s}^2)}\right] g_1 = \overline{m_1} - \bar{\sigma}\sqrt{\frac{\omega_2}{\omega_1}}, g_2 = \overline{m_2} - \overline{m_1} \\ \bar{s} = \frac{1}{n}\sum_{i=1}^{n}\frac{(x_i-\overline{m_1})^3}{\sigma^3} = \frac{\overline{m_1}+2\overline{m_1}^3-3\overline{m_1}\,\overline{m_2}}{\sigma^3} \end{gathered} \tag{7}$$

Consequently, the edge position in the actual image is determined as follows:

$$k = np_1 - \frac{1}{2} = \frac{n}{2}\sqrt{\frac{s}{4+\bar{s}^2}} + (n-1)/2 \tag{8}$$

2.1.2. Bilinear Interpolation (BLI) Method

Bilinear interpolation, an advancement of linear interpolation in two dimensions, facilitates the estimation of pixel values at undefined points within an image. This method approximates the value of a new pixel point by utilizing the values of its four neighboring pixels. It is particularly effective for tasks like image scaling, rotation, and sub-pixel edge detection [38].

Consider a pixel point with a sub-pixel location designated as (x, y). To employ bilinear interpolation, it is necessary to identify the four nearest integer pixel locations surrounding this point. Typically, these locations are represented as (x_1, y_1), (x_1, y_2), (x_2, y_1), (x_2, y_2), where x_1, x_2 are the nearest integers to x, obtained by rounding down and up, respectively, and similarly for y_1, y_2 with respect to y.

Initially, linear interpolation is conducted in the x-direction, which entails calculating the values at points (x, y_1) and (x, y_2).

$$\begin{aligned} f(x,y_1) &= f(x_1,y_1) + (x - x_1) \cdot (f(x_2,y_1) - f(x_1,y_1)) \\ f(x,y_2) &= f(x_1,y_2) + (x - x_1) \cdot (f(x_2,y_2) - f(x_1,y_2)) \end{aligned} \tag{9}$$

A linear interpolation is then conducted in the direction where the values of (x, y_1) and (x, y_2) have already been calculated.

$$f(x,y) = f(x,y_1) + (y - y_1) \cdot (f(x,y_2) - f(x,y_1)) \tag{10}$$

The interpolation results from the two aforementioned directions are combined to derive the final interpolation expression, incorporating the respective weights of these two directions:

$$\begin{aligned} f(x,y) = {} & (1-t)(1-u)f(x_1,y_1) + t(1-u)f(x_2,y_1) \\ & +(1-t)uf(x_1,y_2) + tuf(x_2,y_2) \end{aligned} \tag{11}$$

where t, u represent the differences in distances along the x-axis and y-axis, respectively, defined as $t = x - x_1$ and $u = y - y_1$, where x_1, y_1 correspond to the coordinates of the lower left pixel point.

2.1.3. Hyperbolic Tangent Fitting (HTF) Method

Angela Cantatore et al. proposed the hyperbolic tangent fitting method as an edge model. This approach aligns the actual edge gray-level distribution with the model to enhance sub-pixel edge detection accuracy. The hyperbolic tangent function, an S-type function, is generally expressed as follows:

$$f(x) = A \cdot \tanh(B(x - C)) + D \tag{12}$$

where A is the curve's magnitude, B is the slope, C is the curve's center position (i.e., the edge position), and D is the offset.

In sub-pixel edge detection, a set of grayscale values $I(x)$ from edge pixels is analyzed, where x indicates the pixel position. The objective is to determine the optimal values of A, B, C, and D that best fit these data. Initially, a rough edge location is identified using initial edge detection, and the surrounding gray values are extracted. Subsequently, a fitting window encompassing a series of pixel points near the edge is chosen in the vicinity of this initial edge location. The method employs least squares to determine the optimal parameters A, B, C, and D, thereby minimizing the discrepancy between the fitted function $f(x)$ and the actual gray value $I(x)$. This is accomplished by addressing the following optimization problem:

$$\min_{A,B,C,D} \sum_x [I(x) - (A \cdot \tanh(B(x - C)) + D)]^2 \tag{13}$$

The optimization problem can be solved using a nonlinear least-squares algorithm, and in this paper the Levenberg–Marquardt algorithm is used to solve it.

2.1.4. Simulation Experiment

This comparative analysis evaluates the three methods discussed above through simulation. Firstly, a slanted-edge image, measuring 30 pixels in both length and width, is selected. The results of applying these three sub-pixel edge detection methods are illustrated in Figure 4.

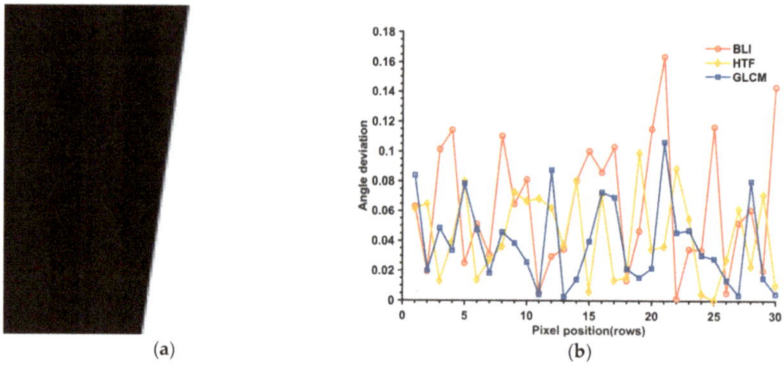

Figure 4. The calculated results for three sub-pixel edge detection methods are as follows: (**a**) the slanted-edge image; (**b**) the deviation between the calculated and theoretical edge positions.

Without considering random noise, Table 1 shows that all three sub-pixel edge detection accuracies are high, with the interpolation method having an error of less than 0.06 pixels, while the GLCM and fitted methods have significantly higher accuracies than the interpolation method, with an error of less than 0.03 pixels.

Table 1. Mean and standard deviation of sub-pixel edge deviations calculated by the three methods.

Indicator \ Method	GLCM	BLI	HTF
Angle deviation MEAN	0.063718	0.044824	0.038874
Angle deviation SD	0.043189	0.02798	0.028086

Gaussian noise, with a noise standard deviation ranging from 0.001 to 0.010 in increments of 0.001, was added to the simulated image for blurring purposes. Subsequently, the root-mean-square error (RMSE) between the measured and theoretical values of the GLCM, BLI, and HTF detection methods was calculated under varying noise conditions. The results of these calculations are depicted in Figure 5.

The introduction of varying levels of Gaussian noise reduces the accuracy of the three methods under consideration. Specifically, the accuracy of the BLI method is significantly compromised by random noise, with edge detection errors escalating rapidly from 0.06 pixels to 0.7 pixels post-noise addition. Conversely, the GLCM and HTF methods exhibit more resilience to random noise, with errors incrementally rising to only 0.1 pixels. Notably, the GLCM method demonstrates commendable accuracy, comparable to the fitting method. However, its shorter processing time makes it particularly suited to real-time detection requirements. Consequently, this paper opts for the GLCM method, balancing accuracy with operational efficiency.

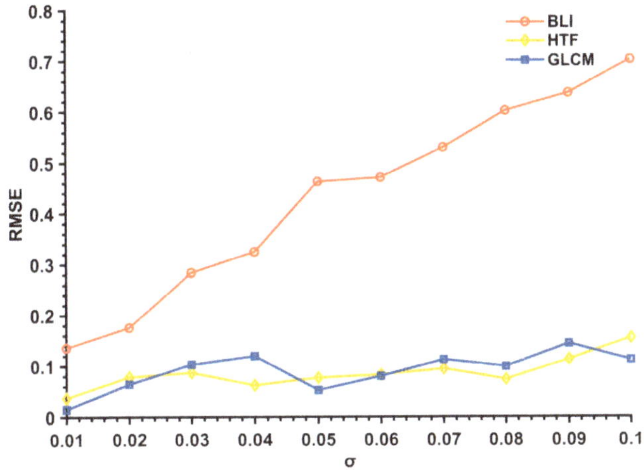

Figure 5. RMSE of three methods for detecting edges under different noise conditions.

2.2. ESF Processing Methods

Extracting the ESF curve is a crucial step in ensuring the computational accuracy of the slanted-edge method. The optical imaging system's conversion of optical signals into electrical signals on the detector results in a down-sampling phenomenon [39]. This varies based on pixel size and leads to the edge image data manifesting as a discrete sequence of points rather than a continuous curve. Consequently, part of the ESF curve information is missing, which directly impacts the accuracy of the MTF. To address this issue, a multi-line edge alignment method is employed to create a more densely sampled ESF curve.

Firstly, the pixels within the edge image's ROI are projected. This projection can be executed in two ways: horizontally or perpendicular to the slanted edge, as illustrated in Figure 6. While these projection methods yield similar results when the edge angle is small, they diverge significantly as the edge angle increases.

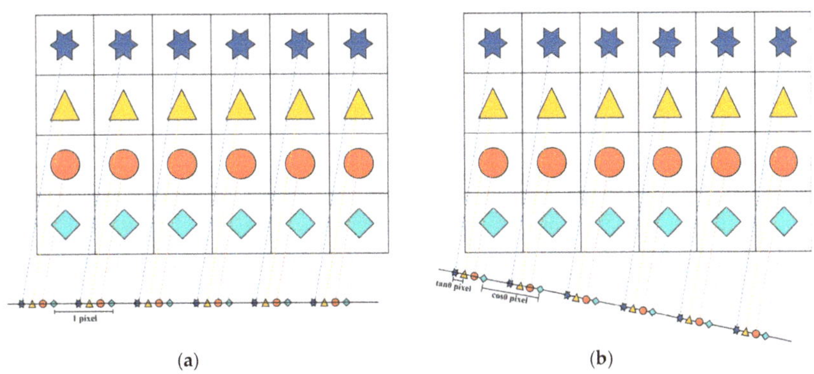

Figure 6. Comparison of different projection methods: (**a**) horizontal direction; (**b**) vertical slanted-edge direction.

In this paper, the ROI image is projected orthogonally to its edge. The subsequent one-dimensional (1D) pixel array requires sub-pixel oversampling, where the bin width is a critical parameter. In the methods outlined in ISO 12233, ISO-cos, and OMNI, a 4x OSR is utilized. The primary distinction among these methods lies in the variation in bin width relative to the edge's inclination angle. Specifically, ISO 12233 and OMNI employ a constant bin width of 1/4 pixel, independent of the angle, whereas ISO-cos adjusts the width based on the neighboring pixel dimensions at the projection level, resulting in a bin width of $(\cos\theta)/4$ pixel. At any edge angle, when $\tan\theta = 1/4$, the ISO-cos method aligns the edge image's pixel projection intervals with the oversampling bin width. However, for angles exceeding 14.04°, where the projection interval of each pixel column fails to align with the bin width, resulting in a gap larger than the bin width, the assumptions about the random phases of the imaging system's sampling points are violated. This misalignment leads to inaccuracies in the MTF estimation. Since $\cos\theta \leq 1$, the bin width in the OMNI method is always greater than or equal to that in ISO-cos at a fixed OSR. This reduces the likelihood of empty bins in edge image projections at larger edge angles. Nonetheless, in scenarios involving smaller angles, the OMNI method's accuracy for sub-pixel oversampling binning does not match that of the ISO-cos method.

Masaoka [25] proposed the OMNI-sine method, an enhancement of the OMNI approach, which adapts the bin width in response to the edge angle. In this method, the sampling interval for each row of the ROI aligns with the bin width at intervals of $\cos\theta$ pixel, and for each column, it aligns at $\tan\theta$ pixel intervals. However, challenges arise when the slanted-edge angle θ approaches 0°, causing the bin width to diminish towards zero and potentially leading to an infinitely large OSR. When the angle is less than 4°, the OSR can exceed 14°. While, theoretically, a higher OSR correlates with greater computational accuracy, practically, an excessively high rate increases computational cost with marginal gains in accuracy. Conversely, a very low rate leads to data underutilization and compromised accuracy, particularly in accurately determining the system MTF at the Nyquist frequency. To mitigate these issues, the bin width is confined within $1/n$ bin to minimize the occurrence of empty bins. The OSR, denoted as ν_{bin}, is defined in Equation (14) and ensures computational efficiency comparable to the ISO method across all angles.

$$\begin{aligned} \theta_{\text{sym}} &= \frac{\arccos(\cos 4\theta)}{4} \\ \nu_{\text{bin}} &= n_{\text{bin}} \cdot 2^{[\log_2(\sin\theta_{\text{sym}})] - \log_2(\sin\theta_{\text{sym}})} \end{aligned} \tag{14}$$

This paper introduces an optimal OSR adaptive model, which is developed by analyzing various ESF processing methods. The model segments the edge angle interval to determine the optimal OSR for each specific angle. The segmentation is as follows: for an edge angle θ in the range [0°, 5.711°), the optimal OSR is set to 5; for θ in [5.711°, 18.435°), it is determined by the cotangent of θ; and for θ in [18.435°, 45°), the optimal OSR is fixed at 3.

In practical applications, optical systems often exhibit aberrations that lead to distortion in edge images, thereby impeding accurate characterization of the ESF distribution in actual images. To maximize the utility of image data while minimizing the influence of random noise, it is essential to reprocess the ESF. This paper describes the application of a fifth-order filter for initial fitting of the ESF, followed by the use of a Savitzky–Golay filter for subsequent smoothing.

2.3. LSF Processing Methods

The ESF can be transformed into the LSF by differentiating it from the discrete ESF. This process occurs post-noise reduction [40], as outlined in Equation (15), where the actual edge distribution overlays the noise interference.

$$\begin{aligned} f'_{ESF}(x) &= f_{ESF}(x) + f_{noise}(x) \\ F\left\{\frac{d}{dx} f'_{ESF}(x)\right\} &= F_{LSF}(j2\pi) + j2\pi F_{noise}(j2\pi) \end{aligned} \tag{15}$$

Due to the spatial domain differentiation of the ESF, noise increasingly impacts the MTF values of the system. The noise coefficient escalates with rising frequencies, indicating a more pronounced noise effect. Therefore, it is necessary to smooth the LSF to mitigate this noise impact. For smoothing, Hamming windows are typically employed. This paper compares the efficacy of the Tukey and Hamming window functions in both time and frequency domains, as depicted in Figure 7.

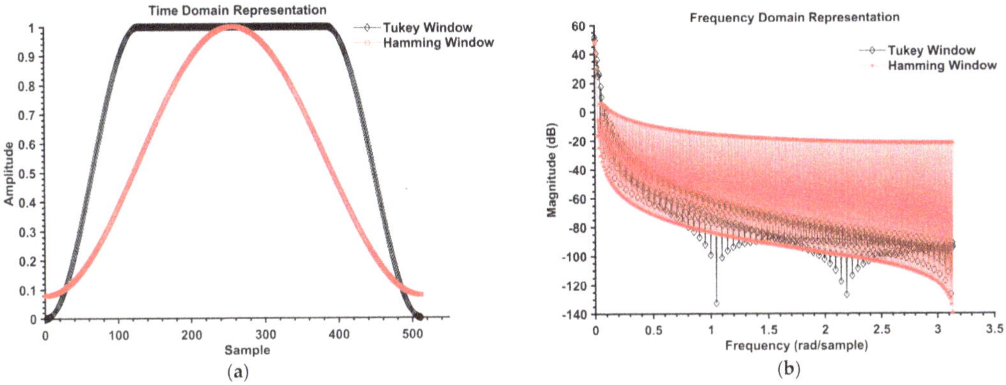

Figure 7. Hamming window and Tukey window frequency and time–domain response curves: (**a**) the time-domain response curve; (**b**) the frequency-domain response curve.

Figure 7 illustrates that in the time domain, the Tukey window (represented by a black line) features smooth transitions at the window's extremities. This characteristic diminishes signal abruptness, thereby aiding in the reduction in spectral leakage. In contrast, the Hamming window (represented by a red line) is generally smoother but exhibits more pronounced jumps at the window's onset and conclusion, potentially leading to increased spectral leakage [41]. In the frequency domain, the Tukey window typically exhibits lower sidelobe magnitudes (peaks outside the main peak) compared to the Hamming window, suggesting enhanced efficacy in minimizing spectral leakage. Moreover, the main sidelobes (central highest peaks) of the Hamming window [42] are marginally wider than those of the Tukey window, indicating a slight inferiority in frequency resolution. These distinctions imply that the Tukey window might be preferable for applications requiring a balance between time–frequency characteristics, reduced spectral leakage, and controlled sidelobe magnitude. Particularly in scenarios demanding precise frequency characteristic measurement or analysis, the Tukey window's attributes may offer superior performance. Therefore, this paper employs the Tukey window for smoothing purposes.

2.4. Processing Steps of the Algorithm Based on the Slanted-Edge Optimal OSR Adaptive Model

According to the comparative analysis of each key link processing method of the slanted-edge method above, the process of the optimal OSR adaptive model based on the slanted edge is shown in Figure 8.

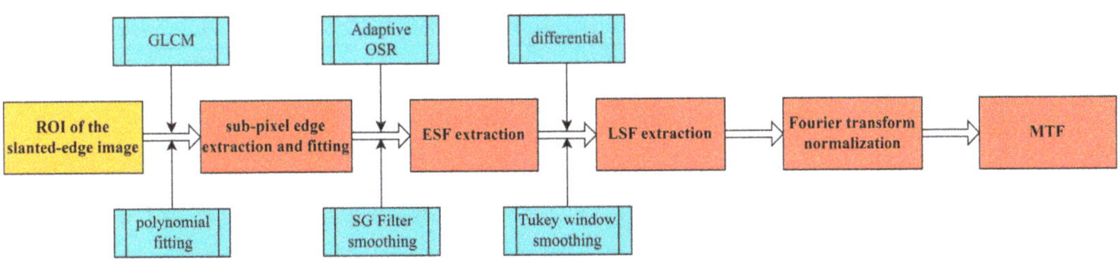

Figure 8. Processing steps of the algorithm in this paper.

2.5. Simulation Experiment Analysis

To precisely assess the accuracy of the algorithm, the RMSE between the test and theoretical MTF curves across various frequencies is utilized as the evaluation metric. This method of accuracy calculation, in contrast to the commonly employed absolute error measurement of MTF values at the Nyquist frequency, encompasses a broader range of spatial frequencies. It more effectively captures the congruence between the computed and theoretical MTF curves, thereby providing a more comprehensive representation of the algorithm's accuracy.

In line with imaging system theory, the response of an imaging system to spatial frequency can be simulated using a mathematical function, provided the system's parameters are known. These parameters include the aperture diameter D, focal length f, and wavelength λ. Furthermore, Equation (16) defines the cutoff frequency f_{cut} in the context of the diffraction limit.

$$f_{\text{cut}} = \frac{1}{\lambda}\frac{D}{f} \tag{16}$$

The OTF in the diffraction limit can be expressed in Equation (17) as:

$$\text{OTF}_{\text{diff}}(f) = \frac{2}{\pi}\left(\arccos\left(\frac{f}{f_{\text{cut}}}\right) - \frac{f}{f_{\text{cut}}}\sqrt{1-\left(\frac{f}{f_{\text{cut}}}\right)^2}\right) \tag{17}$$

The imaging system's image sensor size is designated as 8 μm, with an aperture diameter of 30 mm, a focal length of 25 mm, and a diffraction wavelength of 10 μm. Utilizing these parameters, the system's OTF is derived in accordance with Equation (17). Subsequently, the PSF is computed through an inverse Fourier transform of the OTF, simulating the system's response to a point light source. The cumulative sum of these PSFs yields the ESF, which represents the system's response to an idealized edge. This ESF is then projected onto a two-dimensional grid, determined by setting the edge angle and the image dimensions (H and V). The ESF effectively models the imaging system's response to the slanted edge, culminating in the generation of slanted-edge images at angles of 5°, 10°, 14°, and 26°, as shown in Figure 9.

Figure 9. Images of slanted edge with different angles: (**a**) slanted-edge angle of 5°; (**b**) slanted-edge angle of 10°; (**c**) slanted-edge angle of 14°; (**d**) slanted-edge angle of 26°.

In this section, edge images are generated using software, after which Gaussian noise with a variance of 0.005 is introduced to the images. Subsequently, the ISO-cos, OMNI-sine, and the adaptive OSR method proposed in this paper are employed to calculate the MTF of the edge images at various angles. Simulations are conducted to validate the accuracy of the algorithm presented in this paper, as illustrated in Figure 10.

Figure 10 clearly demonstrates that the MTF curves calculated using the algorithm proposed in this paper align more closely with the theoretical values compared to the other two methods. Specifically, Figure 10a indicates that the OMNI-sine method is prone to MTF aliasing at high frequencies with slanted-edge angles. Similarly, the ISO-cos method exhibits aliasing in MTF curves at high frequencies when the slanted-edge angle is substantial. The system MTF value at the Nyquist frequency, as depicted in Figure 10d, is 0.1586. Additionally, a comparison of the RMSE between the MTFs and the system values for each method across different frequencies is presented in Table 2.

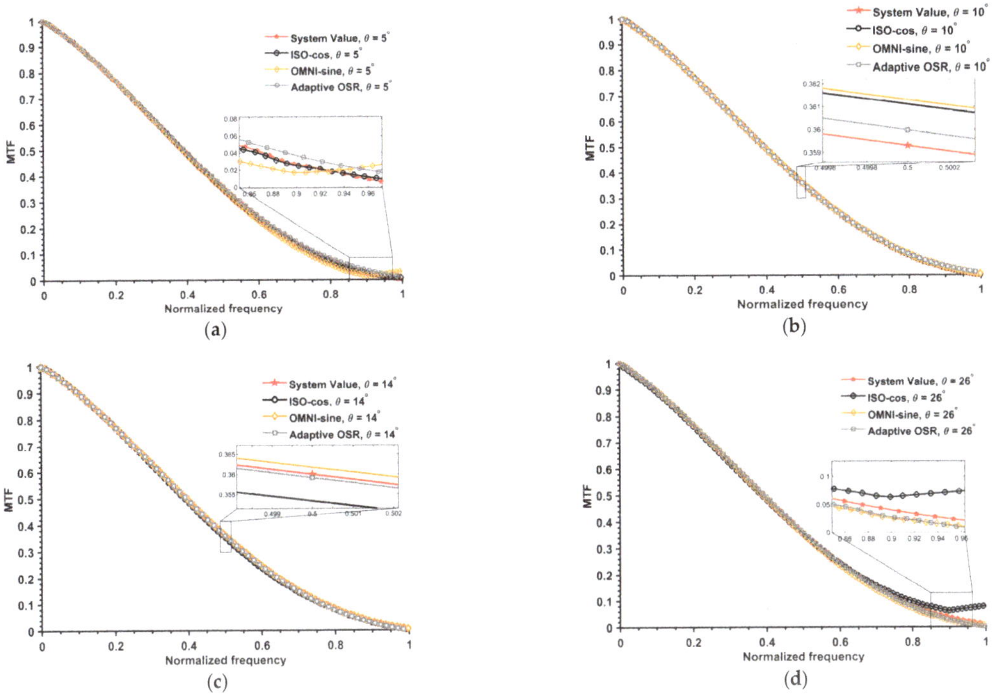

Figure 10. MTF calculation curves at various slanted-edge angles for the three algorithms: (**a**) slanted-edge angle of 5°; (**b**) slanted-edge angle of 10°; (**c**) slanted-edge angle of 14°; (**d**) slanted-edge angle of 26°.

Table 2. RMSE of MTF calculation by different methods.

Angle \ Method	ISO-cos	OMNI-sine	Algorithm in This Paper
5°	0.0495	0.0808	0.0527
10°	0.0301	0.0303	0.0276
14°	0.0337	0.0405	0.0319
26°	0.0905	0.0446	0.0478

It can be seen that the MTF value measured by the algorithm in this paper has the smallest mean square error with the theoretical value, which is closer to the real result.

3. Analysis of Factors Influencing the Results

The slanted edge of the imaging system represents a continuous input signal. When this signal is incident on the image detector element, it remains continuous. However, the sampling process discretizes this continuous signal, transforming it into a step-like distribution on the detector, as illustrated in Figure 11a. This distribution, highlighted by a red circle in the figure, defines the structure known as a "sampling step". The embodiment of the sampling step varies with different tilt angles, requiring different numbers of image lines, as demonstrated in Figure 11b,c.

According to the mathematical relationship, the number of image elements, denoted as y, required to represent a slanted edge with a width of x pixels is given by Equation (18):

$$y \geq x\lceil \cot\theta \rceil \tag{18}$$

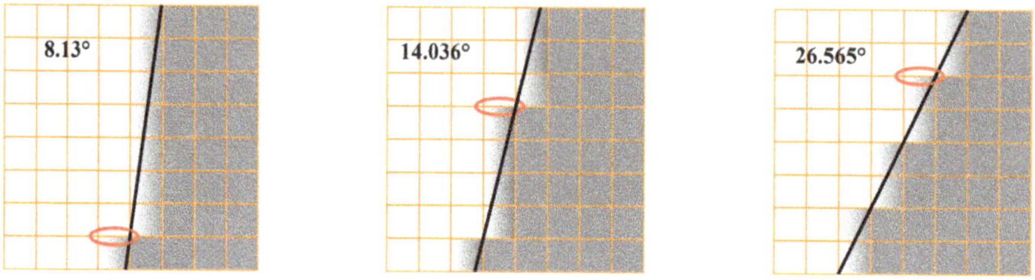

(a) (b) (c)

Figure 11. Schematic diagram of slanted-edge discretization: (**a**) slanted-edge angle of 8.13°; (**b**) slanted-edge angle of 14.036°; (**c**) slanted-edge angle of 26.565°.

Figure 12 depicts the minimum number of pixels along the height of the slanted edge necessary to represent the slanted-edge angle, varying from 1° to 45°, when the width of the slanted edge is fixed at 1 pixel.

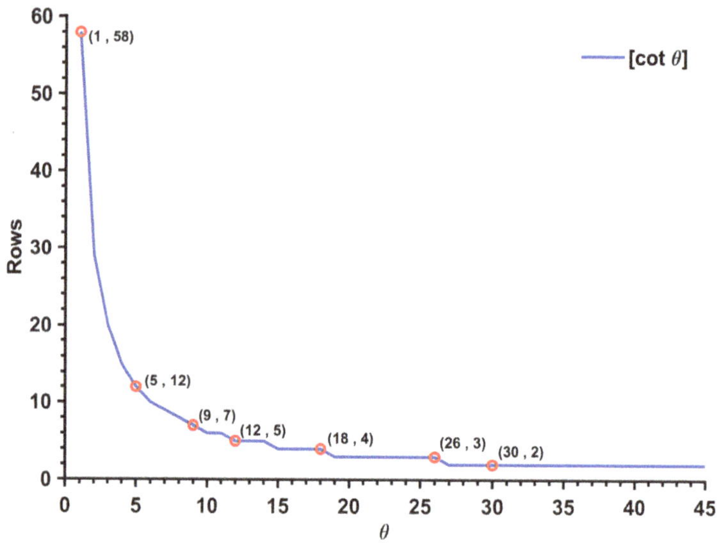

Figure 12. Relationship between slanted-edge angle and number of pixels.

3.1. *Effect of Slanted-Edge Angle on Slanted-Edge Method*

From Figure 11, it is evident that a smaller slanted-edge angle necessitates a greater number of pixels to represent the angle, leading to an increase in the image size in the height direction of the slanted edge. To explore the impact of the slanted-edge method, particularly how the slanted-edge angle, constrained by the height of the edge, affects the method's results, the imaging process of the imaging system was simulated as per Section 2.4. This simulation involved using a computer to generate 100 $\times$ 100-pixel images of a slanted edge with angles increasing from 1° to 45°. The grayscale values for the light and dark areas were set at 255 and 0, respectively. The slanted-edge method was then applied to calculate the MTF of the system at the Nyquist frequency. This paper's analysis includes comparing the results of the proposed algorithm with the theoretical values to evaluate the effectiveness of the slanted-edge method. The comparison of the theoretical and actual calculated values is presented in Figure 12.

Figure 13b demonstrates that the relative error escalates notably when the slanted-edge angle is less than 6°, suggesting that the solution results become highly volatile at smaller angles. This is attributed to the inadequacy of the actual image's slanted-edge height in fully capturing the nuances of smaller angles. Conversely, when the angle exceeds 12°, the relative error also increases as the angle widens. This rise in error is due to the increased width of the edge under a constant height, which introduces additional noise into the solution results. Therefore, the optimal range for the slanted-edge angle is between 6° and 12°.

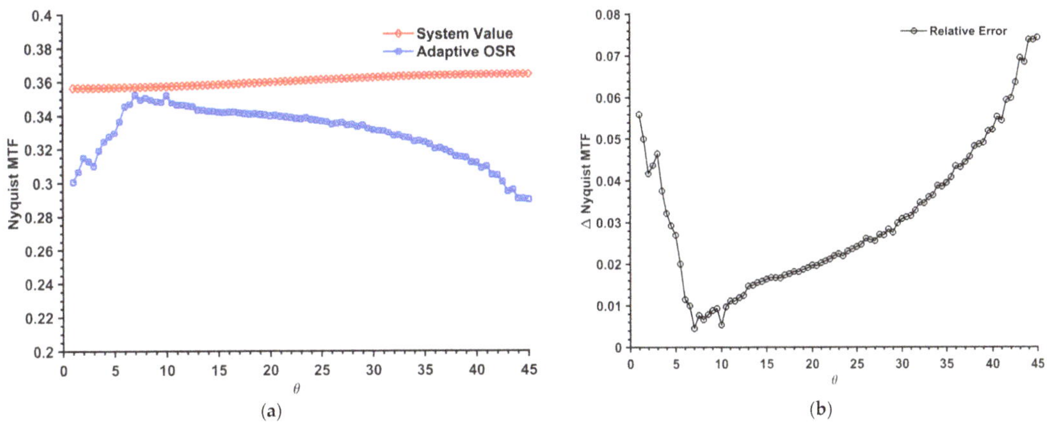

Figure 13. Nyquist MTF for different angles: (**a**) numerical curve at Nyquist frequency; (**b**) relative error curve.

When the slanted-edge angle ranges from 6° to 12°, the relative error between the calculated and theoretical values remains below 0.01. This suggests that within this angle range, the slanted-edge method demonstrates optimal stability, making it the most ideal range for the slanted-edge angle.

3.2. Effect of ROI in Slanted-Edge Images on the Slanted-Edge Method

This section analyzes the impact of slanted-edge height on the slanted-edge method under a specific slanted-edge angle of 7°. Examination of Figure 10 reveals that the continuous signal becomes discretized following the sampling process. Subsequently, the signal's representation on the detector resembles a stair-step distribution. The height of the slanted edge, corresponding to a varying number of steps for a constant angle, introduces result uncertainty. With a slanted-edge angle fixed at 7°, the height fluctuates within a 9-pixel range. Table 3 presents the varied slanted-edge heights for each step. Consequently, simulation experiments were conducted to assess the relationship between slanted-edge height and step distribution.

Table 3. Height of slanted edge for different steps.

Steps	1	2	3	4	5	6	7	8
Rows	2~18	11~27	20~36	29~45	38~54	47~63	56~72	65~81

Figure 14 vividly illustrates the convergence of data, highlighting the correlation between the number of steps in the slanted edge and the precision of experimental results. As the number of steps increases, there is a notable improvement in convergence, with the root-mean-square error remaining below 0.1 when the step count exceeds three. Therefore, an image of the slanted edge should include a minimum of three step structures. Additionally, considering the relationship between the number of slanted-edge steps and their height, as detailed in Table 3, the slanted edge's height within the ROI should be at least 30 pixels for an ideal slanted angle range of $\theta \in [6°, 12°]$.

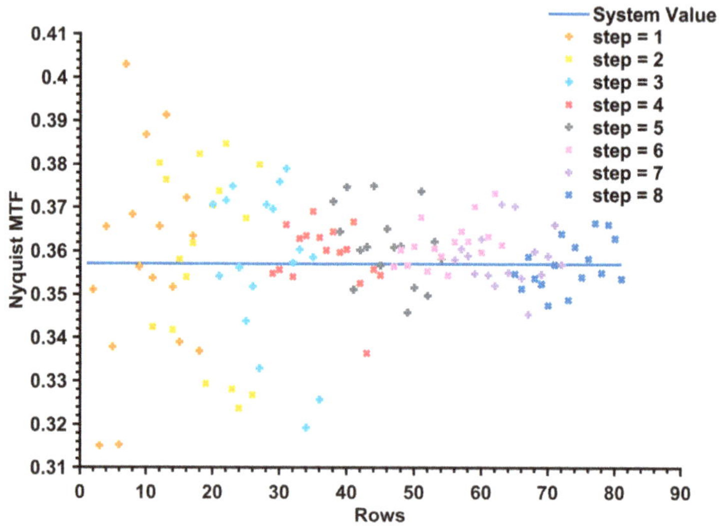

Figure 14. Nyquist MTF versus sampling step curve.

3.3. Effect of Contrast and SNR of Slanted-Edge Images on the Slanted-Edge Method

In this section, we examine the impact of image contrast and SNR on MTF measurements using a slanted-edge angle of 7°. Low edge contrast can hinder the visual distinction of edges, thereby affecting the stability of MTF results. We analyze images with contrasts of 0.95, 0.85, 0.75, 0.66, 0.55, 0.45, 0.35, 0.25, 0.15, and 0.05. Additionally, various noise levels are introduced to these images to achieve slanted-edge images with SNRs of 100, 90, 80, 70, 60, 50, 40, 30, 20, and 10. Using the algorithms proposed in this paper, we explore how image contrast and SNR influence the slanted-edge method. The simulation results are presented in Figure 15.

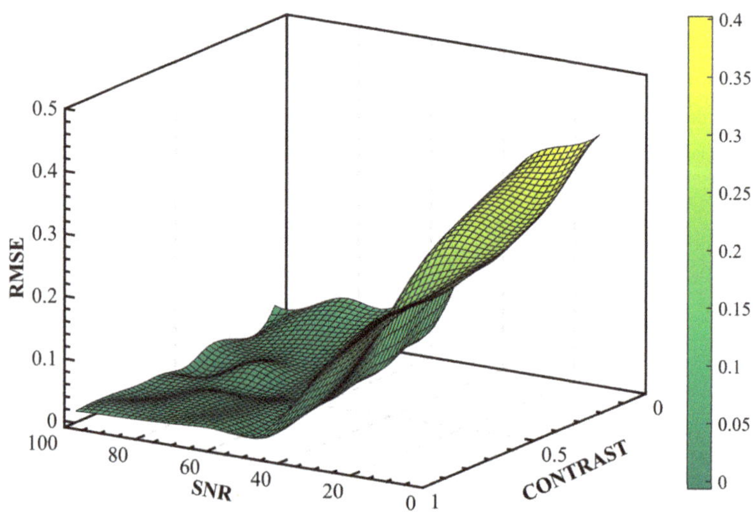

Figure 15. Nyquist MTF at different contrast and SNR.

The presence of noise in the edge image inevitably leads to fluctuations in the MTF curve. As illustrated in Figure 15, random noise significantly impacts MTF calculations. A decrease in the image's SNR results in an increase in the RMSE, leading to a rapid decline in calculation accuracy. Specifically, when the SNR falls below 30 dB, the calculated RMSE of the MTF exceeds 0.2, markedly increasing the likelihood of MTF aliasing. Consequently, the reliability of the measured MTF under these conditions is considerably diminished. In the absence of noise interference, the contrast of the images on either side of the edge marginally affects the accuracy of the MTF calculations. This is evident from Figure 15, which demonstrates that even when the contrast of the gray values on both sides of the edge is reduced to 0.3, the MTF calculation still retains high accuracy. However, as random noise increases, images with higher contrast demonstrate a superior ability to resist noise interference, thereby maintaining higher calculation accuracy.

Consequently, for optimal measurement results, it is essential to maintain an image contrast of at least 0.3 and ensure that the image's SNR exceeds 30.

4. Laboratory Test Results and Analysis

To validate the accuracy and reliability of the testing methodology proposed in this paper, the MTF test experiment was designed for a laboratory CMOS camera, and an experimental verification platform was constructed. This platform comprises an integrating sphere light source, a slanted-edge target, an optical lens, a CMOS camera, a three-dimensional rotary platform, and an image acquisition device. The integrating sphere provides uniform illumination for the slanted-edge target. The light signal, after traversing the optical lens, is captured by the CMOS camera and subsequently converted into a grayscale image through signal acquisition and processing. This image represents the target as captured by the camera's imaging system and is depicted in Figure 16, showcasing the test equipment. The MTF of the optical system under test is computed from the slanted-edge target image using the proposed method. The resulting transfer function values are then employed to evaluate the accuracy and stability of the focal plane position and the MTF values obtained through this test method.

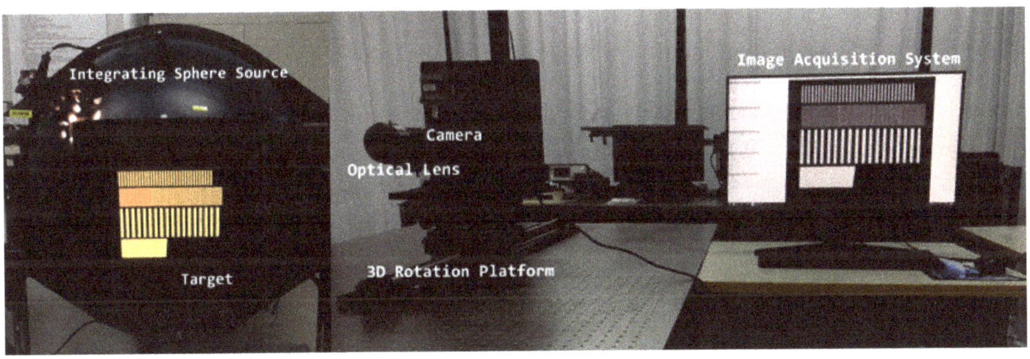

Figure 16. MTF laboratory test system.

The experimental setup included an integrating sphere source, a slanted-edge target, an optical lens, a CMOS camera, a 3D rotation platform, and an image acquisition device, all mounted on an optical vibration isolation platform. The 3D rotation platform was meticulously aligned to ensure the optical path was coaxial, enabling the CMOS camera to fully capture the slanted-edge target image. We maintained a precise 7° angle between the slanted-edge target and the detector. Additionally, the 3D rotation platform facilitated image acquisition at 0.1 mm intervals within a 1 mm depth range beyond the camera's focal length, capturing 20 frames at each position (see Figure 17a). Using the slanted-edge method outlined in this paper, we calculated the MTF values at various positions. These values were then used to construct through-focus MTF curves, as depicted in Figure 17b. The accuracy of our proposed algorithms was validated against reference MTFs derived from three-bar target images in a controlled laboratory environment.

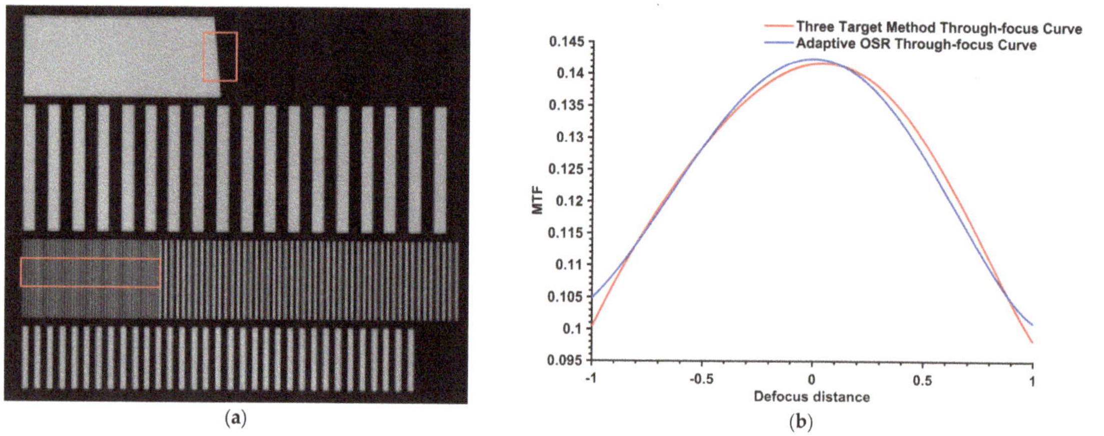

Figure 17. Target image and test results: (**a**) target, red rectangular areas are the ROI; (**b**) MTF through−focus curve.

The curve displayed in the figure closely aligns with the through-focus curve derived using the three-bar method. This congruence provides experimental evidence that the slanted-edge method introduced in this paper is a viable approach for calculating the MTF.

Subsequently, we altered the angle between the CMOS camera and the slanted-edge target. The 3D rotary stage was adjusted incrementally, starting from 1°. Slanted-edge target images were captured at every 0.5° increment, collecting 20 frames at each angle, up to a maximum of 45°, as illustrated in Figure 18. This procedure resulted in the acquisition of a total of 1760 slanted-edge target images. To mitigate the effects of air disturbance, the 20 frames obtained at each angle were aggregated and averaged, producing a representative slanted-edge image for each angle. We then computed the MTF values at the Nyquist frequency for the 88 distinct sets of slanted-edge images, facilitating the assessment and analysis of the impact of the slanted-edge angle on the accuracy of the slanted-edge method.

Figure 19 presents the results, illustrating that the trend of MTF values at the Nyquist frequency for various slanted-edge images, as measured in the laboratory, aligns with the simulated results. This alignment corroborates the efficacy of the optimal OSR-based adaptive modeling algorithm introduced in our study. Notably, the MTF relative error remains within 1.5% for slanted-edge angles ranging from 6° to 12°. The stability is particularly pronounced at a 7° slanted-edge angle, suggesting that the slanted-edge algorithm demonstrates optimal stability within this angle range. Additionally, this finding affirms the reliability and precision of the constraints specified in the slanted-edge method proposed in this paper.

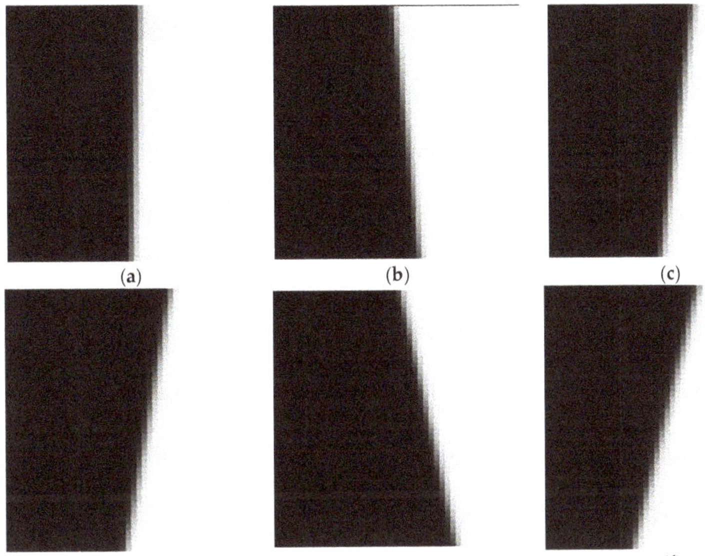

Figure 18. Images of slanted-edge targets at different angles: (a) slanted-edge angle of 3°; (b) slanted-edge angle of 7°; (c) slanted-edge angle of 8°; (d) slanted-edge angle of 10°; (e) slanted-edge angle of 14°; (f) slanted-edge angle of 20°.

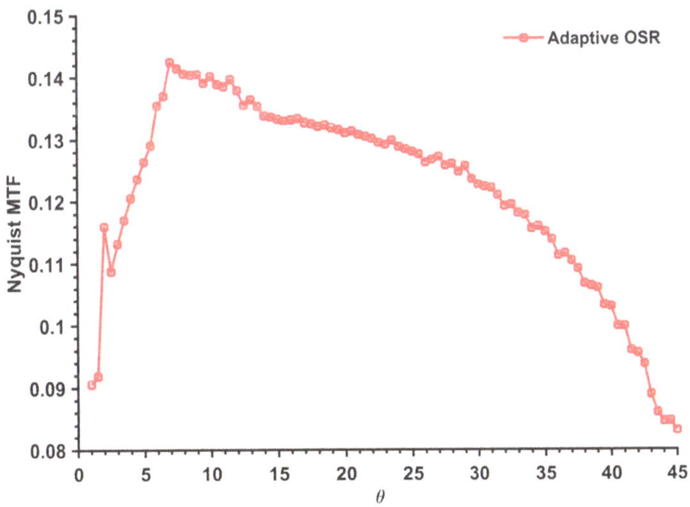

Figure 19. Nyquist MTF at different angles in laboratory testing.

5. Conclusions

In this paper, we address the limitations of existing sub-pixel edge detection and ESF curve extraction algorithms, specifically their low accuracy and high error rates. Through simulation experiments, we generate ideal computer-generated slanted-edge images and introduce varying levels of noise to assess the algorithms' performance. We focus on comparing the accuracy and stability of different methods for calculating the angle in the slanted-edge technique. Furthermore, we examine the impact of the OSR on the method's measurement accuracy and propose an adaptive model based on the optimal OSR. This model adjusts the sampling rate according to the edge angle, enhancing its applicability across a broader range of angles. Additionally, we employ higher-order polynomial edge fitting to mitigate the effects of spatial aberrations and image noise more effectively.

This paper subsequently analyzes the image factors influencing the slanted-edge method's calculations. Specifically, it analyzes how the slanted-edge angle, ROI, contrast, and SNR impact the method's accuracy. These factors are quantitatively assessed as constraints of the slanted-edge method through simulation experiments. The simulation results show that when the slanted-edge angle is between 6° and 12°, the relative error between the calculated value and the theoretical value is within 0.01, which indicates that when the slanted-edge angle is between 6° and 12°, the slanted-edge method has the best stability in the solving results, and it is a more ideal range of the slanted-edge angle. The ROI of slanted-edge height is at least 30 pixels, and the slanted-edge width is selected based on the principle of including as large a bright and dark area as possible. The SNR of the slanted-edge image significantly influences the results of MTF calculations. As the SNR decreases, there is a rapid decline in calculation accuracy. Conversely, contrast has a minimal impact on MTF calculation accuracy, with results maintaining a high level of precision even when the contrast ratio decreases to 0.3. To ensure optimal calculation results, it is imperative to maintain an SNR greater than 30dB for the slanted-edge image, along with a contrast ratio exceeding 0.3.

At present, the selection of the ROI for slanted-edge images remains a manual process in this paper. Future advancements may involve the integration of neural networks or feature recognition techniques to intelligently automate the selection of the effective ROI. Factors such as atmosphere and motion blur that cause MTF attenuation in photoelectric imaging systems can be added to the subsequent analysis to make it more universally applicable, further improve the algorithm, and increase the computational accuracy. In this paper, we use both simulation and laboratory experiments for algorithm validation and accuracy analysis, which can be followed by on-orbit experiments using real on-orbit image data for further evaluation.

Author Contributions: Conceptualization, J.Y. and Y.Z.; methodology, J.Y.; software, J.Y. and B.Q.; validation, J.Y., X.B. and W.W.; formal analysis, H.L.; data curation, J.Y.; writing—original draft preparation, J.Y.; writing—review and editing, Y.Z., B.Q. and X.B. All authors have read and agreed to the published version of the manuscript.

Funding: This research received no external funding.

Institutional Review Board Statement: Not applicable.

Informed Consent Statement: Not applicable.

Data Availability Statement: Data is contained within the article.

Conflicts of Interest: The authors declare no conflict of interest.

References

1. Kabir, S.; Leigh, L.; Helder, D. Vicarious Methodologies to Assess and Improve the Quality of the Optical Remote Sensing Images: A Critical Review. *Remote Sens.* **2020**, *12*, 4029. [CrossRef]
2. Xu, L.; Yan, C.; Gu, Z.; Li, M.; Li, C. Analysis of Dynamic Modulation Transfer Function for Complex Image Motion. *Appl. Sci.* **2019**, *9*, 5142. [CrossRef]
3. Gu, Y.; Shen, X.; He, G. MTF Estimation via BP Neural Networks and Markov Model for Space Optical Camera. *J. Frankl. Inst. B* **2013**, *350*, 3100–3115. [CrossRef]
4. Boreman, G.D.; Yang, S. Modulation Transfer Function Measurement Using Three- and Four-Bar Targets. *Appl. Opt.* **1995**, *34*, 8050–8052. [CrossRef] [PubMed]
5. Najafi, S.; Madanipour, K. Measurement of the Modulation Transfer Function of a Charge-Coupled Device Array by the Combination of the Self-Imaging Effect and Slanted Edge Method. *Appl. Opt.* **2013**, *52*, 4724–4727. [CrossRef] [PubMed]
6. Kamali-Asl, A.; Sarkar, S.; Shahriari, M.; Agha-Hosseini, H. Slit Slat Collimator Optimization with Respect to MTF. *Appl. Radiat. Isot.* **2005**, *62*, 461–468. [CrossRef] [PubMed]
7. Otón, J.; Sorzano, C.O.S.; Marabini, R.; Pereiro, E.; Carazo, J.M. Measurement of the Modulation Transfer Function of an X-ray Microscope Based on Multiple Fourier Orders Analysis of a Siemens Star. *Opt. Express* **2015**, *23*, 9567–9572. [CrossRef] [PubMed]

8. Sanli, U.T.; Rodgers, G.; Zdora, M.-C.; Qi, P.; Garrevoet, J.; Falch, K.V.; Müller, B.; David, C.; Vila-Comamala, J. Apochromatic X-ray Focusing. *Light Sci. Appl.* **2023**, *12*, 107. [CrossRef]
9. Viallefont-Robinet, F.; Helder, D.; Fraisse, R.; Newbury, A.; van den Bergh, F.; Lee, D.; Saunier, S. Comparison of MTF Measurements Using Edge Method: Towards Reference Data Set. *Opt. Express* **2018**, *26*, 33625. [CrossRef]
10. Nishi, K. Does the Slanted-Edge Method Provide the True Value of Spatial Frequency Response? *J. Opt. Soc. Am. A JOSAA* **2023**, *40*, 259–269. [CrossRef]
11. Ryan, R.; Baldridge, B.; Schowengerdt, R.A.; Choi, T.; Helder, D.L.; Blonski, S. IKONOS Spatial Resolution and Image Interpretability Characterization. *Remote Sens. Environ.* **2003**, *88*, 37–52. [CrossRef]
12. Toutin, T. Comparison of Stereo-Extracted DTM from Different High-Resolution Sensors: SPOT-5, EROS-a, IKONOS-II, and QuickBird. *IEEE Trans. Geosci. Remote Sens.* **2004**, *42*, 2121–2129. [CrossRef]
13. Baraldi, A.; Durieux, L.; Simonetti, D.; Conchedda, G.; Holecz, F.; Blonda, P. Automatic Spectral Rule-Based Preliminary Classification of Radiometrically Calibrated SPOT-4/-5/IRS, AVHRR/MSG, AATSR, IKONOS/QuickBird/OrbView/GeoEye, and DMC/SPOT-1/-2 Imagery—Part II: Classification Accuracy Assessment. *IEEE Trans. Geosci. Remote Sens.* **2010**, *48*, 1326–1354. [CrossRef]
14. Han, L.; Gao, K.; Dou, Z.; Zhu, Z.; Wang, H.; Fu, X. On-Orbit MTF Estimation for GF-4 Satellite Using Spatial Multisampling on a New Target. *IEEE Geosci. Remote Sens. Lett.* **2020**, *17*, 17–21. [CrossRef]
15. Crespi, M.; De Vendictis, L. A Procedure for High Resolution Satellite Imagery Quality Assessment. *Sensors* **2009**, *9*, 3289–3313. [CrossRef] [PubMed]
16. *ISO 12233:2023*; Photography-Electronic Still Picture Imaging-Resolution and Spatial Frequency Responses, Document ISO 12233:2023. Multiple. Distributed through American National Standards Institute (ANSI): Washington, DC, USA, 2023.
17. Kobayashi, S. Nonlinear Model for an Optical Read-Only-Memory Disk Readout Channel Based on an Edge-Spread Function. *Appl. Opt.* **2002**, *41*, 2679–2685. [CrossRef] [PubMed]
18. Li, L.; Cao, J.; Wei, S.; Jiang, Y.; Shen, X. Improved On-Orbit MTF Measurement Method Based on Point Source Arrays. *Remote Sens.* **2023**, *15*, 4028. [CrossRef]
19. Lv, S.; Kemao, Q. Modeling the Measurement Precision of Fringe Projection Profilometry. *Light. Sci. Appl.* **2023**, *12*, 257. [CrossRef]
20. Cho, H.-W.; Yoon, H.-J.; Yoon, J.-C. Analysis of Crack Image Recognition Characteristics in Concrete Structures Depending on the Illumination and Image Acquisition Distance through Outdoor Experiments. *Sensors* **2016**, *16*, 1646. [CrossRef]
21. Masaoka, K.; Yamashita, T.; Nishida, Y.; Sugawara, M. Modified Slanted-Edge Method and Multidirectional Modulation Transfer Function Estimation. *Opt. Express* **2014**, *22*, 6040. [CrossRef]
22. Roland, J.K.M. *A Study of Slanted-Edge MTF Stability and Repeatability*; Larabi, M.-C., Triantaphillidou, S., Eds.; SPIE: San Francisco, CA, USA, 2015; p. 93960L.
23. Tzannes, A.P.; Mooney, J.M. Measurement of the Modulation Transfer Function of Infrared Cameras. *Opt. Eng.* **1995**, *34*, 1808–1817. [CrossRef]
24. Hwang, H.; Choi, Y.-W.; Kwak, S.; Kim, M.; Park, W. MTF Assessment of High Resolution Satellite Images Using ISO 12233 Slanted-Edge Method. In Proceedings of the Image and Signal Processing for Remote Sensing XIV, Cardiff, UK, 15–18 September 2008; SPIE: Bellingham, WA, USA, 2008; Volume 7109, pp. 34–42.
25. Masaoka, K. Edge-Based Modulation Transfer Function Measurement Method Using a Variable Oversampling Ratio. *Opt. Express* **2021**, *29*, 37628. [CrossRef] [PubMed]
26. Burns, P.D.; Williams, D.; Griffith, J.; Hall, H.; Cahall, S. Application of ISO Standard Methods to Optical Design for Image Capture. *Electron. Imaging* **2020**, *2020*, 240–241. [CrossRef]
27. Liu, S.; Liu, K.; Kewei, E.; Wang, T.; Li, Z.; Yao, B. Correction of the Error Induced by Obscurations of Ritchey-Chretien Collimators for High-Resolution Space Camera MTF Measured with the ISO 12233 Slanted-Edge Method. *Optik* **2021**, *235*, 166653. [CrossRef]
28. Wen, G.; Li, S.; Wang, L.; Chen, X.; Sun, Z.; Liang, Y.; Jin, X.; Xing, Y.; Jiu, Y.; Tang, Y.; et al. High-Fidelity Structured Illumination Microscopy by Point-Spread-Function Engineering. *Light Sci. Appl.* **2021**, *10*, 70. [CrossRef]
29. Zhang, Y.; Xian, H. Optical Transfer Function for Incomplete Circular Pupil. *Optik* **2019**, *198*, 162957. [CrossRef]
30. Hu, Z.; Huang, X.; Yang, Z.; Qiu, J.; Song, Z.; Zhang, J.; Dong, G. Reversible 3D Optical Data Storage and Information Encryption in Photo-Modulated Transparent Glass Medium. *Light Sci. Appl.* **2021**, *10*, 140. [CrossRef]
31. Yang, X.; Liang, L.; Li, F.; Tian, Q.; Lu, X.; Xin, L.; Guo, Y.; Dong, W. Hyper-Temporal Data Based Modulation Transfer Functions Compensation for Geostationary Remote Sensing Satellites. *IEEE Trans. Geosci. Remote Sens.* **2022**, *60*, 1–10. [CrossRef]
32. Asnani, S.; Presti, F.L.; Amato, L.; Montrucchio, B. MTFCalculator—A Mobile Application for Measuring the Modulation Transfer Function of Built-in Cameras of Smartphones Using ISO 12233 Slanted-Edge Method. In Proceedings of the 2021 IEEE 6th International Conference on Signal and Image Processing (ICSIP), Nanjing, China, 22–24 October 2021; pp. 1273–1278.
33. Pan, Y.; Cohen, E.; Karimi, E.; Gover, A.; Schönenberger, N.; Chlouba, T.; Wang, K.; Nehemia, S.; Hommelhoff, P.; Kaminer, I.; et al. Weak Measurements and Quantum-to-Classical Transitions in Free Electron–Photon Interactions. *Light Sci. Appl.* **2023**, *12*, 267. [CrossRef]
34. Xiong, J.; Hsiang, E.-L.; He, Z.; Zhan, T.; Wu, S.-T. Augmented Reality and Virtual Reality Displays: Emerging Technologies and Future Perspectives. *Light Sci. Appl.* **2021**, *10*, 216. [CrossRef]
35. Roberti de Siqueira, F.; Robson Schwartz, W.; Pedrini, H. Multi-Scale Gray Level Co-Occurrence Matrices for Texture Description. *Neurocomputing* **2013**, *120*, 336–345. [CrossRef]

36. Gribbon, K.T.; Bailey, D.G. A Novel Approach to Real-Time Bilinear Interpolation. In Proceedings of the DELTA 2004: Second IEEE International Workshop on Electronic Design, Test and Applications, Perth, WA, Australia, 28–30 January 2004; pp. 126–131.
37. Namin, A.H.; Leboeuf, K.; Muscedere, R.; Wu, H.; Ahmadi, M. Efficient Hardware Implementation of the Hyperbolic Tangent Sigmoid Function. In Proceedings of the 2009 IEEE International Symposium on Circuits and Systems, Taipei, Taiwan, 24–27 May 2009; pp. 2117–2120.
38. Chen, H.; Huang, L.; Liu, T.; Ozcan, A. Fourier Imager Network (FIN): A Deep Neural Network for Hologram Reconstruction with Superior External Generalization. *Light Sci. Appl.* **2022**, *11*, 254. [CrossRef] [PubMed]
39. Xie, X.; Fan, H.; Wang, A.; Zou, N.; Zhang, Y. Regularized Slanted-Edge Method for Measuring the Modulation Transfer Function of Imaging Systems. *Appl. Opt.* **2018**, *57*, 6552. [CrossRef] [PubMed]
40. Zhang, S.; Wang, F.; Wu, X.; Gao, K. MTF Measurement by Slanted-Edge Method Based on Improved Zernike Moments. *Senssors* **2023**, *23*, 509. [CrossRef]
41. Burns, P.D.; Masaoka, K.; Parulski, K.; Wueller, D. Updated Camera Spatial Frequency Response for ISO 12233. *Electron. Imaging* **2022**, *34*, IQSP-357. [CrossRef]
42. Bojkovic, Z.S.; Bakmaz, B.M.; Bakmaz, M.R. Hamming Window to the Digital World. *Proc. IEEE* **2017**, *105*, 1185–1190. [CrossRef]

applied sciences

MDPI

Article

Global Time-Varying Path Planning Method Based on Tunable Bezier Curves

Longfei Jia [1,*], Si Zeng [1], Lei Feng [2], Bohan Lv [1], Zhiyuan Yu [1] and Yuping Huang [1]

1 Laboratory of Aerospace Servo Actuation and Transmission, Beijing Institute of Precision Mechatronics and Controls, Beijing 100076, China
2 Beijing Institute of Tracking and Telecommunication Technology, Beijing 100094, China
* Correspondence: longfei.jia@lasat.com

Abstract: In this paper, a novel global time-varying path planning (GTVP) method is proposed. In the method, real-time paths can be generated based on tunable Bezier curves, which can realize obstacle avoidance of manipulators. First, finite feature points are extracted to represent the obstacle information according to the shape information and position information of the obstacle. Then, the feature points of the obstacle are converted into the feature points of the curve, according to the scale coefficient and the center point of amplification. Furthermore, a Bezier curve representing the motion path at this moment is generated to realize real-time adjustment of the path. In addition, the 5-degree Bezier curve planning method consider the start direction and the end direction is used in the path planning to avoid the situation of abrupt change with oscillation of the trajectory. Finally, the GTVP method is applied to multi-obstacle environment to realize global time-varying dynamic path planning. Through theoretical derivation and simulation, it can be proved that the path planned by the GTVP method can meet the performance requirements of global regulation, real-time change and multi-obstacle avoidance simultaneously.

Keywords: Bezier curve; path planning; obstacle avoidance; global time-varying; dynamic obstacle; real-time

Citation: Jia, L.; Zeng, S.; Feng, L.; Lv, B.; Yu, Z.; Huang, Y. Global Time-Varying Path Planning Method Based on Tunable Bezier Curves. *Appl. Sci.* **2023**, *13*, 13334. https://doi.org/10.3390/app132413334

Academic Editors: Qi Song and Qinglei Zhao

Received: 31 October 2023
Revised: 11 December 2023
Accepted: 12 December 2023
Published: 18 December 2023

1. Introduction

Path planning is a mapping from perceptual space to behavioral space and the planning method is one of the research hotspots at present. There are a variety of path planning methods commonly used, such as the potential energy method [1], heuristic search algorithm [2], Dijkstra algorithm [3], LPA* algorithm (Life Planning A*) [4], Floyd algorithm [5], PRM algorithm [6], RRT algorithm [7], unit division method [8] and intelligent algorithm [9–11]. However, these path planning algorithms cannot satisfy global adjustment, real-time change and multi-obstacle avoidance at the same time.

The planned motion path can be divided into two categories: segmented paths and continuous paths. Segmental paths include the linear path, circular path, segmental function path, etc. Continuous paths includes the B-spline curve, spline function, polynomial function, Dubins curve [12], clothoid curve [13], etc. The above methods have the characteristics of optimizing velocity and acceleration curves, but the planned trajectory cannot change with dynamic obstacles.

Bezier curve is a parametric polynomial curve family with adjustability, continuity and smoothness, which has been widely used in path planning. Wang [14] combines gliding motion with the three-dimensional path planning method of robot dolphins to propose segmented Bezier curves, which can be implemented to hybrid underwater robots. Zhang [15] proposed a path planning method based on the combination of a jump point search and Bezier curve, which adopts improved heuristic functions based on distance and direction to reduce costs and generate optimal trajectories based on Bezier curves. Zafer [16] proposed a novel method based on Bezier curves to address excessive nodes and spikes in path planning in a given environment by using network mapping. In order to reduce the computational cost of motion planning, Arslan [17] adopts the parameter matching reduction method to make multiple low-order Bezier segments approximate the high-order Bezier curve adaptively. This method can be implemented in the trajectory optimization of non-holonomic constrained mobile robots. Song [18] proposed an improved particle swarm optimization algorithm to plan the smooth path of mobile robots in order to meet the requirements of continuous curvature derivative continuity, combined with the continuous high-order Bezier curve, so as to solve the local optimal solution and premature convergence problems. Blazi [19] proposed a new parameterization method of motion primitive based on Bezier curves, which is suitable for path planning applications of wheeled mobile robots. In this paper, the analytical solution of the motion primitive of a 3-order Bezier curve is given under the given boundary conditions that guarantee the continuous curvature of the combined spline path. Bulut [20] proposed the use of quintic triangular Bezier curves with two shape parameters and C3 continuity for path planning. When there is an obstacle, the predetermined path can be adjusted only by the shape parameter in this method. Xu [21] proposed a new smooth path planning method for mobile robots based on quadratic Bezier transition curve and improved particle swarm optimization algorithm. Simulation results demonstrate the effectiveness and superiority of this method combined with quadratic Bezier transition curve and improved PSO-AWDV algorithm.

Researchers have proposed a number of path planning methods for multi-obstacle environments. Deng [22] proposed a multi-obstacle path planning and optimization method for multi-obstacle avoidance. This method uses the convex hull to optimize obstacles, so as to obtain the base point set and generate the corresponding extension point set. The multi-objective D* Lite algorithm is utilized to design the distance and smoothness of the path planner to obtain a reasonably optimized path in a complex environment. Finally, the third Bezier curve is used to smooth the path.

To solve the dynamic obstacle avoidance problem, many methods have been proposed. However, the traditional method [23] can realize the local dynamic obstacle avoidance of a mobile car, but it cannot be applied to the global dynamic obstacle avoidance of a manipulator. Scoccia [24] used the optimal fitting interpolation of a Bezier curve to smooth the trajectory and improved the obstacle avoidance ability of the robot in the dynamic environment by considering the speed of obstacles. In order to optimize the distance between the start point and the target point, the improved genetic algorithm is used to explore the Bezier curve control points, and the optimal smooth path is selected to minimize the total distance between the start point and the end point [25]. Kang [26] proposed a new collision cost prediction network (CCPN) that adopts a real-time updated sensor data occupation grid to estimate collision costs and avoid robot collisions with static and dynamic obstacles. Minnetoglu [27] proposed an effective real-time path planning algorithm based on the geometry applied to three-dimensional environments, which adopts a three-dimensional potential field to generate the intermediate point that characterizes the path of the robot with less degrees of freedom and significantly improves the maneuverability of the manipulator to avoid obstacles.

The researchers propose a variety of local real-time path planning methods for single obstacles, moving vehicles, or remotely piloted aircraft. However, the global real-time path planning method of the snake manipulator is lacking. Therefore, a global time-varying path planning method based on a Bezier curve (GTVP) is proposed. The GTVP method generates the real-time motion trajectory of the manipulator according to the real-time data of dynamic obstacles and then obtains the trajectory of the center point of each joint of the manipulator according to the repeated path method, which skillfully combines the trajectory planning of joint space with the trajectory planning of Cartesian space.

The layout of this paper is as follows. Section 2 introduces the improvement of the proposed method. Sections 3 and 4 describe the trajectory planning process of this method in single-obstacle and multi-obstacle environments. Section 5 carries out theoretical verification of the method. Section 6 carries out simulation verification and Section 7 summarizes this article.

2. Characteristics of GTVP

The high-order Bezier curve has the disadvantages of large computation, oscillations and complex trajectory, so it cannot be applied to dynamic environment. Although the low-order Bezier curve can guarantee the continuity of the path, it cannot provide continuous curvature and arbitrary setting of the second derivative of the characteristic points of the Bezier curve. The more feature points the Bezier curve has, the more flexible the smooth path is, but the more computational the complexity is and vice versa. In conclusion, in the process of path planning, it is necessary to balance the amount of calculation with the flexibility of the trajectory.

The planned path should meet the following conditions:

(1) The manipulator can realize dynamic obstacle avoidance along the path;
(2) Realize G3 continuity and continuous curvature derivative;
(3) Minimize the maximum curvature of smooth paths;
(4) The length of the smooth path is as short as possible under the premise of meeting the basic conditions.

It is difficult to satisfy the requirements of real-time obstacle avoidance by using a traditional intelligent algorithm to optimize the solution. In order to address the problems of real-time path planning and Bezier curves, the GTVP method is improved on the premise of satisfying the elementary criteria. The novelty and contributions of this paper are summarized and listed as following.

(1) Considering obstacles of different positions and shapes, the GTVP method extracts a finite number of feature points to characterize the key information of dynamic obstacles, which reduces the complexity of the obstacle model. In this way, the dynamic obstacle information can be analyzed in real time during path planning.
(2) Before real-time path planning, the GTVP method has formulated the conversion relationship between feature points and paths through equation deduction (Step 5, Step 6, Step 7). In real-time path planning, the corresponding spline curve can be generated by bringing in the real-time data of feature points and the spline curve is the motion path of the snake manipulator, which greatly reduces the calculation time.
(3) There are many inflection points in the path planned by traditional methods and the variation curve of the joint declination angle is unsmooth when the snake manipulator moves along the path. By virtue of the characteristics of Bezier curves, the smooth path can be directly planned by the GTVP method and then the smooth path can be adjusted in real time according to the nodes on the path and the feature points of dynamic obstacles.
(4) The traditional method can be applied to the dynamic obstacle avoidance of a trolley or car, but it cannot be applied to the snake manipulator to avoid obstacles on the global path. Built on the global characteristics of Bezier curves, the GTVP method can adjust the global path in real time by adjusting the obstacle feature points and curve feature points.
(5) The GTVP method extracts real-time information of dynamic obstacles and utilizes feature points to generate the corresponding smooth trajectory curve, which can realize real-time obstacle avoidance of the manipulator. This method avoids numerous unnecessary calculations, improves search efficiency and efficiently solves path planning problems in multi-target conditions or multi-obstacle environments.
(6) The GTVP method can not only adjust the direction of the start point of the path, but also adjust the direction of the end point of the path.
(7) There are several adjustable parameters in the GTVP method: center point of obstacle, number and position of obstacle feature points, location of scaling center point, scaling ratio coefficient, etc. Individual parameters can be selected or adjusted according to the specific application environment, so this method has good environmental adaptability.

3. Transient Path under the Single-Obstacle Environment

3.1. Overall Process

For moving obstacles with different shapes, it is necessary to plan the corresponding time-varying trajectory according to the real-time information of the obstacle, so that the manipulator can ensure that it does not collide with the obstacle and can also move from the starting point to the end point. In order to meet the above requirements, the global time-varying path planning (GTVP) method for dynamic obstacles is proposed in this paper. The pseudo-code for the main function in the method is shown in Algorithm 1.

Algorithm 1 The main function of the time-varying trajectory planning algorithm

```
1: Input: x_start, x_end, d_start, d_end;
2: Obs_Information ← Get_obstacle(t);
3: Central_Point ← Mid(Obs_Information, x_start, x_end);
4: Feature_Points ← Extract(Obs_Information, x_start, x_end);
5: Bezier_Points ← Amplify(Feature_Points,  Central_Point,  x_start, x_end, d_start, d_end);
6: Bezier_Curve ← Bezier_Function(Bezier_Points);
7: Output: Bezier_Curve;
```

The whole process of the algorithm combined with the pseudo-code is explained as follows. First, input the initial data and the moving object model, including the position x_{start} and direction d_{start} of the start point, the position x_{end} and the direction d_{end} of the end point and the model of the snake manipulator (Step 1). Then, obtain the real-time shape and position information of the obstacle (Step 2) and find the center point O_0, which is the preparation for the curve generation (Step 3). Extract feature points of the surface that represents the obstacle information and judge which side of the obstacle the moving object walks from (Step 4). Enlarge the surface of the obstacle according to the formulated scale coefficient and magnification center point and generate the feature points of the Bezier curve according to the specified law (Step 5). Next, generate the Bezier curve according to the characteristic point of the curve (Step 6). Finally, output the real-time Bezier curve (Step 7).

The Bezier curve corresponding to each moment can be obtained through the above process. Next, the process and principle of the GTVP method are described in four examples shown in Figures 1–4. The direction of the x-axis is from the start point x_{start} to the end point x_{end} and the y-axis is perpendicular to the x-axis. The GTVP method is suitable for snake manipulators, redundant manipulators, continuous manipulators, mobile cars, mobile robots, remote control aircraft and other moving objects. In this paper, the process and principle of the GTVP method are described by taking the snake manipulator as an example. The details of each step are described in Sections 3.2–3.6.

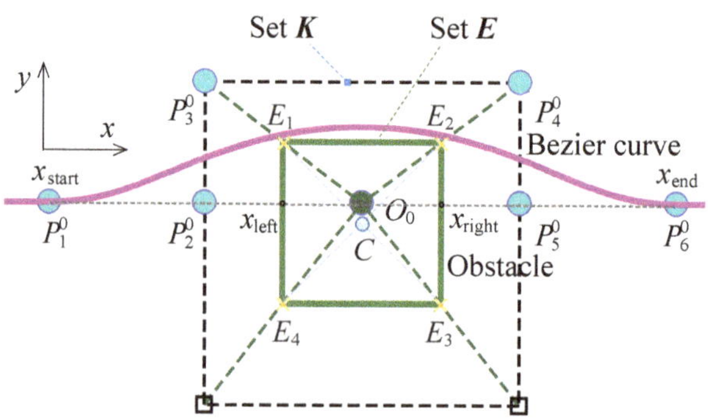

Figure 1. Bezier curve generated for rectangular obstacles.

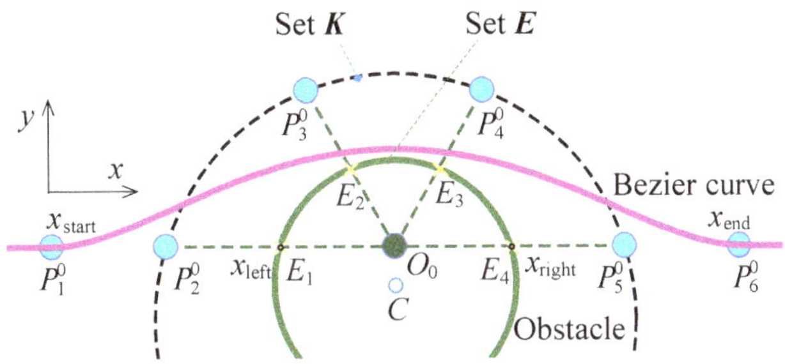

Figure 2. Bezier curve generated for circular obstacles.

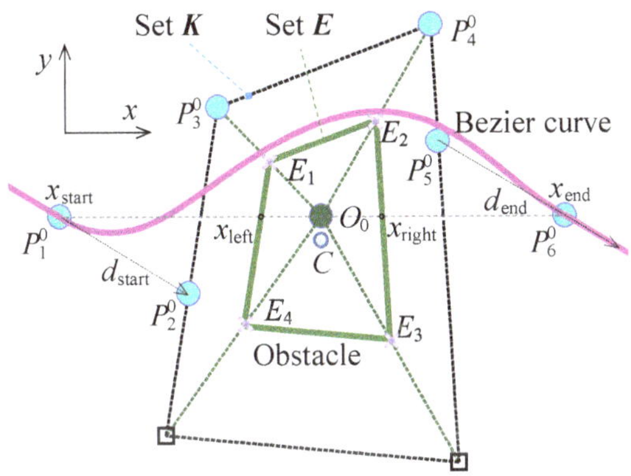

Figure 3. Bezier curve generated for non-parallelogram obstacles.

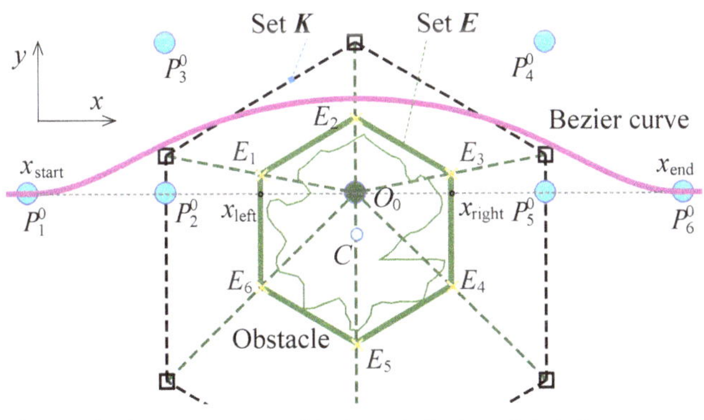

Figure 4. Bezier curve generated for irregularly shaped obstacles.

3.2. Obtaining Initial Information

In Step 1, the kinematic model of the snake manipulator is constructed, including establishment of the D-H coordinate system and analysis of the conversion relationship between the parameters. According to the task requirements, the start point x_{start}, the end point x_{end}, the start point direction d_{start} and the end point direction d_{end} are set corresponding to the motion trajectory of the manipulator. The relationship between the parameters is presented in Equation (1). The d_{start} and d_{end} in Figures 1, 2 and 4 are both 0°, so they are not marked. The d_{start} and d_{end} in Figure 3 are −30°.

$$\begin{cases} d_{\text{start}} = x'_{\text{start}} \\ d_{\text{end}} = x'_{\text{end}} \end{cases} \tag{1}$$

In Step 2, the detailed information of obstacles can be obtained through image processing, construction of sparse maps and so on. The feature information of obstacles (represented by the Obs_Information symbol) can be extracted and only one set of data is needed to store the shape data of obstacles: Obs_Information (1). The data of x, y, z, θ_x, θ_y and θ_z are used for storage of the position data of three-dimensional obstacles: Obs_Information (2:7). Where x, y and z are the distance between the current position and the origin point, θ_x, θ_y and θ_z are the angles of rotation of the initial pose around the axes x, y and z. The position data of two-dimensional obstacles are stored through the data of x, y and z. The data of z, θ_x and θ_y are always zero. Size data of the obstacle are stored through obstacle features: Obs_Information (8:), the size data of obstacles of different shapes occupy different numbers of data. For example, the size data of spherical obstacles are stored through r, the size data of cuboid obstacles are stored through a, b and c. The size data of irregular polyhedral obstacles are stored through the initial position information of each vertex.

The problems of image recognition and segmentation can be solved by existing methods proposed by other researchers. For obstacle recognition, Chen [28] proposed an adaptive object recognition system, which can effectively identify specific targets under complex backgrounds. For the extraction of edge information, Gu [29] used the improved wavelet mode maximum algorithm to extract image edges, which can obtain edge image information with better clarity and connectivity. Yu [30] extracted the boundary of an obstacle from the semantic segmentation result by applying pixel filtering. For irregular obstacles, Bai [31] conducted grid preprocessing and convex preprocessing for concave obstacles, which enhanced the safety of UAV path obstacle avoidance. In order to determine turning points, Dai [32] proposed to use motion coherence to distinguish dynamic and static visual feature points and remove the edges between irrelevant points in the point correlation optimization process.

The shape and size of obstacles do not change with time, but the position of dynamic obstacles changes over time, such as translation, rotation and other movements. Therefore, it is necessary to extract the information of dynamic obstacles in real time to obtain preliminary data for obstacle avoidance.

3.3. *Center Point of Obstacle*

Extract unchanged initial data in Step 1 and the initial data that need to be updated in real time are extracted in Step 2. In Step 3, the center point O_0 is obtained through the two endpoints of the trajectory and the obstacle, which lays the groundwork for the subsequent amplification. The start point x_{start} and the end point x_{end} are connected to generate a straight line. If the line does not intersect the obstacle, there is no need to consider obstacle avoidance and the moving object can move along the line from x_{start} to x_{end}. If part of the line segment is inside the obstacle, the two ends of the line segment are represented by x_{left} and x_{right}, respectively and the midpoint of the line segment is the center point, which is marked O_0. The pseudo-code corresponding to the Mid function that determines the center point O_0 is shown in Algorithm 2.

Algorithm 2 Central_Point ← Mid(Obs_Information, x_{start}, x_{end})

```
Input: Obs_Information, x_start, x_end;
for x_left from x_start to x_end do
    if x_left in obstacle then
        break for
    end if
end for
for x_right from x_end to x_start do
    if x_right in obstacle then
        break for
    end if
end for
if x_left is x_end then
    Central_Point ← Null;
else
    Central_Point ← (x_left+x_right)/2;
end if
Output: Bezier_Curve;
```

After inputting the characteristic information of the obstacle "Obs_Information", x_{start} and x_{end}, let x_{left} move step by step from x_{start} to x_{end}, according to the specified step δ. The calculation equation is as follows.

$$x_{\text{left}} = x_{\text{start}} + n \cdot \delta \tag{2}$$

where n = 1,2,3, ...

Each step determines whether the x_{left} in the step is in the obstacle space. If it is not in the obstacle space, it analyzes whether x_{left} reaches or exceeds the x_{end} point. If so, there is no need to consider obstacle avoidance. If x_{left} does not reach or does not exceed x_{end}, let x_{left} continue moving from x_{start} to x_{end} and cycle again.

If x_{left} is in the obstacle space, let x_{right} move step by step from x_{end} to x_{start}, according to the specified step δ and the calculation equation is as follows.

$$x_{\text{right}} = x_{\text{end}} - n \cdot \delta \tag{3}$$

where n = 1, 2, 3, ...

Each step determines whether the x_{right} in the obstacle space. If not x_{right} continues to move from x_{end} to x_{start} and cycle again. If x_{right} is in the obstacle space, the center point O_0, is calculated as follows.

$$O_0 = \left(x_{\text{right}} + x_{\text{left}}\right)/2 \tag{4}$$

3.4. Feature Points of Obstacles

In Step 4, the feature points of the obstacle are extracted. The turning point of each object is regarded as the feature point on the surface of each object. Several feature points and locations of feature points need to be extracted, which can be set according to the task, mainly related to the following factors.

(1) Obs_Information. The type of obstacle and the number of inflection points in the characteristic information of the obstacle;
(2) The position of the start point x_{start} and the end point x_{end};
(3) Which_Side. Whether the planned trajectory is above or below the obstacle needs to be determined; in other words, which side of the obstacle the planned trajectory bypasses needs to be determined.

The pseudo-code corresponding to the Extract function is shown below.

Line 2–3 in Algorithm 3. The E_i in Figure 1 represents the feature points of the rectangular obstacle, which are the points that characterize the shape and size of the obstacle. The fuzzy center of gravity C is determined according to the above feature points of the obstacle (also known as the inflection point of the obstacle), which can reduce the amount of calculation. Some specially shaped obstacles have no inflection point and several feature points of the obstacle can be set according to the specified rules, as shown in Figure 2. Where four straight lines with adjacent angles of 60° are made through the point O_0 and the intersection point E_i of the straight line, and the intersection point with the circular obstacle is regarded as the feature point of the obstacle. By setting feature points for circular obstacles with the help of the method, the real-time generated Bezier curve never intersects with obstacles. The relevant proof process is shown in Section 5. Three or six feature points E_i can be extracted from triangular obstacles, four feature points E_i can be extracted from quadrangular obstacles, the number of feature points E_i extracted from N-sided obstacles is N and the number of feature points E_i extracted from circular obstacles is n, where n can be set as an integer, such as 3–6, etc. After obtaining these feature points of the obstacle, the blurred center of gravity C of the obstacle can be obtained by taking the average of the above points. The calculation equation is as follows.

$$C = \sum_{i=1}^{w} E_i / w \tag{5}$$

where w is the number of feature points of the obstacle.

Algorithm 3 Feature_Points ← Extract(Obs_Information, x_{start}, x_{end})

1: **Input**: Obs_Information, x_{start}, x_{end};
2: E_i ← Obs_Feature(Obs_Information);
3: C ← Equation (5)(E_i);
4: Which_Side ← Judge_Side(C, x_{start}, x_{end});
5: Feature_Points ← (E_i, Which_Side);
6: **Output**: Feature_Points;

Line 4 in Algorithm 3. By determining which side of the connecting line between the start point x_{start} and the end point x_{end}, the fuzzy center of gravity C is the side of the obstacle the moving object goes through can be determined. The fuzzy center of gravity C, in Figures 1–4 is above the line, so the trajectory only needs to be planned in the upper part of the obstacle.

3.5. Feature Points of Curves

The feature points of the obstacle are extracted in Step 4 and the feature points of the curve are extracted in Step 5. In this step, the obstacle is magnified with the O_0 point as the center point. The enlarged boundary is used to obtain the points P_i^0 by which the Bezier curve is drawn and the line between these points is a Bezier polygon. In this paper, the magnification scale k is set to 2 and the points on the obstacle surface are regarded as set $\boldsymbol{E}$ and the points on the magnified boundary are regarded as set $\boldsymbol{K}$ and then the two sets satisfy the following relationship.

$$\boldsymbol{K} = k \cdot \boldsymbol{E} - O_0 \tag{6}$$

It can be specifically shown in the image that the line segments in Figure 1 satisfy the relationship shown in Equation (7):

$$\begin{cases} \overline{P_2^0O_0} = k \cdot \overline{x_{\text{left}}O_0} \\ \overline{P_3^0O_0} = k \cdot \overline{E_1O_0} \\ \overline{P_4^0O_0} = k \cdot \overline{E_2O_0} \\ \overline{P_5^0O_0} = k \cdot \overline{x_{\text{right}}O_0} \end{cases} \tag{7}$$

The line segments in Figure 2 satisfy the relationship shown in Equation (8):

$$\begin{cases} \overline{P_2^0O_0} = k \cdot \overline{x_{\text{left}}O_0} = k \cdot \overline{E_1O_0} \\ \overline{P_3^0O_0} = k \cdot \overline{E_2O_0} \\ \overline{P_4^0O_0} = k \cdot \overline{E_3O_0} \\ \overline{P_5^0O_0} = k \cdot \overline{x_{\text{right}}O_0} = k \cdot \overline{E_4O_0} \end{cases} \tag{8}$$

The line segments in Figure 3 satisfy the relationship shown in Equation (9):

$$\begin{cases} \overline{P_3^0O_0} = k \cdot \overline{E_1O_0} \\ \overline{P_4^0O_0} = k \cdot \overline{E_2O_0} \\ \angle P_2^0P_1^0 = d_{\text{start}} \\ \angle P_6^0P_5^0 = d_{\text{end}} \end{cases} \tag{9}$$

The points in Figure 4 satisfy the relationship shown in Equation (10):

$$\begin{cases} P_2^0 = (\min(\boldsymbol{K}, x), 0) \\ P_3^0 = (\min(\boldsymbol{K}, x), \max(\boldsymbol{K}, y)) \\ P_4^0 = (\max(\boldsymbol{K}, x), \max(\boldsymbol{K}, y)) \\ P_5^0 = (\max(\boldsymbol{K}, x), 0) \end{cases} \tag{10}$$

where min($\boldsymbol{K}$, x) is the minimum value of the set $\boldsymbol{K}$ in the x-axis direction.

In summary, in addition to finding set $\boldsymbol{K}$ of the enlarged boundary points according to set $\boldsymbol{E}$ and the magnification scale k, the following steps are included in Step 5:

(1) Set the start point x_{start} as the first point in feature points of the curve and the end point x_{end} as the last points in feature points of the curve;
(2) Make a straight line through the x_{start} point in the direction d_{start} (start direction) and find the solution in set $\boldsymbol{K}$ (set $\boldsymbol{K}$ is the set of points on the boundary after the enlarged surface of the obstacle), which is the second point in the characteristic point of the curve;
(3) Make a straight line through the x_{end} point in the direction d_{end} (end direction) and find the solution in set $\boldsymbol{K}$, which is the penultimate point in the feature point of the curve;
(4) Amplify the E_i of the part in which the trajectory planning needs to be developed and the enlarged point is regarded as a feature point of the curve (such as Figures 1–3), or the maximum value point of the coordinate is regarded as a feature point of the curve (such as Figure 4).

3.6. Generating Curve

In Step 4 and Step 5, the feature points of obstacles and the feature points of curves are extracted respectively. In Step 6, a Bezier curve is generated from the feature points of the curve P_i^0. The curves representing the path can be circular arcs, sine and cosine curves, N-polynomial curves, Bezier splines, B-splines, and so on. In this paper, the Bezier curve is taken as an example to describe how to use the feature points of the curve to plan the time-varying trajectory.

Feature points can be obtained according to the obstacle P_1^0-P_n^0. According to these feature points and the scale factor κ, a Bezier curve with n-1 order can be generated. From P_1^0 to P_n^0, a polygon composed of polylines formed by various feature points is referred to as a feature polygon. In Figure 5, the point after the first iteration satisfies the following relationship.

$$\begin{cases} P_1^1 = (1-\kappa)P_1^0 + \kappa P_2^0 \\ P_2^1 = (1-\kappa)P_2^0 + \kappa P_3^0 \\ P_3^1 = (1-\kappa)P_3^0 + \kappa P_4^0 \end{cases} \tag{11}$$

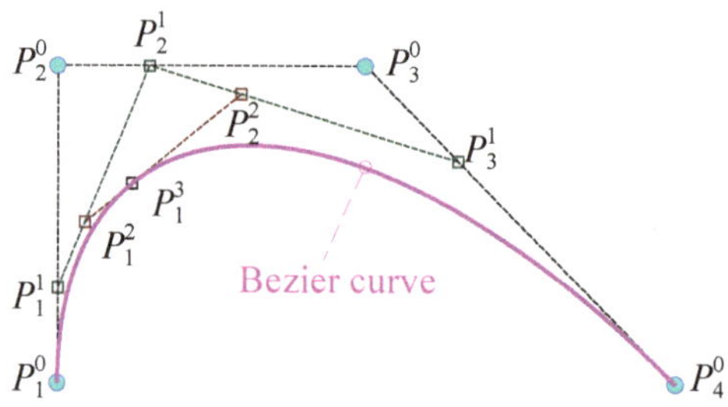

Figure 5. Schematic diagram of the Bezier curve (example of κ = 0.3).

The point after the second iteration satisfies the following relationship.

$$\begin{cases} P_1^2 = (1-\kappa)P_1^1 + \kappa P_2^1 \\ P_2^2 = (1-\kappa)P_2^1 + \kappa P_3^1 \end{cases} \tag{12}$$

After the third iteration, the point on the Bezier curve is obtained, which satisfies the following relationship.

$$p(\kappa) = P_1^3 = (1-\kappa)P_1^2 + \kappa P_2^2 \tag{13}$$

The points and segments in the diagram satisfy the following relationships.

$$P_i^j = (1-\kappa)P_i^{j-1} + \kappa P_{i+1}^{j-1} \tag{14}$$

$$\overline{P_i^{j-1}P_i^j} \Big/ \overline{P_i^j P_{i+1}^{j-1}} = \kappa/(\kappa-1) \tag{15}$$

where P_i^j is the ith point after the jth iteration and $p(\kappa)$ is the function representing the Bezier curve, $\kappa \in [0, 1]$.

When κ changes from 0 to 1, a cubic Bezier curve defined by n = 4 vertices in the graph is generated. By analogy, the nth degree Bezier curve $p(\kappa)$ defined by n+1 vertices can be obtained that satisfies the following equation.

$$p(\kappa) = \sum_{i=1}^{n+1} C_n^{i-1}(1-\kappa)^{n+1-i}\kappa^{i-1}P_i^0 \tag{16}$$

where C_n^{i-1} is the number of combinations expressed in probability theory.

The $p(\kappa)$ corresponding to the Bezier curve can be sorted out as a matrix as follows.

$$p(\kappa) = \boldsymbol{P}\boldsymbol{K}_1(1-\kappa)\boldsymbol{K}_2(\kappa)\boldsymbol{H}^T \tag{17}$$

The matrices $\boldsymbol{P}$, $\boldsymbol{K}_1$, $\boldsymbol{K}_2$ and $\boldsymbol{H}$ in the equation are as follows.

$$\boldsymbol{P} = \left[P_1^0, P_2^0, \cdots, P_n^0\right] \tag{18}$$

$$\boldsymbol{K}_1(1-\kappa) = \left[(1-\kappa)^n, (1-\kappa)^{n-1}, \cdots, (1-\kappa)^0\right] \cdot \boldsymbol{I}_n \quad (19)$$

$$\boldsymbol{K}_2(\kappa) = \left[\kappa^0, \kappa^1, \cdots, \kappa^n\right] \cdot \boldsymbol{I}_n \quad (20)$$

$$\boldsymbol{H}^{\mathrm{T}} = [\lambda_1, \lambda_2, \cdots, \lambda_n] \quad (21)$$

where $\boldsymbol{P}$ is the geometric matrix of feature points; $\boldsymbol{K}_1$ and $\boldsymbol{K}_2$ are diagonal matrices related to parameter κ; $\boldsymbol{H}$ is the weight matrix; $\lambda_i = C_n^{i-1}$ is the weight coefficient.

Deriving the variables $\boldsymbol{K}_1$, $\boldsymbol{K}_2$ and $p(\kappa)$ to κ, the following equations can be obtained.

$$\frac{d\boldsymbol{K}_1}{d\kappa} = \left[-n(1-\kappa)^{n-1}, -(n-1)(1-\kappa)^{n-2}, \cdots, -1, 0\right] \cdot \boldsymbol{I}_n \quad (22)$$

$$\frac{d\boldsymbol{K}_2}{d\kappa} = \left[0, 1, \cdots, (n-1)\kappa^{n-2}, n\kappa^{n-1}\right] \cdot \boldsymbol{I}_n \quad (23)$$

$$\frac{dp(\kappa)}{d\kappa} = \boldsymbol{P}\left[\frac{d\boldsymbol{K}_1}{d\kappa}\boldsymbol{K}_2 + \boldsymbol{K}_1\frac{d\boldsymbol{K}_2}{d\kappa}\right]\boldsymbol{H}^T \quad (24)$$

$$\frac{d^2p(\kappa)}{d\kappa^2} = \boldsymbol{P}\left[\frac{d^2\boldsymbol{K}_1}{d\kappa^2}\boldsymbol{K}_2 + 2\frac{d\boldsymbol{K}_1}{d\kappa}\frac{d\boldsymbol{K}_2}{d\kappa} + \boldsymbol{K}_1\frac{d^2\boldsymbol{K}_2}{d\kappa^2}\right]\boldsymbol{H}^T \quad (25)$$

Just substituting the specific κ value into the above derivation equations, the matrix related to κ can be solved in advance. The final improvement curve $p(\kappa)$ and its first and second derivatives can be obtained by adjusting the weight matrix according to expectations and the curve is the final Bezier curve at this moment.

The curve $p(\kappa)$ in two-dimensional space changes along the x-axis and y-axis directions and the corresponding change functions can be written as $p_x(\kappa)$ and $p_y(\kappa)$. The radius of curvature R corresponding to this curve is calculated as follows.

$$R = \left|\frac{\left[(p'_x)^2 + \left(p'_y\right)^2\right]^{3/2}}{p'_x p''_y - p''_x p'_y}\right| \quad (26)$$

By controlling the parameters in the equation, it is guaranteed that the radius of curvature R of the Bezier curve is always within the specified range. So that the deflection angle of the joint is always within the limit when the moving object (such as a snake manipulator) moves along this path.

4. Dynamic Path under the Multi-Obstacle Environment

Section 3 introduces the process of generating feature points of obstacles, feature points of curves and Bezier curves with the help of pseudo-code for a single dynamic obstacle. In this section, how to plan paths for multiple obstacles in real time is described, as shown in Figure 6.

Step 1: Input the initial data and the model of the moving object: the position x_{start} and direction d_{start} of the start point, the position x_{end} and the direction d_{end} of the end point and the model of the snake manipulator.

Step 2: Obtain real-time shape position information for each obstacle: Obs_Information_j.

Step 3: Find the center point O_{j_0} of each obstacle.

Step 4: Extract the surface feature point E_{j_i} that characterize the information of each obstacle and find the fuzzy center C_j of each obstacle through Equation (27).

$$C_j = \sum_{i=1}^{w} E_{j\_i}/w \quad (27)$$

where j is the number of obstacles and w is the number of obstacle feature points.

Step 5: The magnified surface K_{j_i} of obstacle is obtained by Equation (28) according to the scale factor k, magnification center point O_{j_0} and surface feature point E_{j_i}.

$$K_{j\_i} = k \cdot E_{j\_i} - O_{j\_0} \tag{28}$$

The jth obstacle can generate a set $\boldsymbol{K}_j$ from which the main feature points are filtered to obtain the feature points of the Bezier curve.

In this method, it is not necessary to analyze the feature points that are far away from the obstacle and it is not necessary to consider multiple adjacent feature points repeatedly.

Step 6: According to the feature points of the curve and the principle described in Section 3.6, a smooth transition Bezier curve can be generated.

Step 7: The motion path corresponding to each time is output, so that the moving object can achieve dynamic obstacle avoidance when moving along the path.

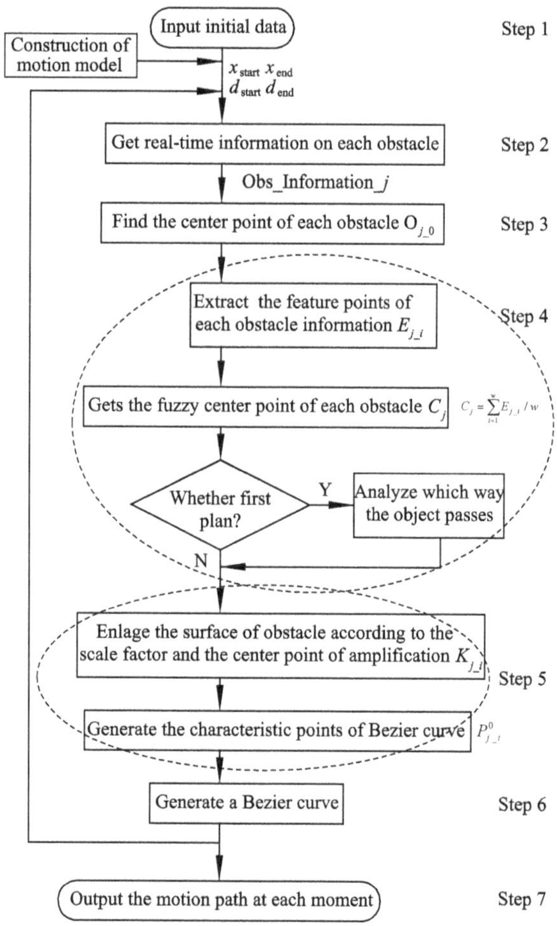

Figure 6. Flowchart of real-time trajectory generation in a multi-obstacle environment.

According to the above steps, the obstacle avoidance curve generated at a certain moment can be obtained when moving in a multi-obstacle environment, as shown in Figure 7. The figure includes L-shaped obstacles, rectangular obstacles, triangular obstacles, circular obstacles and pentagram obstacles. These five obstacles move along the path represented by Equations (29)–(33) and these five motion paths are plotted by curves in Figure 7. The different colored asterisks * in the figure correspond to the feature points of each obstacle. When these five obstacles move, the Bezier curve is generated in real time by this method to ensure that the moving objects do not collide with the obstacles.

$$\begin{cases} x_{\text{circle}}/50 = 9\sin(\pi t/200) + 5 \\ y_{\text{circle}}/50 = -10\cos(\pi t/200) - 2 \end{cases} \tag{29}$$

$$\begin{cases} x_{\text{rectangle}}\Big/200 = \sin(\pi t/50) + 5 \\ y_{\text{rectangle}}\Big/200 = \cos(\pi t/200) + \cos(\pi t/50) + 1 \end{cases} \tag{30}$$

$$\begin{cases} x_{\text{triangle}}\Big/50 = t/100 + 30 \\ y_{\text{triangle}}\Big/50 = -t/10 + 13 \end{cases} \tag{31}$$

$$\begin{cases} x_{\text{star}}/50 = -t/10 + 35 \\ y_{\text{star}}/50 = 3t/20 - 18 \end{cases} \tag{32}$$

$$\begin{cases} x_{\text{L}}/50 = \sin(\pi t/30) + 5 \\ y_{\text{L}}/50 = -t/10 + 11 \end{cases} \tag{33}$$

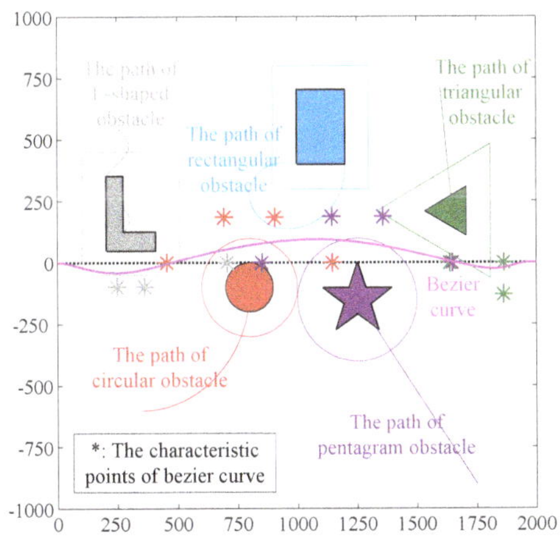

Figure 7. Obstacle avoidance curve in a multi-obstacle environment.

When the five obstacles move in real time, feature points of obstacles, feature points of curves and Bezier curves are generated in real time through the GTVP method.

5. Theoretical Verification

The process of the GTVP method is described in Sections 3 and 4. In these two sections, the effectiveness and practicability of the GTVP method are proved by theoretical derivation. The path planned by this method can ensure that the moving object will never encounter the obstacle and the distance between the moving object and the obstacle can be adjusted by adjusting the magnification factor. Next, a circular obstacle is taken as an example to describe the proof process.

If O_0 is taken as the origin point, the coordinates of the four points E_1, E_2, E_3 and E_4 in Figure 2 are:

$$\begin{cases} E_1: \left(-\sqrt{r^2-y_0^2}, 0\right) \\ E_2: \left(\frac{-1}{4}\xi, \frac{\sqrt{3}}{4}\xi\right), \xi = \sqrt{4r^2-y_0^2} - \sqrt{3}y_0 \\ E_3: \left(\frac{1}{4}\xi, \frac{\sqrt{3}}{4}\xi\right), \xi = \sqrt{4r^2-y_0^2} - \sqrt{3}y_0 \\ E_4: \left(\sqrt{r^2-y_0^2}, 0\right) \end{cases} \tag{34}$$

where r is the dimensional information of the circular obstacle: the radius of the circle; y_0 is the distance between point O_0 (center point) and point C (fuzzy center of gravity).

Through adopting the method in the paper, six feature points can be obtained. And then, the expression of the 5-order Bezier curve can be derived:

$$p(\kappa) = \sum_{i=1}^{N} C_{N-1}^{i-1} (1-\kappa)^{N-i} \kappa^{i-1} P_i^0 \tag{35}$$

where N = 6. Equation (36) can be obtained by expanding the combination number in the above equation.

$$p(\kappa) = \sum_{i=1}^{N} \left[\frac{N!}{(i-1)!(N-i-1)!}\right] (1-\kappa)^{N-i} \kappa^{i-1} P_i^0 \tag{36}$$

Put N = 6 into Equation (36) and expand to obtain Equation (37).

$$\begin{aligned} p(\kappa) = (1-\kappa)^5 P_1^0 + 5(1-\kappa)^4 \kappa P_2^0 + 10(1-\kappa)^3 \kappa^2 P_3^0 + \\ 10(1-\kappa)^2 \kappa^3 P_4^0 + 5(1-\kappa)\kappa^4 P_5^0 + \kappa^5 P_6^0 \end{aligned} \tag{37}$$

Bringing in these 6 points P_1^0 - P_6^0, it can be analyzed that, when κ = 0.5, the distance between the Bezier curve and the obstacle is the closest. Let κ = 0.5 , and yields:

$$\begin{aligned} p(\kappa = 0.5) = (0.5)^5 P_1^0 + 5(0.5)^5 P_2^0 + 10(0.5)^5 P_3^0 + \\ 10(0.5)^5 P_4^0 + 5(0.5)^5 P_5^0 + (0.5)^5 P_6^0 \end{aligned} \tag{38}$$

Ensure that the curve is outside the obstacle, namely:

$$p(\kappa = 0.5) \geq r - y_0 \tag{39}$$

After organization,

$$10 \cdot (0.5)^3 \cdot \frac{\sqrt{3}}{4} \cdot \xi \geq r - y_0 \tag{40}$$

Bringing in ξ, and tidying up, yields:

$$5\sqrt{3}\left(\sqrt{4r^2-y_0^2} - \sqrt{3}y_0\right) \geq 16(r - y_0) \tag{41}$$

Let y_0=$k \cdot r$, where $k \in (0, 1)$. Equation (42) is obtained after simplification.

$$5\sqrt{3}\left(\sqrt{4-k^2} - \sqrt{3}k\right) \geq 16(1-k) \tag{42}$$

After further organization,

$$5\sqrt{3}\left(\sqrt{4-k^2}\right) + k - 16 \geq 0 \tag{43}$$

Let the left of the inequality sign in the above equation be $f(k)$, namely:

$$f(k) = 5\sqrt{3}\left(\sqrt{4-k^2}\right) + k - 16 \quad (44)$$

When $f(k)$ is derived and the increase or decrease of the function between the interval (0,1) is analyzed, it can be obtained that the function keeps as the increase function between the interval $(0, 1/\sqrt{19})$ and the subtraction function between the interval $(1/\sqrt{19}, 1)$. The two minimum points $f(k=0) = 1.32$ and $f(k=1) = 0$ satisfy the condition of greater than or equal to 0. Through the theoretical derivation in this section, it can be proved that the Bezier curve planned by this method never touches obstacles.

6. Simulation

6.1. Feasibility Analysis of Obstacle Avoidance

The experiment was run on a CPU of Inter i5-6500 with 4G of RAM. According to Section 3, the GTVP method can ensure that the manipulator avoids obstacles in a single-obstacle environment. It can be seen from Section 4 that, when the manipulator moves in a multi-obstacle environment, it still keeps a certain distance from the obstacles. The movement process of the manipulator is shown in Figure 8. It can be seen from Figure 8 that complex obstacles are characterized by the basic obstacle model and the corresponding obstacle feature points are generated in the process of motion planning. Whereafter, a continuous path with global dynamic change is generated according to the dynamic feature points, so that the manipulator can repeat the time-varying following movement along the time-varying path, so as to achieve obstacle avoidance.

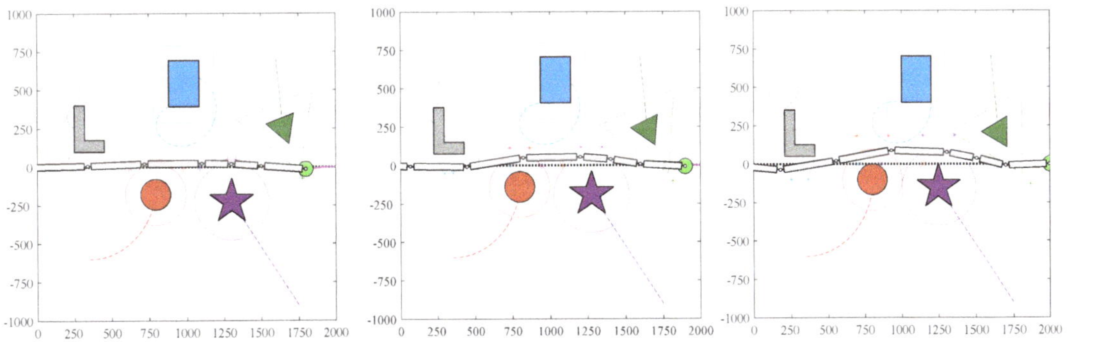

Figure 8. Diagram of the motion process of the manipulator in the environment of multiple obstacles. (**a**) t = 90 s. (**b**) t = 95 s. (**c**) t = 100 s.

During the movement of the manipulator from 90 s to 100 s, the closest distance between the key node of the manipulator and all obstacles can be analyzed to obtain the change curve shown in Figure 9. The curve is the closest distance between each key node on the manipulator and the obstacle, where d_8 is the closest distance between the end point of the manipulator and all obstacles and the remaining d_i corresponds to the closest distance between the start point of the ith link and all obstacles.

As can be seen from the figure, the maximum distance between the manipulator and the obstacle is 15 mm, which is not less than the set minimum distance. Because, when the closest distance between the manipulator and the obstacle is less than the specified value, the real-time path can be quickly adjusted by adjusting the curve feature point of the symbolic path in this method, so that the manipulator can retreat to a safe area. Through the above simulation, it can be proved that the path planned by the proposed method meets the following conditions: "global", "time-varying" and "obstacle avoidance in multi-obstacle environment".

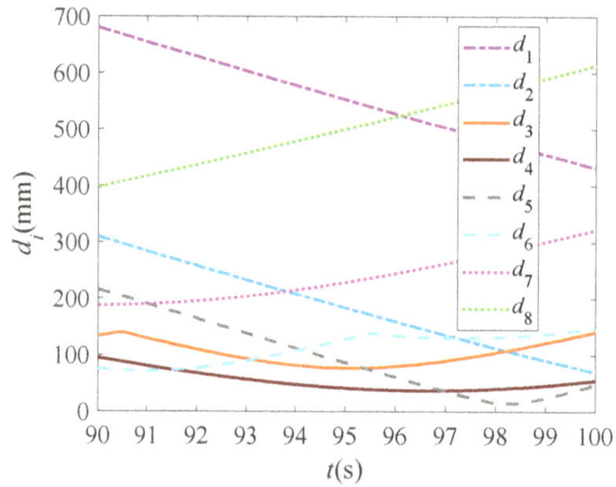

Figure 9. The closest distance between several key nodes and obstacles.

6.2. Comparison of Different Methods

For path planning of moving objects, traditional path planning methods used for obstacle avoidance cannot consider multiple conditions: (1) real-time obstacle avoidance; (2) global obstacle avoidance; and (3) obstacle avoidance under multiple obstacles. Up to now, a variety of path planning methods are proposed, and the comparison between these methods is summarized in Table 1.

Table 1. Features of different algorithms.

Methods	Real-Time	Global	Multi-Obstacle	Types of Path	Application Objects
Method 1 [7]	×	✓	✓	RRT/ Straight path	Manipulator
Method 2 [9]	×	✓	✓	VDSM/ Straight path	Manipulator
Method 3 [3]	×	×	✓	Bezier curves	Wheeled mobile robot
Method 4 [20]	×	×	✓	Bezier curves	Autonomous vehicles
Method 5 [21]	✓	✓	✓	Bezier curves	Mobile robots
Method 6 [22]	✓	×	✓	Bezier curves	Robot
Method 7 [23]	✓	×	×	Bezier curves	Mobile robots
Method 8 [24]	✓	×	×	Bezier curves	Robot
Method 9 [33]	×	✓	✓	RRT	Robot
GTVP	✓	✓	✓	Bezier curve	Manipulator

It can be seen from the table that most methods cannot take into account multiple-obstacle avoidance conditions at the same time. It can be seen from the simulation results that method [7] and method [9] not only cannot realize real-time path planning, but also the simulation time in the static path is longer than that in the present method. Although global real-time path planning can be achieved for multiple obstacles in paper [21], the PSO algorithm is adopted in the algorithm, which takes a certain amount of time to perform iterative operations in the process of solving the key points of the path. However, the path that meets multiple conditions can be solved without the help of an intelligent optimization solving algorithm in this method and the planned path is smoother.

The distance between the robot arm and the obstacle is relatively close from 90 s to 100 s. During this period, path planning was performed through RRT, Q-RRT*[33], MDA+RRT [7] and GTVP to obtain Figures 10–13. In these four pictures, (a)–(f) are the corresponding simulation results of the six moments 90 s, 92 s, 94 s, 96 s, 98 s and 100 s, respectively. In the figure, the black solid part is the obstacle, the red line segment is the planned path and the black dot in Figure 13 is the feature point.

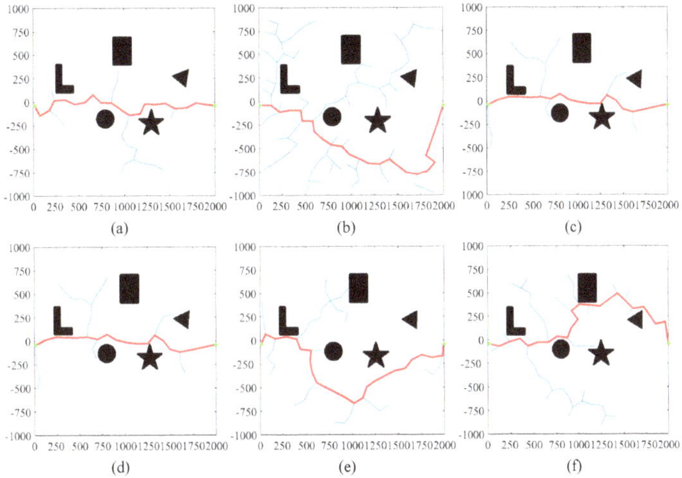

Figure 10. Simulation results obtained by RRT. (**a**) t = 90 s. (**b**) t = 92 s. (**c**) t = 94 s. (**d**) t = 96 s. (**e**) t = 98 s. (**f**) t = 100 s.

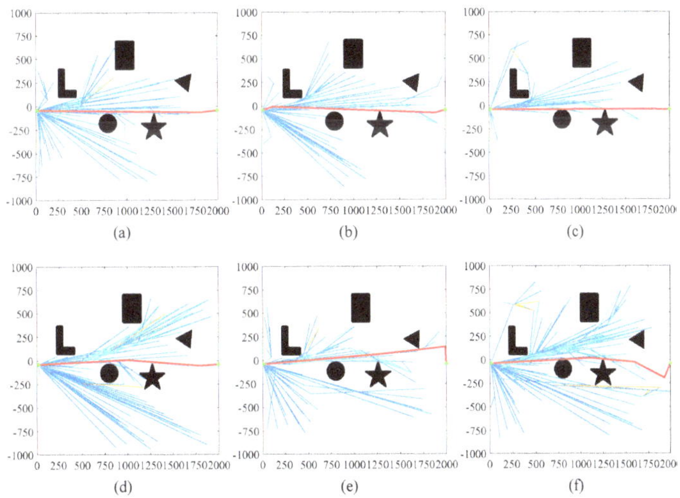

Figure 11. Simulation results obtained by Q-RRT* [33]. (**a**) t = 90 s. (**b**) t = 92 s. (**c**) t = 94 s. (**d**) t = 96 s. (**e**) t = 98 s. (**f**) t = 100 s.

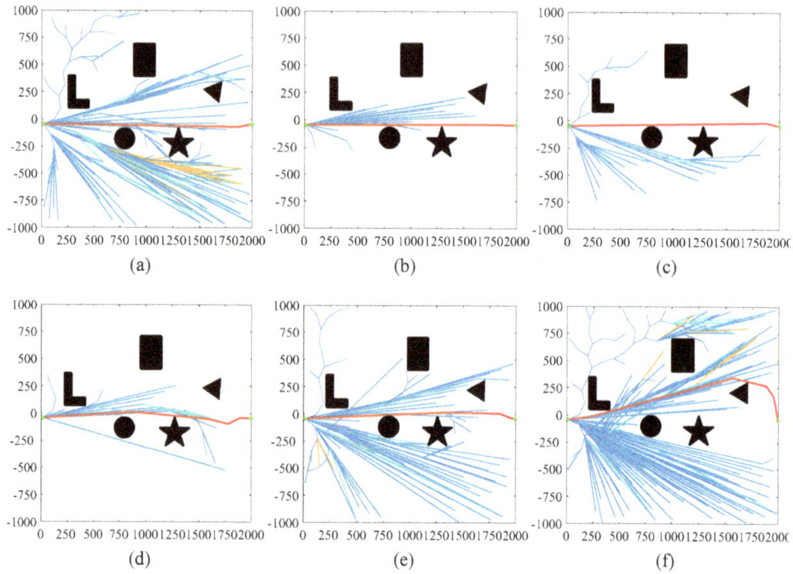

Figure 12. Simulation results obtained by MDA+RRT. (**a**) $t = 90$ s. (**b**) $t = 92$ s. (**c**) $t = 94$ s. (**d**) $t = 96$ s. (**e**) $t = 98$ s. (**f**) $t = 100$ s.

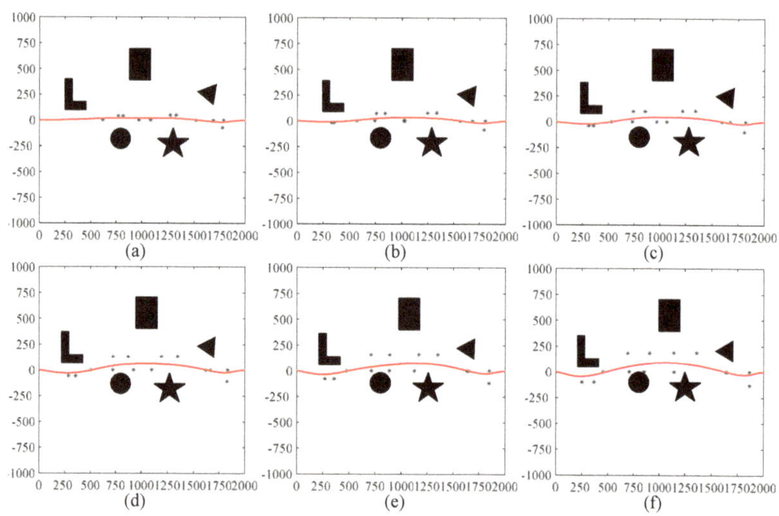

Figure 13. Simulation results obtained by GTVP. (**a**) t=90s. (**b**) $t = 92$ s. (**c**) $t = 94$ s. (**d**) $t = 96$ s. (**e**) $t = 98$ s. (**f**) $t = 100$ s.

As can be seen from Figure 10, the path planned by the RRT algorithm requires a small number of nodes, but the total length of each planned path is long and there are many turning points. When the manipulator moves along this path, the joint angle will exceed the limit of the deflection angle. As can be seen from Figures 11 and 12, the path length planned by the Q-RRT* algorithm or MDA+RRT algorithm is short, but the distance between the path and obstacles is close, and the planned path has uncertainty at any time.

As can be seen from Figures 10–12, the path planned by other path planning methods has the following characteristics: (1) the distance between the path and the obstacle is relatively close, which cannot meet the conditions of obstacle avoidance by the manipulator;

(2) the path is not smooth; and (3) there is a big difference between the paths planned at each time. All in all, in the movement process of the manipulator, other methods are difficult to ensure real-time, global and obstacle avoidance at the same time.

As can be seen from Figure 13, feature points change with the change of obstacle position, representing the key information of the obstacle position. It can be seen from the figure that the path planned by the GTVP method is not the shortest path, but the path has smooth characteristics. The GTVP method is used to plan the path in the middle of the obstacles, so that the path is as far away from all obstacles as possible. For real-time path planning, the path planning of each moment with the GTVP method is related to the path planning result of the previous moment, thus reducing the overall simulation time and realizing real-time path planning. In a word, the GTVP method adjusts the path corresponding to each moment through the feature points of obstacles, so as to achieve the requirements of ensuring real-time, global and obstacle avoidance at the same time.

Each method is used to plan the path in the environment of t = 100 s and the boxplots shown in Figure 14 shows the simulation time statistics for 50 simulations. As shown in the figure, the average times required for simulation using RRT, Q-RRT*, MDA+RRT and GTVP are 0.008 s, 0.5992 s, 1.6136 s and 0.0934 s, respectively.

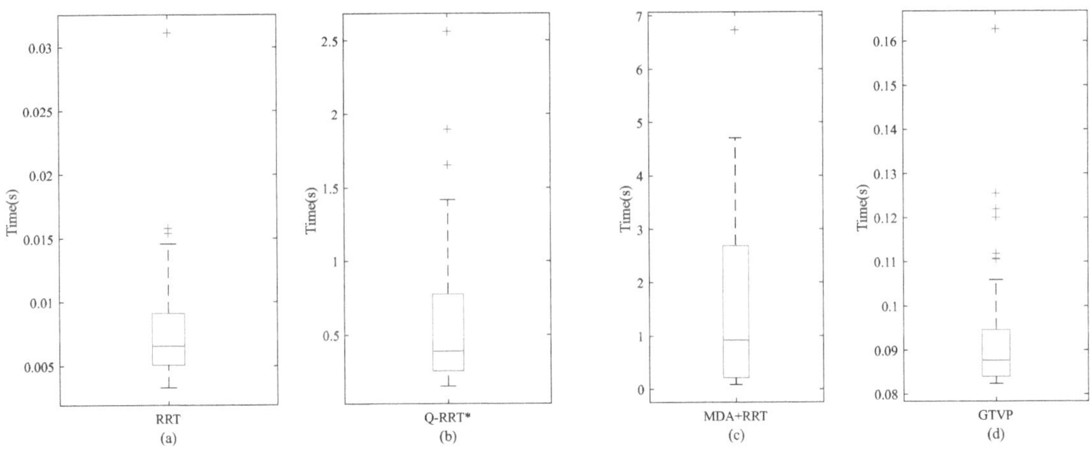

Figure 14. The computational time of each method. (**a**) RRT. (**b**) Q-RRT*. (**c**) MDA+RRT. (**d**) GTVP.

As shown in the figure, path planning using the RRT algorithm takes the shortest time, but the planned path is not smooth , unfeasible and the path changes greatly at adjacent moments. The path planning time of the GTVP method is much less than the other two methods and the path obtained is smooth and continuous in real time.

6.3. Comparison of Bezier Curves under Different Orders

The high-order Bezier curve has some disadvantages, such as a large calculation amount, oscillation, complexity of the trajectory and it cannot be applied to dynamic environments. Although the low-order Bezier curve can guarantee the continuity of the path, it cannot provide the continuous curvature and any setting of the second derivative of the Bezier curve.In short, it is necessary to balance the computational load of trajectory planning with the flexibility of the trajectory. In this paper, the characteristics of Bezier curves under different orders are compared and analyzed. Finally, the GTVP method selects the 5-order Bezier curve with a global real-time obstacle avoidance function and the calculation amount of the 5-order Bezier curve is moderate.

Taking the time-varying environment with five obstacles mentioned in this paper as the simulation environment, a set of Bezier curves is obtained every 0.1 s in the process of

0 s to 100 s. The total simulation time t_n required for real-time path planning using Bezier curves of different orders is summarized in Table 2.

Table 2. Characteristics of methods under different orders.

Methods	Simulation Time t_n(s)	Global Obstacle Avoidance	Characteristics
Bezier curves (n = 9)	71.552	✓	Trajectory mutation
Bezier curves (n = 7)	44.491	✓	Feasible
Bezier curves (n = 5)	20.313	✓	Feasible
Bezier curves (n = 3)	-	×	Infeasible

The GTVP method in this paper considers the starting direction , the ending direction and there are 6 points representing the characteristics of obstacles. Therefore, the degrees of the Bezier curve is at least 5-order, while the 3-order Bezier curve cannot meet the requirements. It can be seen from the table that the simulation time increases with the rise of the order. However, the generated path is more complex when the order of Bezier curve is 9 and the trajectory mutation with oscillation will occur. Moreover, the simulation time is as long as 71.552 s, which cannot guarantee real-time planning. In summary, the degrees of the Bezier curve with the GTVP method can be 5-order or 7-order. This paper preferentially selects the 5-order Bezier curve planning method to ensure real-time requirements.

7. Conclusions

In this paper, a novel path planning method, GTVP, is proposed. In this method, the center point is obtained according to the real-time shape and position information of the obstacle and feature points representing the obstacle information are extracted. Then the obstacle surface is amplified by the scale coefficient to generate the center point and Bezier curve feature point. Finally, the curve corresponding to the real-time motion path of the manipulator is output. The GTVP method is applied to trajectory planning in single obstacle or multi-obstacle environment and each process of the method is described in detail. The simulation results demonstrate that the path planned with the GTVP method can meet various conditions at the same time: (1) real-time obstacle avoidance; (2) global obstacle avoidance; and (3) obstacle avoidance under multiple obstacles. When the manipulator moves in a multi-obstacle environment, the closest distance between the manipulator and the obstacle is 15 mm, which is greater than the set minimum distance. In addition, compared with other path planning algorithms, the GTVP method can plan real-time smooth paths that meet multiple conditions without the help of an intelligent optimization algorithm.

In the future, we can continue to improve the GTVP method in the following directions: (1) analyze how to adjust multiple adjustable parameters better in the GTVP method; (2) the GTVP method is combined with other intelligent optimization algorithms to plan optimal trajectory planning based on time, space, speed and other goals; and (3) the GTVP method can be applied to smooth obstacle avoidance in a three-dimensional environment.

Author Contributions: methodology, L.J.; software, B.L.; writing—original draft, L.F. and Y.H.; writing—review & editing, S.Z. and Z.Y. All authors have read and agreed to the published version of the manuscript.

Funding: This research was funded by the National Key Research and Development Program of China under Grant 2022YFB4700800.

Institutional Review Board Statement: Not applicable.

Informed Consent Statement: Not applicable.

Data Availability Statement: Publicly available datasets were analyzed in this study. This data can be found here: https://github.com/LongfeiJia/GTVP_method_code, accessed on 10 December 2023.

Conflicts of Interest: The authors declare no conflict of interest.

References

1. Huang, Y.; Jia, L.; Chen, J.; Zheng, J.; Guo, Y.; Tao, Y. A Novel Path Planning Algorithm Considering the Maximum Deflection Angle of Joint. *IEEE Access* **2021**, *9*, 115777–115787. [CrossRef]
2. Li, H.; Qian, L.; Mei, H.; Wang, X.; Guo, Z.-Y. An Efficient Maritime Route Planning Method Based on an Improved A* with an Adaptive Heuristic Function and Parallel Computing Structure. *Appl. Sci.* **2023**, *13*, 10873. [CrossRef]
3. Alshammrei, S.; Boubaker, S.; Kolsi, L. Improved Dijkstra Algorithm for Mobile Robot Path Planning and Obstacle Avoidance. *Comput. Mater. Contin.* **2022**, *72*, 5939–5954. [CrossRef]
4. Li, D.; Yin, W.; Wong, W.E.; Jian, M.; Chau, M. Quality-Oriented Hybrid Path Planning Based on A* and Q-Learning for Unmanned Aerial Vehicle. *IEEE Access* **2022**, *10*, 7664–7674. [CrossRef]
5. Ismail, H. The Floyd-Warshall All-Pairs Shortest Paths Algorithm for Disconnected and Very Sparse Graphs. *Softw. Pract. Exp.* **2023**, *53*, 1287–1303. [CrossRef]
6. Cao, Y.; Cheng, X.; Mu, J. Concentrated Coverage Path Planning Algorithm of UAV Formation for Aerial Photography. *IEEE Sens. J.* **2022**, *22*, 11098–11111. [CrossRef]
7. Jia, L.; Huang, Y.; Chen, T.; Guo, Y.; Yin, Y.; Chen, J. MDA+RRT: A General Approach for Resolving the Problem of Angle Constraint for Hyper-Redundant Manipulator. *Expert Syst. Appl.* **2022**, *193*, 116379. [CrossRef]
8. Ma, Y.; Zhao, Y.; Li, Z.; Bi, H.; Wang, J.; Malekian, R.; Sotelo, M.A. CCIBA*: An Improved BA* Based Collaborative Coverage Path Planning Method for Multiple Unmanned Surface Mapping Vehicles. *IEEE Trans. Intell. Transp. Syst.* **2022**, *23*, 19578–19588. [CrossRef]
9. Jia, L.; Yu, Z.; Zhou, H.; Pan, Z.; Ou, Y.; Guo, Y.; Huang, Y. Variable Dimensional Scaling Method: A Novel Method for Path Planning and Inverse Kinematics. *Machines* **2022**, *10*, 1030. [CrossRef]
10. Yang, X.; Wu, F.; Li, R.; Yang, D.; Li, M.; He, A. Real-Time Path Planning for Obstacle Avoidance in Intelligent Driving Sightseeing Cars Using Spatial Perception. *Appl. Sci.* **2023**, *13*, 11183. [CrossRef]
11. Raj, R.; Kos, A. An Optimized Energy and Time Constraints-Based Path Planning for the Navigation of Mobile Robots Using an Intelligent Particle Swarm Optimization Technique. *Appl. Sci.* **2023**, *13*, 9667. [CrossRef]
12. Yang, D.; Li, D.; Sun, H. 2D Dubins Path in Environments with Obstacle. *Math. Probl. Eng.* **2013**, *2013*, 291372. [CrossRef]
13. McCrae, J.; Singh, K. Sketching Piecewise Clothoid Curves. *Comput. Graph.* **2009**, *33*, 452–461. [CrossRef]
14. Wang, J.; Wu, Z.; Tan, M.; Yu, J. 3-D Path Planning with Multiple Motions for a Gliding Robotic Dolphin. *IEEE Trans. Syst. Man Cybern. Syst.* **2021**, *51*, 2904–2915. [CrossRef]
15. Zhang, B.; Zhu, D. A New Method on Motion Planning for Mobile Robots Using Jump Point Search and Bezier Curves. *Int. J. Adv. Robot. Syst.* **2021**, *18*, 172988142110192. [CrossRef]
16. Durakli, Z.; Nabiyev, V. A New Approach Based on Bezier Curves to Solve Path Planning Problems for Mobile Robots. *J. Comput. Sci.* **2022**, *58*, 101540. [CrossRef]
17. Arslan, O.; Tiemessen, A. Adaptive Bézier Degree Reduction and Splitting for Computationally Efficient Motion Planning. *IEEE Trans. Robot.* **2022**, *38*, 3655–3674. [CrossRef]
18. Song, B.; Wang, Z.; Zou, L. An Improved PSO Algorithm for Smooth Path Planning of Mobile Robots Using Continuous High-Degree Bezier Curve. *Appl. Soft Comput.* **2021**, *100*, 106960. [CrossRef]
19. Blazic, S.; Klancar, G. Effective Parametrization of Low Order Bezier Motion Primitives for Continuous-Curvature Path-Planning Applications. *Electronics* **2022**, *11*, 1709. [CrossRef]
20. Bulut, V. Path Planning for Autonomous Ground Vehicles Based on Quintic Trigonometric Bezier Curve. *J. Braz. Soc. Mech. Sci. Eng.* **2021**, *43* , 104. [CrossRef]
21. Xu, L.; Cao, M.; Song, B. A New Approach to Smooth Path Planning of Mobile Robot Based on Quartic Bezier Transition Curve and Improved PSO Algorithm. *Neurocomputing* **2022**, *473*, 98–106. [CrossRef]
22. Deng, X.; Li, R.; Zhao, L.; Wang, K.; Gui, X. Multi-Obstacle Path Planning and Optimization for Mobile Robot. *Expert Syst. Appl.* **2021**, *183*, 115445. [CrossRef]
23. Renny Simba, K.; Uchiyama, N.; Sano, S. Real-Time Smooth Trajectory Generation for Nonholonomic Mobile Robots Using Bézier Curves. *Robot. Comput. Integr. Manuf.* **2016**, *41*, 31–42. [CrossRef]
24. Scoccia, C.; Palmieri, G.; Palpacelli, M.C.; Callegari, M. A Collision Avoidance Strategy for Redundant Manipulators in Dynamically Variable Environments: On-Line Perturbations of Off-Line Generated Trajectories. *Machines* **2021**, *9*, 30. [CrossRef]
25. Elhoseny, M.; Tharwat, A.; Hassanien, A.E. Bezier Curve Based Path Planning in a Dynamic Field Using Modified Genetic Algorithm. *J. Comput. Sci.* **2018**, *25*, 339–350. [CrossRef]
26. Kang, M.; Cho, Y.; Yoon, S.-E. RCIK: Real-Time Collision-Free Inverse Kinematics Using a Collision-Cost Prediction Network. *IEEE Robot. Autom. Lett.* **2022**, *7*, 610–617. [CrossRef]
27. Minnetoglu, O.; Conkur, E.S. Tight Maneuvering for Path Planning of Hyper-Redundant Manipulators in Three-Dimensional Environments. *Appl. Sci.* **2022**, *12*, 8882. [CrossRef]
28. Chen, W.-C.; Lin, C.-L.; Chen, Y.-Y.; Cheng, H.-H. Quadcopter Drone for Vision-Based Autonomous Target Following. *Aerospace* **2023**, *10*, 82. [CrossRef]
29. Gu, Y.; Lv, J.; Bo, J.; Zhao, B.; Zheng, K.; Zhao, Y.; Tao, J.; Qin, Y.; Wang, W.; Liang, J. An Improved Wavelet Modulus Algorithm Based on Fusion of Light Intensity and Degree of Polarization. *Appl. Sci.* **2022**, *12*, 3558. [CrossRef]

30. Yu, E.; Ryu, B.-S. Recognizing Trained and Untrained Obstacles around a Port Transfer Crane Using an Image Segmentation Model and Coordinate Mapping between the Ground and Image. *Sensors* **2023**, *23*, 5982. [CrossRef]
31. Bai, X.; Jiang, H.; Cui, J.; Lu, K.; Chen, P.; Zhang, M. UAV Path Planning Based on Improved A* and DWA Algorithms. *Int. J. Aerosp. Eng.* **2021**, *2021*, 4511252. [CrossRef]
32. Dai, W.; Zhang, Y.; Li, P.; Fang, Z.; Scherer, S. RGB-D SLAM in Dynamic Environments Using Point Correlations. *IEEE Trans. Pattern Anal. Mach. Intell.* **2022**, *44*, 373–389. [CrossRef]
33. Jeong, I.-B.; Lee, S.-J.; Kim, J.-H. Quick-RRT*: Triangular Inequality-Based Implementation of RRT* with Improved Initial Solution and Convergence Rate. *Expert Syst. Appl.* **2019**, *123*, 82–90. [CrossRef]

applied sciences

MDPI

Article

Low-Cost Data-Driven Robot Collision Localization Using a Sparse Modular Point Matrix

Haoyu Lin, Pengkun Quan, Zhuo Liang, Dongbo Wei and Shichun Di *

School of Mechatronics Engineering, Harbin Institute of Technology, Harbin 150001, China; linhaoyu@hit.edu.cn (H.L.); quanpengkun@hit.edu.cn (P.Q.); liangzhuo@hit.edu.cn (Z.L.); weidb@hit.edu.cn (D.W.)

* Correspondence: dishichun@hit.edu.cn; Tel.: +86-139-046-05946

Abstract: In the context of automatic charging for electric vehicles, collision localization for the end-effector of robots not only serves as a crucial visual complement but also provides essential foundations for subsequent response design. In this scenario, data-driven collision localization methods are considered an ideal choice. However, due to the typically high demands on the data scale associated with such methods, they may significantly increase the construction cost of models. To mitigate this issue to some extent, in this paper, we propose a novel approach for robot collision localization based on a sparse modular point matrix (SMPM) in the context of automatic charging for electric vehicles. This method, building upon the use of collision point matrix templates, strategically introduces sparsity to the sub-regions of the templates, aiming to reduce the scale of data collection. Additionally, we delve into the exploration of data-driven models adapted to SMPMs. We design a feature extractor that combines a convolutional neural network (CNN) with an echo state network (ESN) to perform adaptive feature extraction on collision vibration signals. Simultaneously, by incorporating a support vector machine (SVM) as a classifier, the model is capable of accurately estimating the specific region in which the collision occurs. The experimental results demonstrate that the proposed collision localization method maintains a collision localization accuracy of 91.27% and a collision localization RMSE of 1.46 mm, despite a 48.15% reduction in data scale.

Keywords: automatic charging; data-driven collision localization; sparse modular point matrix; convolutional neural network; echo state network; support vector machine

Citation: Lin, H.; Quan, P.; Liang, Z.; Wei, D.; Di, S. Low-Cost Data-Driven Robot Collision Localization Using a Sparse Modular Point Matrix. *Appl. Sci.* **2024**, *14*, 2131. https://doi.org/10.3390/app14052131

Academic Editor: Paolino Di Felice

Received: 5 February 2024
Revised: 25 February 2024
Accepted: 1 March 2024
Published: 4 March 2024

1. Introduction

In the domain of robot-assisted automatic electric vehicle charging, the connection between the charger and the charging port relies on precise visual positioning [1]. However, the visual positioning system may be subject to disturbances in unstructured environments, such as changes in lighting conditions, leading to instances of positioning failure. Visual positioning failure typically results in three scenarios: in cases of minimal localization deviation, the charger carried by the robot's end-effector is able to connect to the charging port, but may experience jamming. In such situations, impedance control implemented on the robot can effectively suppress jamming [2]. When the positioning deviation is substantial, the charger on the robot's end-effector may fail to make contact with the charging port during the connection process, potentially causing contact with other parts of the electric vehicle's body. In this case, implementing a collision classification protection system on the robotic arm ensures the safety of the vehicle and the robot. When the positioning deviation falls between the aforementioned scenarios, meaning that the charger can make contact with the charging port but cannot smoothly insert due to the presence of a visual positioning deviation, imparting a collision localization capability to the robotic arm can effectively correct the deviation caused by the visual positioning failure, serving as a supplementary localization strategy in the event of a visual failure [3,4].

In the exploration of model-based collision localization methods, J. Vorndamme et al. achieved collision localization for humanoid robots by constructing a generalized momentum observer to calculate joint torques and estimate joint accelerations [5]. This method enables point estimation in single-contact situations using only onboard sensors. Additional force/torque sensors are introduced only when estimating multi-contact positions. M. Iskandar et al. developed a momentum-based external force estimation framework for robot collision localization [6]. This approach includes joint-level residual estimation and uncoupled force–torque estimation in Cartesian space, eliminating the need for acceleration estimation and consequently mitigating the introduction of noise associated with acceleration estimation. D. Zurlo et al. addressed the problem of difficulty in achieving high-precision collision localization solely by relying on a generalized momentum observer to a certain extent by combining the generalized momentum observer method with a particle filtering strategy [7].

In the pursuit of achieving high-precision collision localization, artificial-skin-based methods are generally considered a more favorable option. P. Piacenza et al. utilized low-cost optical components installed along the edges of the perception region to achieve higher accuracy in contact localization by measuring the impact of touch on the passage of light through elastic material [8]. X. Fan et al. designed a set of ultrasound sensors deployable on the surface of a robotic arm to achieve high-precision contact localization and analyze contact pressure [9]. P. Mittendorfer et al. achieved interactive touch in different parts of a humanoid robot by employing self-organizing, multimodal artificial skin [10]. X. Li et al. developed a tactile sensor composed of overlapping air chambers, leveraging the spatiotemporal continuity of learning contact positions to achieve high-precision and high-resolution collision localization [11].

With the rapid advancement of artificial intelligence technology, supervised learning strategies have become widely utilized to address collision localization problems in robotics. These methods are commonly referred to as data-driven collision localization approaches. D. Popov et al. employed onboard sensors to collect collision data from robots, utilizing neural network methods to learn from the relevant data, thereby achieving collision localization at the centimeter level [12]. X. Ha et al. utilized information from multi-core fiber Bragg grating sensors, combined with a k-nearest neighbor (KNN) model to fit a free-space curvature model, successfully estimating collision positions for continuum robots [13]. F. Min et al. mounted accelerometers on the joint near the base and end-effector of a robotic arm to capture vibration signals during collisions. They performed reasonable feature extraction on the collision vibration signals and, in conjunction with an artificial neural network (ANN), successfully achieved the localization of the contact points [14]. W. McMahan et al. mounted four accelerometers on a single robotic arm to form an accelerometer array, capturing collision vibration data. They employed a support vector machine (SVM) to learn from the vibration information of different collision positions, thereby achieving collision localization with an error in the centimeter range [15].

In the realm of robot-assisted automatic electric vehicle charging, the end-effector, which is incapable of establishing direct physical contact with the charging port, exclusively interfaces with it through the intermediary of the carried charger. This situation may introduce unknown disturbances in signal measurements within the model-based method, posing challenges for achieving high accuracy in collision localization with model-based methods. Additionally, during the plug-in process, the forces generated during collisions typically act along the robot. As discussed in [5], model-based methods face increased difficulty in handling collision issues when external forces act along the robot. In addition, due to the frequent contacts and substantial contact forces inherent in the plug-in process, this demanding operational environment will significantly diminish the lifespan of artificial skin. Simultaneously, the deployment of artificial skin encounters certain challenges, serving as a constraint that restricts its application in addressing this issue. Considering data-driven collision localization methods, these approaches heavily depend on formulating rules for gathering data tailored to specific scenarios.

In our previous research, we introduced a collision point matrix template specifically designed for millimeter-level collision localization in the scenario of automatic charging for electric vehicles [4]. The collision point matrix template consists of collision points spaced at 1 mm intervals on a plane. By pre-setting the collision point matrix template on the surface of the charging port and then colliding with each collision point using the charger carried by the automatic charging robot, we can obtain collision vibration information suitable for collision localization. Utilizing a collision point matrix template composed of densely distributed collision points, the collision localization problem can be transformed into a classification problem, with collision information corresponding to different points in the template. To enhance the generalization ability of the collision localization method, it is necessary to consider the adaptation of the collision localization method to variations in joint configurations during the data collection process. Therefore, it is generally required to collect collision information under as many different joint configurations as possible. As the accuracy of collision localization in this method depends on the dense distribution of collision points in the template, the cost of data collection is typically very high.

To alleviate the significant burden of data collection associated with such an approach, we propose a data-driven collision localization method based on a sparse modular point matrix (SMPM). Unlike the earlier collision point matrix template, the SMPM efficiently reduces the density of collision point distribution, thereby reducing the scale and associated costs of building the collision dataset to some extent. The main contributions of this paper are as follows:

1. Building upon the collision point matrix template, the SMPM is first introduced to achieve local sparsity of the template, thereby reducing the data scale required for the data-driven collision localization method;
2. Comparative experiments are conducted by constructing SMPMs of various forms and degrees of sparsity, exploring the optimal way to build SMPMs effectively while maintaining high collision localization performance with a reduced data scale;
3. A data-driven collision localization method combining a convolutional neural network (CNN), an echo state network (ESN), and a support vector machine (SVM) is proposed to enable the SMPM to achieve optimal performance in collision localization.

2. Materials and Methods

2.1. Dataset Description

The SMPM proposed in this study was constructed based on the collision point matrix template introduced in ref. [4]. To investigate the effectiveness of the proposed SMPM, the data used in this study were consistent with our previous work [4]. Specifically, the datasets comprised vibration signals encompassing 3-axis acceleration and 3-axis angular velocity, collected by the IMU mounted on the charger at a sampling frequency of 1500 Hz. An AUBO-i5 robot, a commercially available general-purpose 6-DOF robotic arm, was employed as the automatic charging equipment. It was connected to the charger via a flexible wrist, as depicted in Figure 1. During the data collection process, the end-effector robot moves in a linear motion at a speed of 15 mm/s to execute the collision. Each collision point in the collision point matrix template attached to the charging port undergoes five collisions in the same pattern to minimize the impact of robot positioning errors on the results. Additionally, we considered the impact of different joint configurations on the collision localization results, thereby constructing three independent collision vibration signal datasets named D1, D2, and D3. Each dataset corresponds to three sets of distinct joint configurations, with each dataset containing 4335 samples. For more details, please refer to the table entitled "Joint configuration of the datasets" in ref. [4].

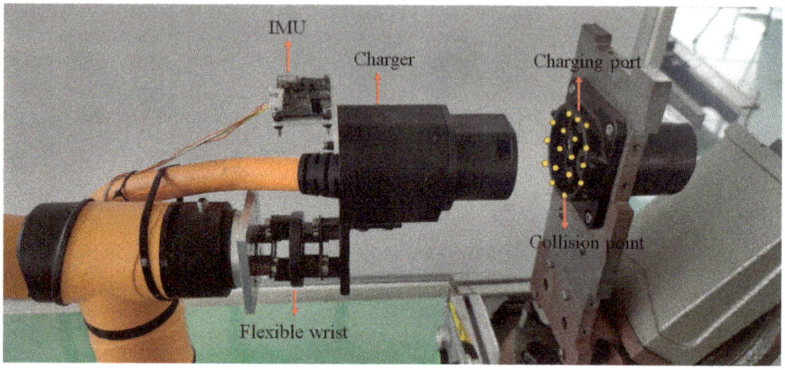

Figure 1. Illustration of the automatic charging equipment.

2.2. SMPM Method

In our previous work, we observed that when using the collision point matrix template, the estimated positions of collision points are prone to confusion with their neighboring collision points. This implies that it is possible to estimate the occurrence of collisions at a particular collision point by leveraging the collision vibration information from its neighboring points. Building upon this idea, we propose a local modularization and sparsification approach for the collision point matrix template, as illustrated in Figure 1. The collision point matrix template mentioned here is identical to the one in ref. [4]. Collision points are defined as intersection points between the central axis of the charger and the plane in which the charging port is located. The template comprises collision points with 1 mm spacing, arranged in 17 rows and 17 columns, with its center located at the intersection of the central axis of the charging port and the plane in which the charging port is situated. In practical applications, the template can be scaled without altering the spacing between collision points. Modularization is achieved by exploiting the similarity in collision vibration signals between the estimated collision points and their neighboring points, eliminating the need to collect data for the estimated collision points during the data collection process. We refer to the estimated collision points that do not appear in the dataset as "zero-shot points", while the collision points used to estimate zero-shot points, requiring collection in the dataset, are defined as "fully observable points". In the process of implementing local modularization, we consider the information from fully observable points nearest to the zero-shot points to estimate collisions occurring at the zero-shot points. This process leads to the formation of a modular point matrix (MPM), as illustrated in the figure, comprising a central zero-shot point and its surrounding eight fully observable points. The MPM utilized results in a 1/9 reduction in the quantity of data collected, compared to the original collision point matrix template.

Furthermore, it is crucial to consider whether utilizing all eight fully observable points is necessary for accurately estimating a collision occurrence at a zero-shot point. In theory, the collision vibration information obtained from these eight fully observable points may contain redundancy when estimating a zero-shot point. If this hypothesis holds true, eliminating the redundant fully observable points could further reduce the scale of the collision dataset. Taking this into consideration, we propose three sparsification methods for the MPM, as illustrated in Figure 2. The first sparsification method involves removing one of the fully observable points from the MPM. This approach results in eight sparse modular point matrices (SMPMs), denoted as Cell 1-1, Cell 1-2, . . ., Cell 1-8, obtained by sequentially removing one fully observable point in clockwise direction starting from the top left corner. The second sparsification method involves removing two fully observable points from the MPM. During this removal process, we consider two extreme cases: one that maximally preserves the zero-shot point's farthest adjacent points (resulting in Cell 2-1 and

Cell 2-2), and another that maximally preserves the zero-shot point's nearest adjacent points (resulting in Cell 2-3 and Cell 2-4). The third sparsification method involves removing four fully observable points from the MPM, specifically resulting in Cell 3-1, which excludes all of the nearest adjacent points, and Cell 3-2, which excludes all of the farthest adjacent points. The collision localization effects arising from the different sparsification methods are explained in detail in the experimental section.

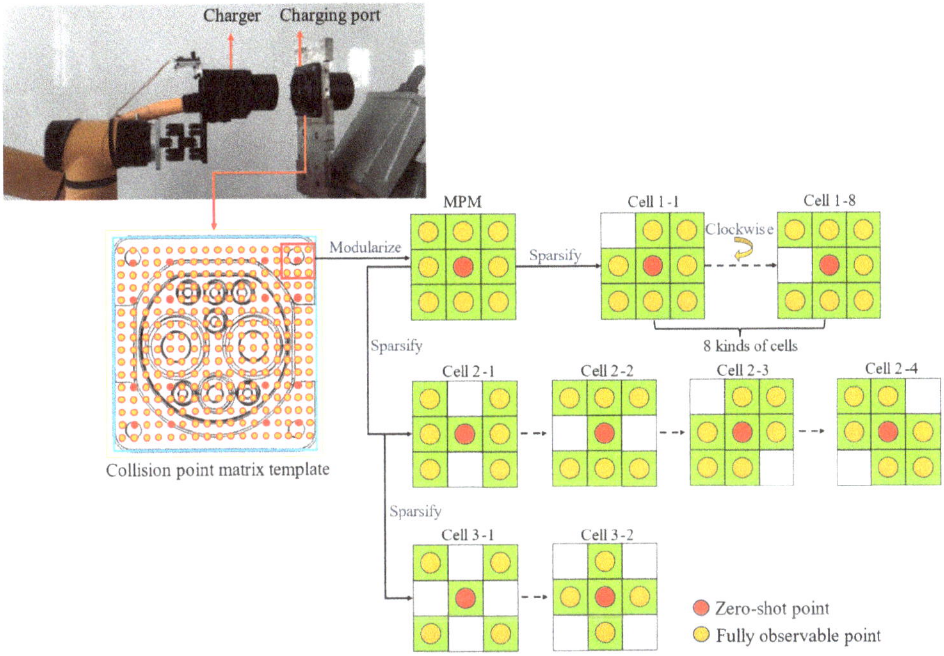

Figure 2. Different sparsification forms of SMPMs.

2.3. *Collision Localization Model*

In our previous work, we explored collision localization models based on a CNN [3] and an ESN [4], respectively. Drawing inspiration from these two approaches, we proposed a collision localization model that integrates a CNN and an ESN as feature extractors. In this model, the CNN demonstrates a propensity for capturing salient features along the line of sight, making it a prevalent choice for feature engineering. Meanwhile, the ESN exhibits the capability to unfold in accordance with the temporal sequence, finding widespread applications in time series analysis. To enhance the ultimate localization performance, we also integrate an SVM model as the final region classifier. Capitalizing on distinctive attributes of the CNN and ESN, we formulate a collision localization model based on a CNN-ESN-SVM (CE-SVM) architecture.

2.3.1. CNN

The CNN, a representative deep learning method, is known for its efficacy in processing time-series and image signals [16,17]. A typical CNN structure comprises two main components: the convolutional layer and the pooling layer. In the convolutional layer, the convolution operation is applied between the input features and convolution kernels, resulting in the generation of new features. Following convolution, the obtained results typically undergo non-linear processing, often facilitated by activation functions. Commonly employed activation functions include Sigmoid, tanh, and ReLU [18]. Based on previous research results, ReLU activation functions are a suitable choice for collision localization problems.

The pooling layer serves two primary functions: dimensionality reduction and mitigating overfitting. There are two main types of pooling methods: average pooling and maximum pooling. In average pooling, the operation involves taking the average of the convolution-derived features as the output, while in maximum pooling, the operation involves selecting the maximum value from the convolution-derived features as the output. In this research, we adopted the same pooling method as in our previous work, specifically utilizing the maximum pooling approach.

2.3.2. ESN

An echo state network (ESN) is a type of recurrent neural network proposed by Jaeger et al. [19], consisting of three main components: an input layer, a reservoir, and an output layer. The reservoir is essentially a randomly connected recurrent network of a certain size, where neurons form a dense structure through random connections. These connections are predetermined and remain unaltered during training. A basic ESN model is illustrated in Figure 3.

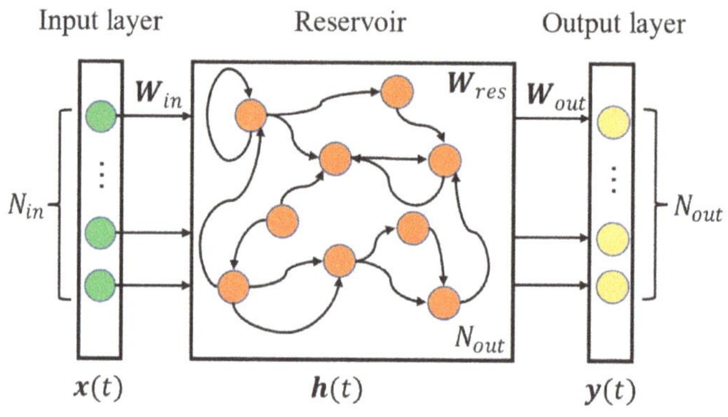

Figure 3. The structure of the ESN.

Let N_{in}, N_{res}, and N_{out} represent the numbers of neurons in the input layer, reservoir, and output layer, respectively. The matrices W_{in}, W_{res}, and W_{out} denote the weight matrices from the input layer to the reservoir, within the reservoir, and from the reservoir to the output layer, respectively. W_{in} and W_{res} are randomly initialized and remain fixed throughout the training process. Only W_{out} undergoes adjustments during the learning process. The specific ESN model can be expressed as follows:

$$h(t) = \varepsilon tanh(\boldsymbol{W_{in}}\boldsymbol{x}(t+1) + \boldsymbol{W_{res}}\boldsymbol{h}(t) + \boldsymbol{W_{out}}\boldsymbol{y}(t)) \tag{1}$$

where $\tanh(\cdot)$ represents the non-linear activation function of the reservoir and $\varepsilon \in (0, 1]$ is the leakage rate. $\boldsymbol{x}(t)$, $\boldsymbol{h}(t)$, and $\boldsymbol{y}(t)$ denote the input vector, the state vector of the reservoir, and the output vector, respectively. Compared to conventional RNNs, the training process of the ESN is simpler, only involving parameter adjustments in the output layer. The entire network does not require the complex process of backpropagation. Furthermore, due to the randomness and dense connections in the reservoir, this structure facilitates enhanced generalization capabilities, enabling the network to capture the non-linear dynamics of input signals effectively. This property contributes to the ESN's strong performance in handling time-series tasks.

In terms of hyperparameter settings, since W_{in} and W_{res} are generated through random initialization, it is essential to predefine the range for their random initialization before training. The appropriate values for these two weight matrices were adopted from Ref. [20].

Additionally, following our previous work [4], the leakage rate ε, spectral radius ρ, and N_{res} were set. Specifically, the hyperparameters of the ESN used in this paper are presented in Table 1.

Table 1. Key parameters of ESN.

Parameters	Symbolic Representations	Values
Weight matrices from the input layer to the reservoir	W_{in}	$[-0.5, 0.5]$
Weight matrices within the reservoir	W_{res}	$[-0.5, 0.5]$
Leakage rate	ε	0.5
Spectral radius	ρ	1
Numbers of neurons in the reservoir	N_{res}	64

2.3.3. Framework of CE-SVM

The framework of the proposed CE-SVM is illustrated in Figure 4, consisting primarily of a feature extractor and a classifier. The collision vibration signals employed are 3-axis acceleration and 3-axis angular velocity signals collected by the IMU mounted on the charger. After normalization, these signals serve as inputs to the model. The definition of the input data length follows the concept of the "effective period" from our previous work, where a segment with rich information meeting collision localization requirements is extracted from the initial data length, as detailed in ref. [4]. As discussed in ref. [4], an effective period with 290 sampling points already contains sufficient information for collision localization. Therefore, this paper also sets the effective period to 290 sampling points. The input layer of the feature extractor is followed by two CNN layers, each composed of a convolutional layer and a maxpooling layer. In the diagram, Conv2D denotes a 2D convolution layer, and MaxPooling2D denotes a 2D pooling layer. Post Conv2D, batch normalization is applied to ensure the data's generalization ability. Subsequently, a non-linear ReLU activation function is used to process the features, enhancing the model's capacity for effective non-linear information processing. Notably, the Conv2D structure employed in this study differs from that of [3]. While the previous work involved symmetric 3×3 convolutional kernels, in this study, we adopt asymmetric kernels to maximally preserve temporally reasonable features extracted by the CNN for subsequent processing by the ESN layer. The convolution kernel size in the temporal direction is significantly larger than that in the different axis dimensions. To effectively transmit temporal information to the ESN layer, the time-distributed technique [21] is employed for the flattening layer connecting the CNN layers and the ESN layer. To enable the SVM to effectively utilize the features extracted by the ESN, the features need to undergo flattening processing after the ESN process. Simultaneously, a fully connected layer is employed to reduce the dimensionality of the features to prevent the curse of dimensionality. In the feature extractor training process, Softmax is used as the final classifier. Based on feedback from the Softmax layer's estimation results, the weights of different components in the feature extractor are adjusted. In the training of the classifier SVM for collision localization, the weights of the pre-trained feature extractor are fixed and used solely for feature extraction. The SVM is then constructed based on the features extracted by the pre-trained feature extractor.

Detailed hyperparameters for the feature extractor and classifier are provided in Table 2. The hyperparameters of the ESN and SVM are taken from [4], while the hyperparameters of the CNN remain consistent with those outlined in ref. [3], with the exception of the convolutional kernel aspect. Since the padding method is not utilized during the initial convolutional computation, it should ensure that the length of the input can be evenly divisible by the convolutional length along the temporal dimension. Furthermore, to maintain an appropriate convolutional size along the temporal dimension, we specify the kernel size in this direction as 10. Meanwhile, the kernel size remains consistent with ref. [3] in the other direction. Our experiments utilized a Windows-based system with the following specifications: Processor: Intel (R) Core (TM) i7-10700K CPU @ 3.80 GHz, Memory: 31.9 GiB, GPU: NVIDIA GeForce RTX 3080.

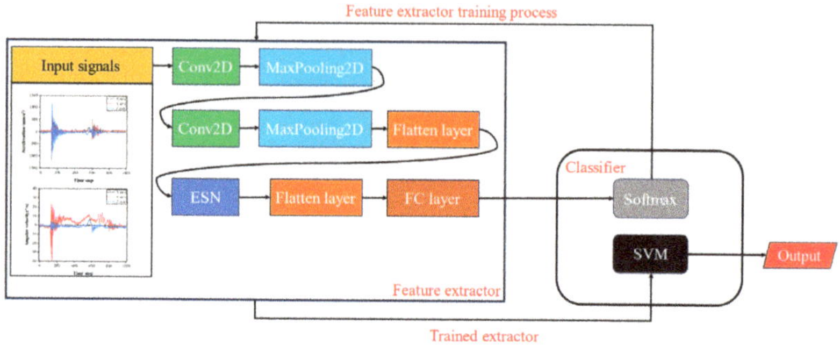

Figure 4. Structure of the CE-SVM.

Table 2. Hyperparameters of the CE-SVM.

Type	Name of Parameter	Values
Conv2D	Number of filters Kernel size Stride	64 (10, 3) 1
Batch normalization	-	-
ReLU	-	-
Maxpooling	Pool size Stride Padding	(2, 2) 1 same
Conv2D	Number of filters Kernel size Stride Padding	64 (10, 3) 2 same
Batch normalization	-	-
ReLU	-	-
Maxpooling	Pool size Stride Padding	(2, 2) 1 same
Time-distributed flattening	-	-
ESN	N_{res} ε ρ	64 0.5 1
FC	Number of hidden units	512
SVM	Regularization parameter Kernel function	100 rbf

3. Results and Discussion

To explore the effectiveness of the SMPM in reducing the required data scale for collision localization model construction, the experimental design of this research mainly consists of two parts. The first part aims to analyze the SMPM structure under discrete distributions and select the optimal structure based on the structural analysis results. The second part aims to investigate the effectiveness of the proposed collision localization method when employing the optimal SMPM layout across the entire collision point template. In the first part of the experiment, various SMPM structures are predefined based on the characteristics of vibration signals corresponding to collision points. Subsequently, utilizing multiple data-driven models, SMPMs with diverse structures across the discrete distribution of collision point templates are evaluated, leading to the selection of the optimal SMPM structure. In the second part of the experiment, the optimal SMPM is deployed throughout the entire collision point template with varying degrees of sparsity. This deployment allows for testing the performance of the optimal SMPM in reducing the necessary data scale for constructing the collision localization model while maintaining collision localization performance. By integrating the test outcomes, the collision localization model that best complements the SMPM is also identified.

3.1. Optimal SMPM Structure

In conducting a comprehensive structural analysis and optimization of the SMPM across the entire collision point matrix template, significant computational costs are incurred. In this study, we mitigate these computational challenges by decomposing the SMPM optimization problem into distinct local regions. This subdivision results in a substantial reduction in the workload for the optimal SMPM structure selection method. The proposed optimal SMPM structure selection method consists of two main steps: firstly, various forms of sparsification are applied to the MPM distributed in the four-corner region. Collision localization is then performed on the SMPMs using multiple models. Based on the accuracy of the localization results, SMPMs with superior performance are initially identified. Subsequently, the position of the SMPM relative to the collision point matrix template is adjusted, and further collision localization using multiple models is conducted on the initially screened SMPMs to select those with the optimal structure. In terms of model selection, the proposed CE-SVM method from this study is employed, along with the DCNN-SVM method introduced in [3], as well as the ESN-SVM, LSTM-SVM, and GRU-SVM methods mentioned in [4]. Additionally, based on our previous findings, the effective handling of collision point localization in the testing set when a particular collision point in the collision point matrix template is present in both the training and testing sets has been validated. Therefore, in the SMPM selection process, greater attention can be directed towards evaluating the performance of collision localization for zero-shot points. Consequently, in the SMPM selection process, we employ all points in datasets D1, D2, and D3 that meet the definition of zero-shot points as the testing set. Simultaneously, all points in datasets D1, D2, and D3 that conform to the definition of fully observable points are utilized as the training set.

The distribution of MPMs in four corner regions of the collision point matrix template are illustrated in Figure 5. To introduce a certain level of similarity interference, four MPMs are set in each region, and are tightly connected to form a square area. These MPMs are labeled for collision localization. Specifically, we defined the regions in the four corners as I, II, III, and IV. Subsequently, we assigned numerical labels from 1 to 4 to the MPMs within each region. Then, different forms of sparsification were applied to the MPMs, as shown in Figure 2. Based on these various sparsification forms, collision localization tests were conducted to preliminarily identify preferable sparsification forms.

Building upon the aforementioned preferable configurations, we further displaced the SMPM to occupy different positions on the collision point matrix template. As illustrated in Figure 6, there are three types of movements: vertical (up and down), horizontal (left and right), and toward the center. A collision localization test was conducted each time the SMPM was shifted by a distance equivalent to one collision point. Based on the results of these tests, we refined the selection of the optimal SMPM structure. To facilitate the explanation, we defined the following situations: horizontal movement by one collision point as LR1 and by two collision points as LR2; vertical movement by one collision point as UD1 and by two collision points as UD2; and movement toward the center by one collision point as CT1 and by two collision points as CT2.

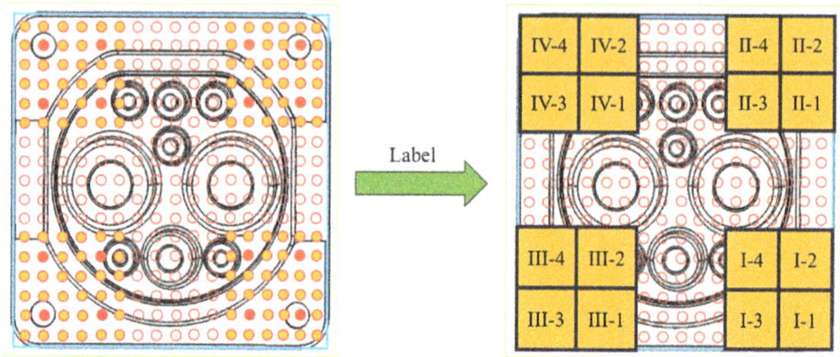

Figure 5. Illustration of SMPMs at four corners of the collision point matrix template.

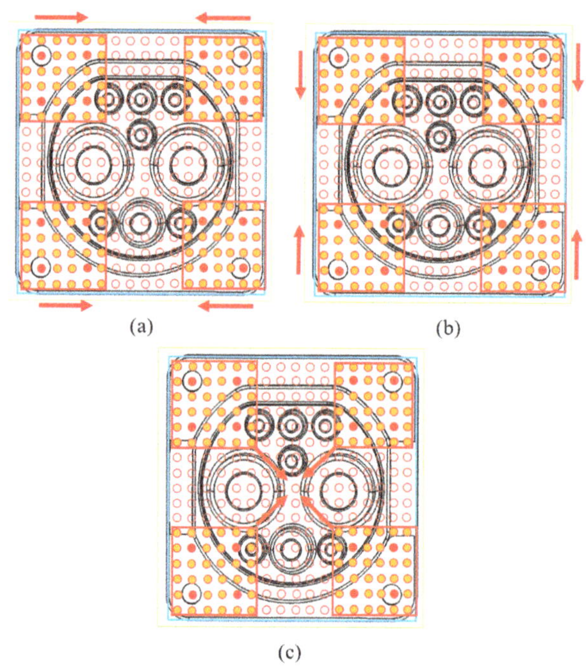

Figure 6. Illustration of SMPMs with different distributions. (**a**) Case with horizontal movement; (**b**) case with vertical movement; (**c**) case with movement toward the center.

As shown in Figure 7, the average collision localization accuracy results for the SMPM positioned at the corners of the collision point matrix template are presented. The results labeled "models with CNN" represent the average collision localization accuracy of fusion models incorporating convolutional modules, specifically the DCNN-SVM and CE-SVM methods. Conversely, "models without CNN" correspond to the average collision localization accuracy of models excluding convolutional modules, including ESN-SVM, LSTM-SVM, and GRU-SVM. From the graph, it is evident that fusion models with convolutional modules significantly outperform those relying solely on recursive neural networks for handling collision localization when applied in conjunction with an SMPM. When employing collision localization methods with convolutional modules, the accuracy of SMPMs (Cell 1-1 to Cell 1-8) after removing single points is slightly higher overall than the accuracy achieved by the MPM. In contrast, for collision results obtained using collision localization models without convolutional layers, single-point removal SMPMs are comparatively disadvantaged. This suggests that, when estimating the localization of collisions at zero-shot points, choosing an appropriate model enables the achievement of accuracy levels, even with sample size reduction, equivalent to or higher than those achieved without reducing the sample size. In the case of double-point and quadruple-point removal SMPMs, it is notably observed that when using models with convolutional modules, Cell 2-3, Cell 2-4, and Cell 3-2 achieve significantly higher average collision localization accuracy compared to single-point removal SMPMs. Especially, Cell 3-2 consistently achieves the highest collision accuracy across different sparsification forms. This implies that certain points in the MPM provide redundant or even disruptive information for collision localization. Furthermore, it can be observed that, in the single-point removal SMPM, using models with convolutional modules results in higher collision accuracy for Cell 1-1, Cell 1-3, Cell 1-5, and Cell 1-7 compared to their adjacent counterparts, Cell 1-2, Cell 1-4, Cell 1-6, and Cell 1-8. Similarly, in the double-point removal SMPM, Cell 2-3 and Cell 2-4 achieve significantly higher collision localization accuracy than Cell 2-1 and Cell 2-2, while in the quadruple-point removal SMPM, Cell 3-2 demonstrates markedly higher collision localization accuracy than Cell 3-1. This phenomenon indicates that the sparsification method removing the farthest adjacent points of zero-shot points in the MPM is more effective than removing the nearest adjacent points. Moreover, the standout performance of Cell 3-2 suggests that information from the farthest adjacent points in the MPM may lead to confusion in different regions, resulting in a decrease in collision localization accuracy.

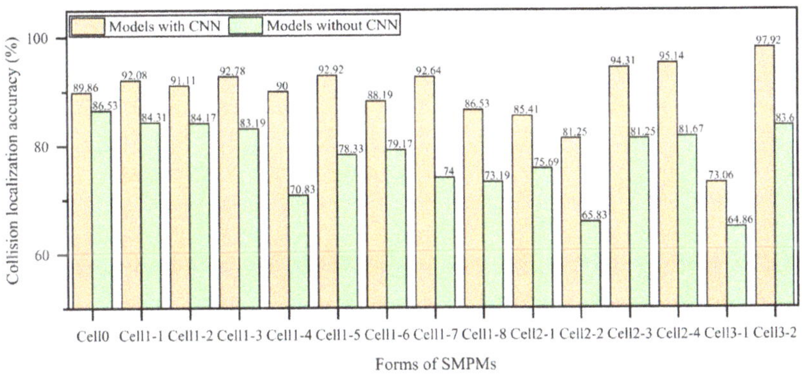

Figure 7. Collision localization accuracy with different forms of SMPMs at four corners.

Considering the significant advantages of collision localization models with convolutional modules when combined with an SMPM, we focused solely on utilizing collision localization models with convolutional modules when analyzing the zero-shot point localization situation of the SMPM at different positions within the collision point matrix template. In the selection of SMPM structures, we experimentally chose SMPM structures that exhibited clear advantages at the corners of the collision point matrix template, specifically those removing the farthest adjacent points: Cell 1-1, Cell 1-3, Cell 1-5, Cell 1-7, Cell 2-3, Cell 2-4, and Cell 3-2. For the purposes of comparison with cases without any sparsification, Cell 0 was introduced as a control experiment. Figure 8 depicts the collision localization results based on SMPMs positioned at different locations. The best performance in UD1 and UD2 is observed with Cell 1-5 and Cell 2-3, in LR1 and LR2 with Cell 0 and Cell 1-3, and in CT1 and CT2 with Cell 1-7 and Cell 3-2. Comparing the results of SMPMs with those of the MPM, it is evident that the vibration signals acquired at the farthest neighboring points may indeed contain information that could degrade the localization model's performance. From the average accuracy results of different movement point numbers in UD, LR, and CT, the relative differences in average accuracy for SMPMs removing single points (Cell 1-1, Cell 1-3, Cell 1-5, and Cell 1-7) compared to the MPM are 1.29%, −0.2%, and −0.02%; for SMPMs removing double points (Cell 2-3 and Cell 2-4), they are −0.03%, −1.18%, and −1.56%;

and for SMPMs removing four points (Cell 3-2), they are 2.16%, −0.7%, and 0.62%. This indicates that information contained in some of the farthest adjacent points is not always redundant. However, even with the removal of these points, collision localization accuracy does not significantly decrease compared to using the MPM, suggesting that removing the farthest neighboring points is effective in reducing the dataset size while maintaining high collision localization accuracy. Furthermore, it is noteworthy that the use of the Cell 3-2 sparsification form consistently demonstrates excellent collision localization performance for SMPMs positioned at different locations. This sparsification form, compared to others, minimizes the required data collection scale to the greatest extent. Hence, we consider the Cell 3-2 sparsification form of the SMPM the optimal choice.

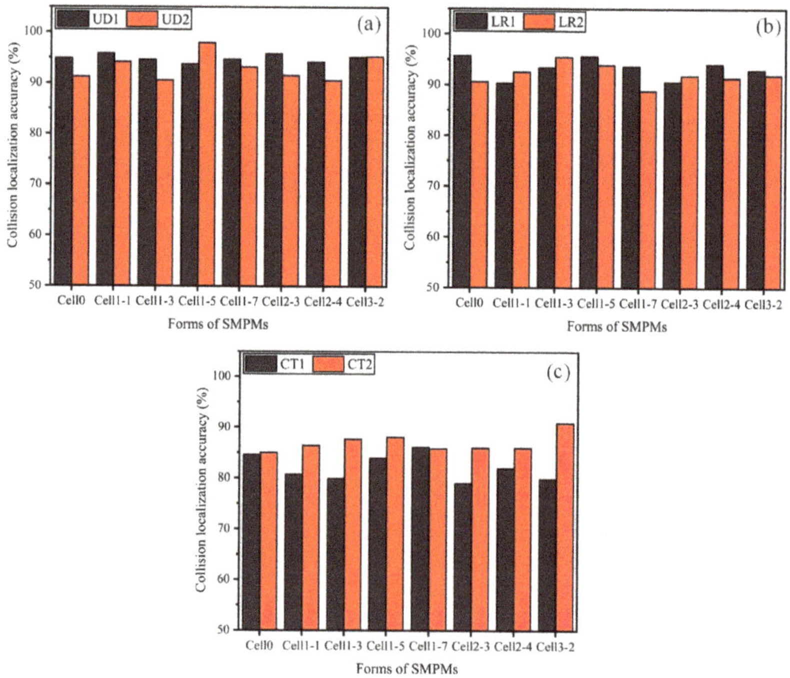

Figure 8. Collision localization results using SMPMs with different distributions. (**a**) UD; (**b**) LR; (**c**) CT.

3.2. Collision Localization Results across the Entire Template

When applying the SMPM with the Cell 3-2 form for collision localization at the charging port, the encountered challenge is not solely limited to accurately identifying collisions occurring at zero-shot points. Rather, it extends to efficiently locating collisions within the entire collision point matrix template. Therefore, it is imperative for the Cell 3-2 SMPM to be comprehensively deployed across the entire collision point matrix template. In order to investigate the feasibility of the proposed sparsification method across the entire domain, a thorough analysis of the complete collision point matrix template area is required under different robot joint angles. The training datasets used for this purpose are D1 and D2, while the testing dataset is D3, isolated from D1 and D2. Additionally, the collisions occurring at the locations of removed fully observable points also need to be addressed. Despite the theoretical capability of SMPMs to effectively locate collisions at both zero-shot and fully observable points, there is currently a lack of adequate information for handling collisions at removed fully observable points. Hence, we introduce the concept of partially observable points, i.e., collision data sampled with a demand lower than that of fully observable points but greater than zero. In conjunction with training datasets D1 and D2, the specific forms of SMPMs containing locally observable points used in the experiments are illustrated in Figure 9. To ensure the comprehensive deployment of SMPMs across the entire collision point matrix template, a departure from the approach outlined in [7] is taken. Specifically, the outermost points of the collision point matrix template are disregarded, and analysis is conducted only on the inner 15 rows and 15 columns. Building upon the Cell 3-2 form, partially observable points are positioned at the farthest adjacent point of the MPM. The samples include 30 instances of fully observable points, 0 instances of zero-shot points, and N instances of partially observable points. To investigate the impact of different sparsity levels of partially observable points on the collision localization results, five different SMPMs are defined, with N values of 5, 10, 15, 20, and 25, denoted as S1 to S5, respectively. Additionally, for comparison with the case of ample data collection, an MPM (S6) is introduced, and a control experiment (S7) utilizes the collision point matrix template directly without downsampling.

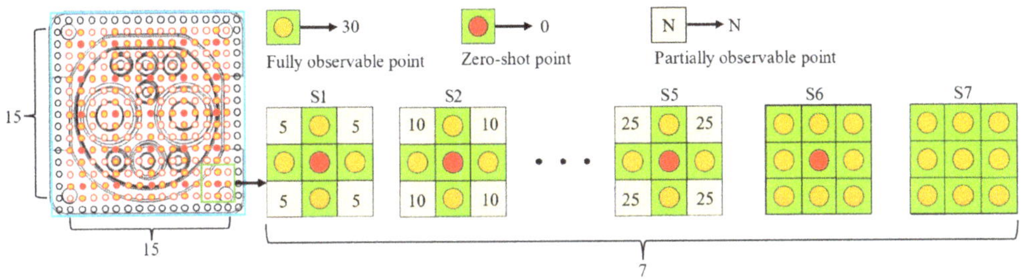

Figure 9. Illustration of SMPMs in the Cell3-2 form with varying degrees of sparsification.

In evaluating the effectiveness of applying SMPMs with varying degrees of sparsity to collision localization, we employ assessment criteria that include collision localization accuracy and root mean square error (RMSE). In this study, we treat the SMPM as a unified entity, with its central position representing the estimated positions of individual points within the SMPM. The results of collision localization accuracy are presented in Table 3, where the data scale of S7 is defined as 100%. In S7, the collision localization accuracy of different models exceeded 96%, with the highest reaching 98.67%. As the data scale decreases, the accuracy of collision localization for each method also declines accordingly. At a data scale of 51.85%, the average accuracy of each method only drops by 8.72%. Notably, the CE-SVM method exhibits the smallest decrease, with a reduction of only 7.22%, maintaining an accuracy above 90%. Furthermore, in cases S1 to S6, the CE-SVM method outperforms other methods, especially in situations with higher sparsity, highlighting the pronounced advantage of CE-SVM. Regarding the deviation in collision localization, the RMSEs of various models for collision localization are presented in Table 4. Due to the utilization of the center position of the SMPM as the estimated location for individual points within the SMPM, additional localization biases are introduced in the RMSEs, even when the collision area is correctly predicted. Therefore, the RMSEs of various models are consistently greater than 1 mm. Without any reduction in data scale, the CNN-SVM model achieves the lowest RMSE in collision localization. However, with the introduction of varying degrees of sparsity, the CE-SVM consistently exhibits a notable advantage. When the data scale is reduced to 51.85%, the RMSE of the CE-SVM method increases by only 0.21 mm, compared to the increase in RMSE for CNN-SVM, which is only 55% of its value. In conclusion, it is evident that the SMPM is capable of maintaining a high level of collision localization performance even in scenarios involving significant data scale reduction. From both the collision localization accuracy and RMSEs perspectives, the SMPM maintains a high level of performance even in scenarios of substantial data scale reduction. Particularly noteworthy is its outstanding collision localization performance when employed in conjunction with the CE-SVM method.

Table 3. Collision localization accuracy achieved using SMPMs with different levels of sparsification.

Case	CE-SVM	CNN-SVM	LSTM-SVM	ENS-SVM	GRU-SVM	Data Scale
S1	**91.27%**	89.29%	88.33%	89.53%	88.73%	51.85%
S2	**94.29%**	91.79%	90.62%	91.54%	90.8%	59.26%
S3	**95%**	94.17%	93.24%	93.27%	93.06%	66.67%
S4	**96.2%**	95%	93.92%	94.2%	93.88%	74.07%
S5	**97.07%**	95.12%	94.44%	95.22%	94.72%	81.48%
S6	**96.73%**	95.74%	94.78%	95.65%	95.28%	88.89%
S7	98.49%	**98.67%**	96.7%	98.64%	98.27%	100%

Table 4. Collision localization RMSEs achieved using SMPMs with different levels of sparsification (mm).

Case	CE-SVM	CNN-SVM	LSTM-SVM	ENS-SVM	GRU-SVM
S1	**1.46**	1.62	1.73	1.63	1.75
S2	**1.4**	1.52	1.59	1.54	1.61
S3	**1.37**	1.46	1.51	1.5	1.49
S4	**1.33**	1.45	1.48	1.47	1.48
S5	**1.34**	1.39	1.54	1.42	1.42
S6	**1.3**	1.39	1.45	1.42	1.44
S7	1.25	**1.24**	1.35	1.26	1.26

4. Conclusions

To achieve higher precision in collision localization, the existing data-driven method for the plug-in process of electric vehicle (EV) automatic charging suffers from high data collection costs. In this study, we propose a novel data-driven approach for robot collision localization specifically tailored to automatic charging scenarios for EVs, effectively mitigating this issue. Our method is grounded in a collision point matrix template and integrates a sparse modular point matrix (SMPM) to reduce the necessary size of the collision dataset for data-driven techniques. By employing an optimized SMPM structure to sparsify the entire template, we achieve a reduction in data scale of 48.15% while maintaining an average localization accuracy of 89.43% and an average RMSE of 1.64 mm. Compared to scenarios without sparsification, the average localization accuracy decreases by only 8.72%, with a minimal increase of 0.37 mm in RMSE for collision localization. Additionally, we exploit the characteristics of convolutional neural network (CNNs) and echo state network (ESNs) to develop an integrated adaptive extractor for dynamic feature extraction from collision vibration signals. Utilizing a support vector machine (SVM) as the classifier, we demonstrate the exceptional performance of the model in addressing collision localization issues when combined with the SMPM. Specifically, even with a 48.15% reduction in data scale, our model achieves an outstanding collision localization accuracy of 91.27% and an RMSE of 1.46 mm for collision localization.

Although our proposed method effectively reduces dataset size while maintaining collision localization performance at a high level, there is still a noticeable decrease in collision localization accuracy when compared to the scenario without any sparsification. In subsequent research, we will focus on exploring whether data augmentation techniques can be employed to generate data for sparsified points, creating a virtual supplement to the dataset. We aim to enhance the performance of the collision localization method further while reducing the need for experimental data acquisition.

Author Contributions: H.L. developed the methodology; H.L. conceived and designed the experiment; Z.L. and P.Q. conducted the data curation and collection; H.L. wrote the original draft of the paper; S.D. and D.W. reviewed and edited the paper. All authors have read and agreed to the published version of the manuscript.

Funding: This research received no external funding.

Institutional Review Board Statement: Not applicable.

Informed Consent Statement: Not applicable.

Data Availability Statement: The raw data supporting the conclusions of this article will be made available by the authors on request.

Conflicts of Interest: The authors declare no conflicts of interest.

References

1. Pan, M.; Sun, C.; Liu, J.; Wang, Y. Automatic Recognition and Location System for Electric Vehicle Charging Port in Complex Environment. *IET Image Process* **2020**, *14*, 2263–2272. [CrossRef]
2. Zhang, H.; Zhu, W.; Huang, Y. A Research on the Control Strategy of Automatic Charging Robot for Electric Vehicles Based on Impedance Control. *J. Phys. Conf. Ser.* **2022**, *2303*, 012085. [CrossRef]
3. Lin, H.; Quan, P.; Liang, Z.; Lou, Y.; Wei, D.; Di, S. Collision Localization and Classification on the End-Effector of a Cable-Driven Manipulator Applied to EV Auto-Charging Based on DCNN–SVM. *Sensors* **2022**, *22*, 3439. [CrossRef]
4. Lin, H.; Quan, P.; Liang, Z.; Lou, Y.; Wei, D.; Di, S. Precision Data-Driven Collision Localization with a Dedicated Matrix Template for Electric Vehicle Automatic Charging. *Electronics* **2024**, *13*, 638. [CrossRef]
5. Vorndamme, J.; Schappler, M.; Haddadin, S. Collision Detection, Isolation and Identification for Humanoids. In Proceedings of the IEEE International Conference on Robotics and Automation (ICRA), Singapore, 29 May–3 June 2017; pp. 4754–4761.
6. Iskandar, M.; Eiberger, O.; Albu-Schaffer, A.; Luca, A.D.; Dietrich, A. Collision Detection, Identification, and Localization on the DLR SARA Robot with Sensing Redundancy. In Proceedings of the IEEE International Conference on Robotics and Automation (ICRA), Xi'an, China, 30 May–5 June 2021.
7. Zurlo, D.; Heitmann, T.; Morlock, M.; De Luca, A. Collision Detection and Contact Point Estimation Using Virtual Joint Torque Sensing Applied to a Cobot. In Proceedings of the IEEE International Conference on Robotics and Automation (ICRA), London, UK, 29 May 2023; pp. 7533–7539.
8. Piacenza, P.; Dang, W.; Hannigan, E.; Espinal, J.; Hussain, I.; Kymissis, I.; Ciocarlie, M. Accurate Contact Localization and Indentation Depth Prediction with an Optics-Based Tactile Sensor. In Proceedings of the IEEE International Conference on Robotics and Automation (ICRA), Singapore, 29 May–3 June 2017.
9. Fan, X.; Lee, D.; Jackel, L.; Howard, R.; Lee, D.; Isler, V. Enabling Low-Cost Full Surface Tactile Skin for Human Robot Interaction. *IEEE Robot. Autom. Lett.* **2022**, *7*, 1800–1807. [CrossRef]
10. Mittendorfer, P.; Yoshida, E.; Cheng, G. Realizing Whole-Body Tactile Interactions with a Self-Organizing, Multi-Modal Artificial Skin on a Humanoid Robot. *Adv. Robot.* **2015**, *29*, 51–67. [CrossRef]
11. Li, X.; Zhang, Y.; Xie, X.; Li, J.; Shi, G. Improving Robotic Tactile Localization Super-Resolution via Spatiotemporal Continuity Learning and Overlapping Air Chambers. *AAAI* **2023**, *37*, 6192–6199. [CrossRef]
12. Popov, D.; Klimchik, A.; Mavridis, N. Collision Detection, Localization & Classification for Industrial Robots with Joint Torque Sensors. In Proceedings of the 26th IEEE International Symposium on Robot and Human Interactive Communication (RO-MAN), Lisbon, Portugal, 28–31 August 2017; pp. 838–843.
13. Ha, X.T.; Wu, D.; Lai, C.-F.; Ourak, M.; Borghesan, G.; Menciassi, A.; Poorten, E.V. Contact Localization of Continuum and Flexible Robot Using Data-Driven Approach. *IEEE Robot. Autom. Lett.* **2022**, *7*, 6910–6917. [CrossRef]
14. Min, F.; Wang, G.; Liu, N. Collision Detection and Identification on Robot Manipulators Based on Vibration Analysis. *Sensors* **2019**, *19*, 1080. [CrossRef]
15. McMahan, W.; Romano, J.M.; Kuchenbecker, K.J. Using Accelerometers to Localize Tactile Contact Events on a Robot Arm. In Proceedings of the Workshop on Advances in Tactile Sensing and Touch-Based Human-Robot Interaction, ACM/IEEE International Conference on Human-Robot Interaction, Boston, MA, USA, 5–8 March 2012.
16. Livieris, I.E.; Pintelas, E.; Pintelas, P. A CNN–LSTM Model for Gold Price Time-Series Forecasting. *Neural Comput. Appl.* **2020**, *32*, 17351–17360. [CrossRef]
17. Sun, Y.; Xue, B.; Zhang, M.; Yen, G.G.; Lv, J. Automatically Designing CNN Architectures Using the Genetic Algorithm for Image Classification. *IEEE Trans. Cybern.* **2020**, *50*, 3840–3854. [CrossRef]
18. Zha, W.; Liu, Y.; Wan, Y.; Luo, R.; Li, D.; Yang, S.; Xu, Y. Forecasting Monthly Gas Field Production Based on the CNN-LSTM Model. *Energy* **2022**, *260*, 124889. [CrossRef]
19. Jaeger, H. Adaptive Nonlinear System Identification with Echo State Networks. In Proceedings of the Advances in Neural Information Processing Systems 15 (NIPS 2002), Vancouver, BC, Canada, 9–14 December 2002.

20. Hua, Z.; Zheng, Z.; Péra, M.-C.; Gao, F. Remaining Useful Life Prediction of PEMFC Systems Based on the Multi-Input Echo State Network. *Appl. Energy* **2020**, *265*, 114791. [CrossRef]
21. Montaha, S.; Azam, S.; Rafid, A.K.M.R.H.; Hasan, M.Z.; Karim, A.; Islam, A. TimeDistributed-CNN-LSTM: A Hybrid Approach Combining CNN and LSTM to Classify Brain Tumor on 3D MRI Scans Performing Ablation Study. *IEEE Access* **2022**, *10*, 60039–60059. [CrossRef]

applied sciences

MDPI

Article

Path Planning Method for Manipulators Based on Improved Twin Delayed Deep Deterministic Policy Gradient and RRT*

Ronggui Cai [1] and Xiao Li [1,2,*]

[1] School of Electronic Engineering and Automation, Guilin University of Electronic Technology, Guilin 541004, China; 21082304004@mails.guet.edu.cn

[2] Key Laboratory of Intelligence Integrated Automation in Guangxi Universities, Guilin 541004, China

* Correspondence: lixiao@guet.edu.cn

Abstract: This paper proposes a path planning framework that combines the experience replay mechanism from deep reinforcement learning (DRL) and rapidly exploring random tree star (RRT*), employing the DRL-RRT* as the path planning method for the manipulator. The iteration of the RRT* is conducted independently in path planning, resulting in a tortuous path and making it challenging to find an optimal path. The setting of reward functions in policy learning based on DRL is very complex and has poor universality, making it difficult to complete the task in complex path planning. Aiming at the insufficient exploration of the current deterministic policy gradient DRL algorithm twin delayed deep deterministic policy gradient (TD3), a stochastic policy was combined with TD3, and the performance was verified on the simulation platform. Furthermore, the improved TD3 was integrated with RRT* for performance analysis in two-dimensional (2D) and three-dimensional (3D) path planning environments. Finally, a six-degree-of-freedom manipulator was used to conduct simulation and experimental research on the manipulator.

Keywords: deep reinforcement learning; twin delayed deep deterministic policy gradient; path planning; six-degree-of-freedom manipulator

check for updates

Citation: Cai, R.; Li, X. Path Planning Method for Manipulators Based on Improved Twin Delayed Deep Deterministic Policy Gradient and RRT*. *Appl. Sci.* **2024**, *14*, 2765. https://doi.org/10.3390/app14072765

Academic Editor: Marco Troncossi

Received: 10 January 2024
Revised: 19 February 2024
Accepted: 26 February 2024
Published: 26 March 2024

1. Introduction

The utilization of multi-degree-of-freedom manipulators is prevalent across various industries including aerospace and industrial manufacturing and so on. The study of path planning problem in complex environments is a crucial aspect in the field of robot control technology. At present, some path planning methods commonly employed include the A* algorithm based on graph traversals [1], the probabilistic roadmap method (PRM) algorithm based on probability sampling [2] and rapidly exploring random tree star (RRT*) utilizing a random sampling technique [3,4]. Informed RRT* uses a heuristic function to guide exploration toward the target region by optimizing the sampling process [5]. RRT*-Smart enhances the optimization speed of paths near obstacle turning points by incorporating path optimization and intelligent sampling techniques within RRT* [6]. The node generation strategy of the Gaussian mixture model RRT* (GMM-RRT*) algorithm utilizes a target-biased policy, resulting in shorter path length [7]. The P-RRT*-connect [8] combines the bidirectional artificial potential field with RRT*, which reduces the time and decreases the number of iterations. The Real-Time RRT* (RT-RRT*) [9] introduces an online tree rewiring strategy, it can find paths to new targets more quickly. These methods exhibit robust environmental exploration abilities, asymptotic optimality and consume fewer computational resources. However, these methods explore the environment randomly and hardly use valuable previous iterative experience to guide sampling. This characteristic may lead to these methods unable to find more optimized solutions.

The introduction of Deep Q-network (DQN) by [10] marked a milestone in the field. DRL has achieved remarkable advancements in many fields [11–13]. DRL empowers agents to conduct autonomous exploration and leverage the experience gained from prior explorations to inform subsequent behavioral decisions. With the advent of DRL, robots are now capable of self-directed learning. The Refs. [14,15] proposed a path planning method of improved learning policy based on different experience depth requirements at different learning stages. Marcin et al used deep deterministic policy gradient (DDPG) [16] and hindsight experience replay (HER) to train the manipulator in a simulated environment [17]. Gu [18] used a policy learning algorithm with deep Q-function to train physical robots. Lin [19] used recurrent neural network and DDPG to predict the collision-free path. Yang [20] proposed a new deep Q-learning method, which was applied to the push and grasp of objects by manipulators. Li [21] proposed a DRL that integrates automatic entropy adjustment. Kim [22] designed a motion planning algorithm that uses TD3 [23] with HER to enhance sample efficiency.

Some advanced DRL algorithms possess certain limitations. Pan [24] used the Boltzmann Softmax operator to estimate the value function, which increased computational costs and involved an amount of parameter adjustment. The policy of DRL is driven by reward during learning, successful learning relies on the design of reward functions and an action selection policy to ensure exploration and exploitation [25]. The universality of the reward function is typically low, and its design poses significant challenges. In the presence of obstacles, employing a general distance-based reward function often leads to policy learning failures [26]. Li [27] layered the DRL model to avoid the construction of complex reward functions. The task or policy model is divided into upper and lower layers to mitigate the coupling between the update formula and the challenging convergence of the reward function [28].

The challenge of applying DRL to path planning tasks remains significant. The current trend is to combine DRL with traditional path planning methods. The traditional path planning methods, such as RRT and PRM, possess robust sampling and search abilities, enabling them to provide reference paths or intermediate waypoints for DRL. In turn, DRL accomplishes the point-to-point task between each pair of nodes so that agents can obtain a smoother path [29]. Sampling path planning methods does not need a reward function in complex environment and has a higher success rate than does the DRL, it can be used to provide a successful experience reference for DRL, and DRL can use these experiences to exert exploration ability and finally complete the learning of the path planning strategy. Gao [30] combined TD3 with PRM, this can decompose the path into multiple local paths, improve development efficiency. Li [21] proposed traditional path planners with DRL to obtain the path in Cartesian space. Florensa [31] decomposed complex problems into multiple subproblems and explored maze paths using dynamic programming. Chiang et al. [32] regarded PRM and RRT as global path planning methods, respectively, and searched for intermediate path points of DRL in indoor navigation. The waypoint selection of some fusion methods is influenced by traditional path planning. To address the above issues, the contributions are outlined as follows:

- In order to improve the ability of DRL algorithm to balance exploration and development, an improved TD3 algorithm was designed and evaluated.
- Aiming at the problems existing in robot path planning, a path planning method is proposed that combines the exploration abilities of sampling-based RRT* and the experience replay mechanism of DRL algorithm.
- The simulation environment of path optimization based on DRL-RRT* is built.

The remaining sections of this paper are organized as follows: Section 2 presents the implementation principle of the proposed CDTD3 algorithm. Section 3 provides a detailed description of the DRL-RRT* path planning algorithm. Section 4 reports the path planning verification in a simulation environment. Section 5 outlines the experimental verification of the manipulator path planning using CDTD3-RRT*. Section 6 concludes the paper.

2. Improved TD3 Algorithm

2.1. Reinforcement Learning

Reinforcement learning (RL) are generally described using the Markov decision process (MDP) [33], and a tuple $M = (S, \mathcal{A}, r, p, \gamma)$ can be used to describe MDP, where S and $\mathcal{A}$ represent a set of states and actions, p is the probability of transition from the current state to the next state, r is the reward given by environmental changes for state transition, and $\gamma \in (0,1)$ is a discount used to determine reward priority. At time step t, for a given state $s \in S$, the agent can obtain r by selecting action $a \in \mathcal{A}$ based on policy μ and transfer to the next state $s_{i+1} \in S$. The goal of the agent is to maximize the discounted return $R_t = \sum_{i=t}^{T} \gamma^{i-t} r(s_i, a_i)$ [23], which can be measured by value function Q shown in Equation (1).

$$Q^{\mu}(s,a) = \mathrm{E}_{\mu}[R_t | s, a] \tag{1}$$

2.2. Algorithm Structure

The actor-critic structure DRL algorithm TD3 [23] for continuous control is advanced. Similar to popular DRL such as soft actor critical (SAC) [34], TD3 uses the double network structure of actor and critic. Both the current actor network $\mu_\theta(s)$ and the current critic network $Q_{\omega_k}(s,a)$ (k = 1, 2) have a corresponding target network, during the implementation, only $\mu_\theta(s)$ and $Q_{\omega_k}(s,a)$ participate in parameter update while the target actor network $\mu_{\theta'}(s)$ and the target critic network $Q_{\omega'_k}(s_{i+1}, a')$ (k = 1, 2) are used to store the parameters of the corresponding target network at the previous time. $\mu_{\theta'}(s)$ and $Q_{\omega'_k}(s_{i+1}, a')$ do not completely copy the parameters of their corresponding original network when storing them; instead, a soft update method shown in Equation (2) is adopted.

$$\begin{cases} \theta' = \tau\theta + (1-\tau)\theta' \\ \omega'_k = \tau\omega_k + (1-\tau)\omega'_k \end{cases} \tag{2}$$

where $\tau \in (0,1)$. θ and θ' represent the network parameters of $\mu_\theta(s)$ and $\mu_{\theta'}(s)$, respectively. ω_k and ω'_k correspond to $Q_{\omega_k}(s,a)$ and $Q_{\omega'_k}(s_{i+1}, a')$, respectively.

TD3 select an action based on policy $\mu_\theta(s_i)$ in the s_i. Due to the deterministic policy adopted by TD3, a certain proportion of random noise is added in the exploration phase to improve the exploration ability of the agent. The output of a_i is shown in Equation (3).

$$a_i = \mu_\theta(s_i) + \delta \tag{3}$$

where $\delta \sim \mathcal{N}(0,\sigma)$ represents a Gaussian distribution with a mean of 0 and variance σ. To limit the action, a is cropped as $a_{low} < a < a_{high}$. The concept of off-policy DRL is to fully utilize previous experience memory, these algorithms usually have a large experience replay buffer $\mathcal{B}$, which is used to store information such as state, action, reward, and next state for each step. When updating the actor network and critic network, the required parameter sequence $\Gamma = (s_i, a_i, r_i, s_{i+1})$ in the update equation will be obtained by sampling from the $\mathcal{B}$, and the objective function can be calculated using Equations (4) and (5).

$$y_i = r_i + \gamma \min_{k=1,2} Q_{\omega'_k}(s_{i+1}, a') \tag{4}$$

$$a' = \mu_{\theta'}(s_{i+1}) + \xi \tag{5}$$

where r_i is the reward, and a' represents the output of the network $\mu_{\theta'}(s_{i+1})$ with noise ξ and $\xi \sim \mathcal{N}(0,\tilde{\sigma})$ denoting a Gaussian distribution, which is clipped to $(-c, c)$ $c > 0$. Adding clipped noise to the actions by $\mu_{\theta'}(s_{i+1})$ is a regularization method that can be used to alleviate overfitting in the output of $Q_{\omega'_k}(s_{i+1}, a')$. In Equation (4), the values of two target networks $Q_{\omega'_k}(s_{i+1}, a')$ is minimized to reduce bias; otherwise, the update of the current critic network can be performed with Equation (6) to minimize the loss $\mathcal{L}_\omega$.

$$\mathcal{L}_\omega = \frac{1}{N}\sum_{i=0}^{N}[y_i - Q_{\omega_k}(s_i, a_i)]^2 \quad (6)$$

where N is the number of batches sampled from $\mathcal{B}$. In deterministic policy gradient RL [35], the policy parameters are updated by calculating the sampled policy gradient. Equation (7) can be used to update the current actor network.

$$\mathcal{L}_\theta = \frac{1}{N}\sum_{i=0}^{N}[\nabla_a Q_{\omega_1}(s_i, a)\Big|_{a=\mu_\theta(s_i)} \nabla_\theta \mu_\theta(s_i)] \quad (7)$$

As shown in Equation (3), although adding a certain proportion of noise in the exploration stage of TD3 can increase a certain exploration ability, the deterministic policy plays a dominant role in the policy learning exploration, which still limits the early exploration of the agent. In order to enhance the exploration ability of the agent, ϵ-greedy is a commonly used policy for balancing exploitation and exploration. The ϵ-greedy policy is shown in Equation (8), which shows that when the agent makes a decision, there is a small probability of positive ϵ will randomly selecting an unknown action, and a probability of $(1-\epsilon)$ will selecting the action with the largest action value among the existing actions.

$$\mu_\theta(s_i) = \begin{cases} \frac{\epsilon}{|A|} + 1 - \epsilon\,, if\ a = \text{argmax}Q(s_i, a_i) \\ \frac{\epsilon}{|A|}\,, \quad if\ a \neq \text{argmax}Q(s_i, a_i) \end{cases} \quad (8)$$

A standard RL algorithm must include exploration and exploitation. Exploration helps the agent fully understand state space and select the other unknown action, and exploitation helps the agent find the optimal action to maximize the expected return at the present moment. Inspired by these methods, TD3 is combined with ϵ-greedy policy. In order to enhance the exploration ability of agent in the early stage and make more stable use of the exploration in the later stage, the decay of ϵ-greedy policy is used and combined with TD3. The ϵ-greedy policy proposed based on cosine decay as shown in Equation (9).

$$\epsilon = \lambda(c + \cos(\frac{\pi t}{i})) \quad (9)$$

where $\lambda \in (0, 0.5)$, c is a constant greater than 1, i is training steps, and t is the current step value. After combining TD3 with the ϵ-greedy policy of cosine decay, the method of action selection shown in Equation (3) can be changed to the form shown in Equation (10).

$$a_i = \begin{cases} \mu_\theta(s_i) + \delta, & if\ \epsilon \leq \text{X} \\ \widetilde{\delta}, & if\ \epsilon > \text{X} \end{cases} \quad (10)$$

where X is the uniform distribution on [0,1); and $\widetilde{\delta} \sim (0, \sigma')$ is a random number conforming to a Gaussian distribution with a mean of 0 and a standard deviation of σ'; it is common to set σ' to a number associated with the maximum action. The flow of the cosine decay TD3 (CDTD3) algorithm is shown in Figure 1.

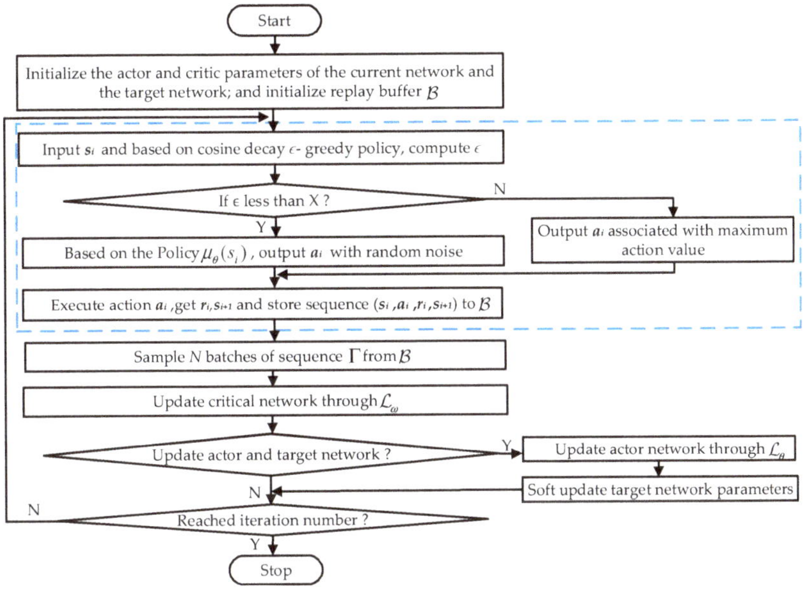

Figure 1. Flowchart of the CDTD3 algorithm.

The performance of CDTD3 was tested in MuJoCo through the OpenAI Gym interface. The original task set was used during testing without modifying the environment and reward function. Except for the individual characteristics of the algorithms, other settings remained the same. Each algorithm was run 10 times under different random seeds. The number of training steps in each task was two million steps. The results are shown in Figure 2, including the results of three different DRL algorithms running on three different robot tasks in MuJoCo. The dimensions of the robot's joints in Figure 2a–c from more to less, and the specified task difficulty ranges from hard to easy. The x-axis is the number of steps, while the y-axis is the average return of ten evaluations per five thousand steps for the current task under two million training steps. The shaded area in the Figure 2 represents the maximum and minimum value intervals with test data smoothed by convolution, while the solid line represents the average of ten experimental results.

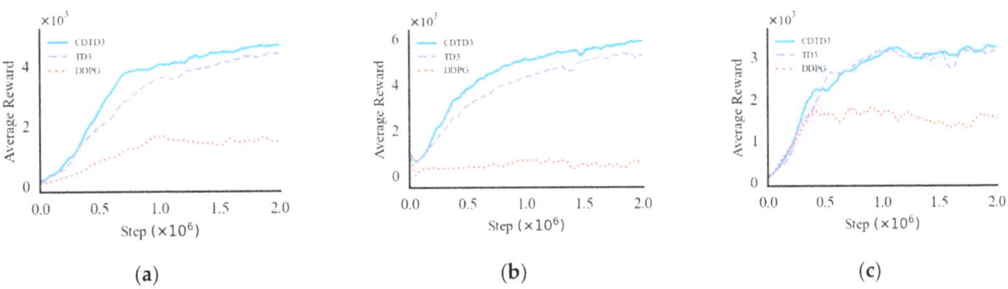

Figure 2. Simulation results of CDTD3 in MuJoCo: (**a**) Walker2d-v4, (**b**) Ant-v4, and (**c**) Hopper-v4.

In Figure 2c, the robust exploratory ability of CDTD3 does not significantly contribute to relatively simple tasks. However, in the scenarios illustrated in Figure 2a,b, CDTD3 exhibits a superior ability to explore actions with higher rewards and to rapidly incorporate them into decision-making process compared to the TD3. CDTD3 has strong exploration ability and adaptability to tasks, which makes policy learning more efficient.

3. The Improved Path Planning Method

RRT* has the characteristics of continuous iteration and path replanning in the environment. However, due to its random nature, each exploration or iteration of RRT* is independent of each other, so the new node positions generated by each iteration may be different, leading to a tortuous and suboptimal path. The reward function in DRL plays a pivotal role in shaping the learning effectiveness of the agent. It is imperative to define an appropriate reward function for each task to guide the agent toward successful task completion. The variation of the task can also easily lead to the failure of the reward function.

Currently, there are many improved variations of RRT*, such as informed RRT* [5], RRT*-Smart [6], and real-time RRT* [9], which have advantages in specific scenarios. However, they still face the problem of independent exploration and iteration. Additionally, introducing target biasing methods in RRT* can yield favorable results, and this approach is relatively easier to implement and debug compared to algorithms such as informed RRT*. Based on these issues, a path planning framework that combines the strengths of DRL and RRT* have been proposed, and is built on the basis of RRT* with target bias.

Performing a path search between two adjacent waypoints is called a dynamic point-to-point (D-P2P) task. The agent defines a spherical range centered at the intermediate waypoint as an explorable region and performs a D-P2P task between the preceding and following waypoint. The initial base path $R^0 = \{r_1{}^0, r_2{}^0, \ldots, r_N{}^0\}$ is obtained by RRT*, N is the total number of waypoints. The length of the original path can be characterized as $l_0^1 = \sum_{n=1}^{N-1} |r_n^1 r_{n+1}^1|$, where $|r_n^1 r_{n+1}^1|$ represents the Euclidean distance between waypoints r_n^1 and r_{n+1}^1.The steps of the DRL-RRT* algorithm is described as follows.

Step 1: Initialization. The number of iterations for RRT* is m, DRL path search task is K times. D-P2P task for each iteration is $J = N - 2$ groups. Using RRT* for planning iteration.

Step 2: Using DRL for path search tasks. When the agent first enters the jth $(1 \le j \le J)$ explorable region in a certain steps, the position c_j^k is retained as the end position of this D-P2P task, and the waypoint replacement task in Step 3 is performed. If the explorable region is not accessed in the specified number of steps, go to Step 4.

Step 3: The procedure of waypoint replacement. The first waypoint replacement is considered, resulting in point c_2^1 being obtained after completion of the D-P2P task; then, the impacts of r_2^1 and c_2^1 on the original path are computed individually. If c_2^1 makes the path shorter, c_2^1 replaces r_2^1, and the length becomes $l_0^1 = \left|r_1^1 c_2^1\right| + \left|c_2^1 r_3^1\right| + \sum_{n=3}^{N-1} \left|r_n^1 r_{n+1}^1\right|$. The waypoints of the path become $R^1 = \{\mathrm{r}_1^1, c_2^1, \ldots, r_N^1\}$; at the same time, c_2^1 also serves as the starting position for the next D-P2P task; otherwise, the original path is unchanged.

Step 4: Termination condition check for a round of DRL search task. If D-P2P task iteration is larger than the iterative number J, break to Step 2.

Step 5: Get the new path R^{k+1}. Assuming that only c_2^1 can shorten the path during the first iteration (k = 1), the path for the second iteration is $R^2 = \{\mathrm{r}_1^2, \mathrm{r}_2^2, \ldots, r_N^2\} = \{\mathrm{r}_1^1, c_2^1, \ldots, r_N^1\}$.

Step 6: Terminating condition check. If the iteration is larger than K, the path explored is retained as $P = R^K$, and the algorithm stops. Otherwise, restart from Step 2.

RRT* with DRL is utilized by Kontoudis [29]; however, the waypoints were unchanged during the process. Reference [30] employed TD3 and PRM as a path planning method, in such cases, the final path is highly influenced by the original path, limiting the effectiveness of the DRL in path planning. In contrast, in the proposed DRL-RRT*, DRL optimizes the underlying path planned by RRT* by automatically adjusting the intermediate nodes of the path. Consequently, potential optimizations can be explored, allowing for direct modification of the underlying path. Reference [21] combined improved DRL with RRT*, and optimized the path by adjusting the intermediate waypoints. In contrast, the proposed method calculates the path replacement as soon as the intermediate waypoints are explored, and the path replacement method is simpler. Base on the planning results of RRT*, the policy learning and experience replay mechanism of DRL are combined to obtain a better path. Figure 3 illustrates the flow of the proposed DRL-RRT* path optimization algorithm.

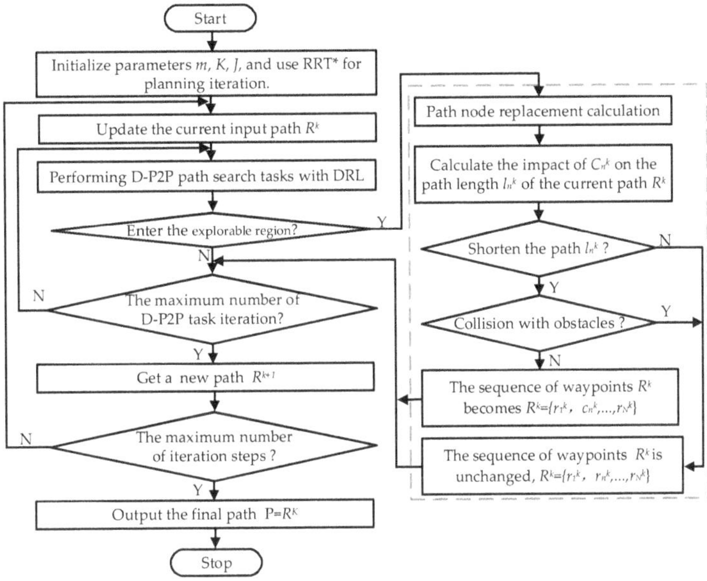

Figure 3. Flowchart of the DRL-RRT* path planning algorithm.

In each iteration of length optimization, the agent using DRL performs a D-P2P path search task. In each task, the reward function is designed as follows:

$$r = \begin{cases} \frac{\alpha}{e^{-\rho d}}, & other \\ 0, & if\ collsion \end{cases} \tag{11}$$

where d is the distance between the agent and the goal. To ensure an appropriate reward value in D-P2P tasks, constants α and ρ are introduced to limit its magnitude. It is crucial to keep $0 < \rho < 1$ to avoid potential issues such as vanishing or exploding gradients.

4. Simulation Analysis

When the DRL-RRT* is used for path optimization simulation in 2D and 3D obstacle environments, the success rate of each path iteration optimization is expressed as follows:

$$\eta = \frac{1}{K}\frac{1}{J}\sum_{k=1}^{K}\sum_{j=1}^{J}\zeta_j^k \times 100\% \tag{12}$$

where K represents the number of path search tasks performed by DRL, J represents the number of groups for D-P2P tasks, and $\zeta_j^k \in \{0,1\}$. If the agent can reach the endpoint of the current task in the maximum step size of the j D-P2P tasks in k times iteration, then $\zeta_j^k = 1$; otherwise, $\zeta_j^k = 0$.

4.1. Analysis of the 2D Simulation Environment

The size of the 2D complex obstacle environment was 100 (cm^2), and the simulation environment was configured with obstacles of diverse shapes is depicted in Figure 4, the lines represent the paths and the gray geometric objects represent obstacles. The starting position was (10, 90), and the target position was (98, 8). In order to fully utilize the exploration ability of RRT*, a target bias strategy is introduced in RRT*, which makes the sampling point equal to the target point with a certain probability p and randomly samples with a probability of $(1-p)$.

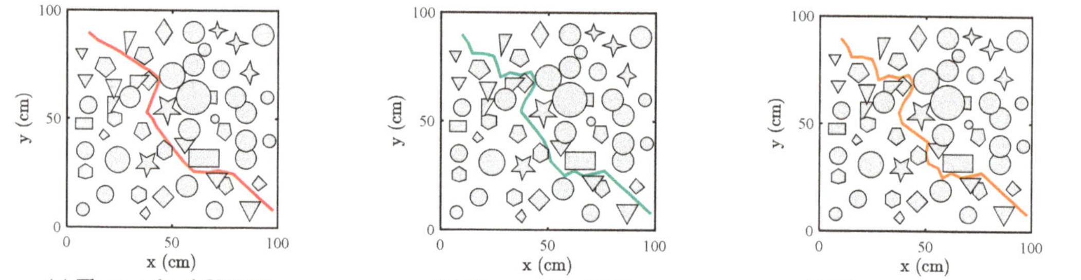

(**a**) The result of CDTD3 (**b**) The result of TD3 (**c**) The result of DDPG

Figure 4. Simulation environment for a complex obstacle space.

RRT* was first used for 10,000 iterations in the 2D simulation scenario. Then, CDTD3, TD3, and DDPG are used for 100 iterations of path optimization experiments on the DRL-RRT* algorithm, respectively. Iteration steps K in each experiment was 200, and the steps in each D-P2P task was 200. Among the 100 times experiments of each algorithm in each environment, different experiments were configured with different random seeds. The resulting path is shown in Figure 4, and the path optimized using the CDTD3 demonstrated improvements in terms of both path length and smoothness, closely resembling a straight line throughout most of the obstacle-free sections.

As DDPG and TD3 use a deterministic policy, they are prone to premature convergence to local optima with fixed actions during the continuous exploration and optimization process, making it difficult to explore better policy. They performed worse than did CDTD3 in terms of path length and success rate. The addition of a random exploration mechanism in CDTD3 enhances the exploration ability of the deterministic policy DRL and prevents premature convergence to local optima which includes fixed action selection. This feature can help the agent to be more inclined to explore the environment in the early stage of the task so as to explore a better policy. In certain obstacle-free spaces, CDTD3 explored and identified more optimal waypoints. By replacing intermediate waypoints through point substitution, the originally curved paths become straighter and the length of the path is shortened.

The results shown in Figure 5a represent the probability density. The x-axis represents the path length, while the y-axis represents the percentage of distribution. The curve represents the distribution of the path lengths from multiple experiments. From the probability density curve, it can be observed that, under the influence of CDTD3, the majority of results were concentrated in the region of the smaller path lengths. The distribution was relatively dense, and the final results exhibited less fluctuation, indicating a higher level of stability. The results of TD3 and DDPG were more dispersed and distributed in the region of larger path lengths. The relationship between the changes in path length obtained by the three algorithms is illustrated in Figure 5b. As the number of iteration steps increases, the path length undergoes continuous optimization and reduction., CDTD3 demonstrates superior ability to explore a more optimal path compared to TD3 and DDPG in most experimental cases, resulting in shorter path lengths and greater consistency with each optimization.

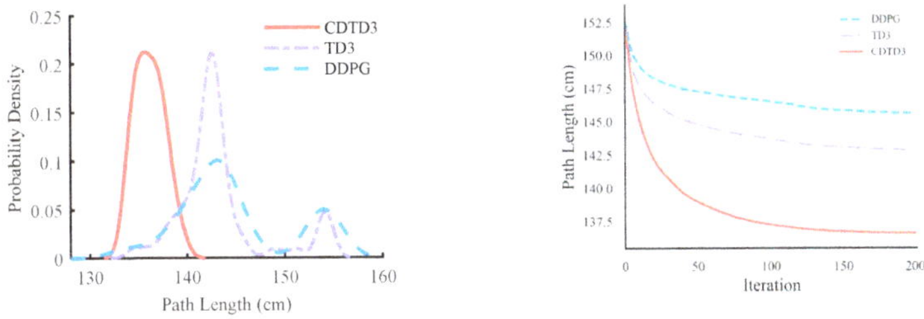

(**a**) (**b**)

Figure 5. Simulation results of the 2D complex obstacle space. (**a**) The probability density of path length. (**b**) Diagram of the length variation during the path optimization process.

The simulations for the maze space and the narrow space environment as shown in Figures 6 and 7 is designed, the lines represent the paths and the purple geometric objects represent obstacles. The starting point and obstacles were set differently. In the narrow space environment depicted in Figure 7, the distance between the two obstacles vertically was less than 10 cm. The obstacle avoidance path connecting the starting point and the target point needed to pass through all the narrow channels. The path explored by RRT* tended to be more winding; however, by using CDTD3 to optimize the path, continuously adjusting the intermediate waypoints, even in the case of a narrow space with a relatively singular path, the optimized path by CDTD3 still demonstrated advantages. It appeared more straight and shorter in the overall path. The probability density statistics of the path length in maze space and narrow space are illustrated in Figure 8, while Figure 9 depicts the changes in the path. Aligning with the outcomes observed in the complex obstacle space, CDTD3 outperformed in these two cases.

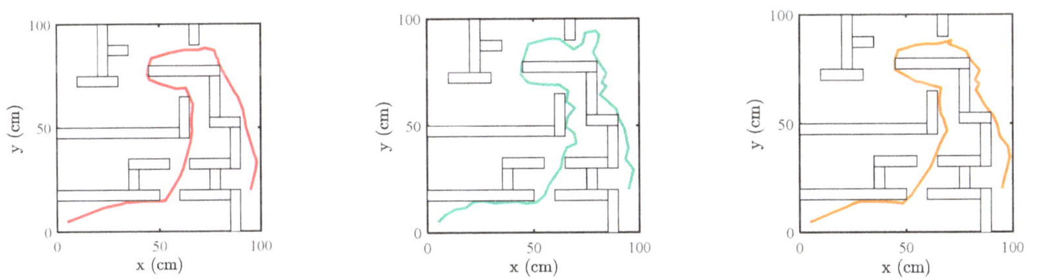

(**a**) The result of CDTD3 (**b**) The result of TD3 (**c**) The result of DDPG

Figure 6. Simulation environment for the maze space.

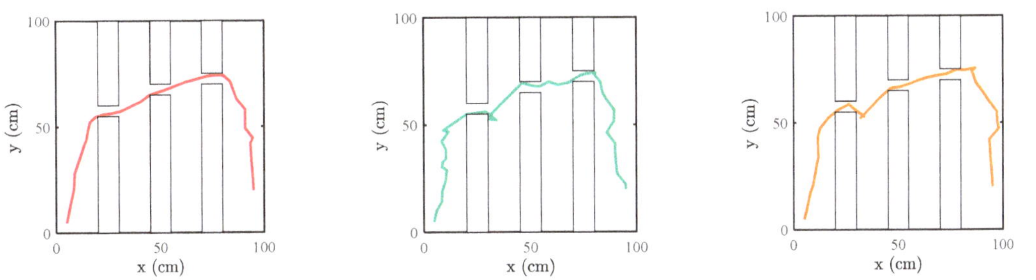

(**a**) The result of CDTD3 (**b**) The result of TD3 (**c**) The result of DDPG

Figure 7. Simulation environment for the narrow space.

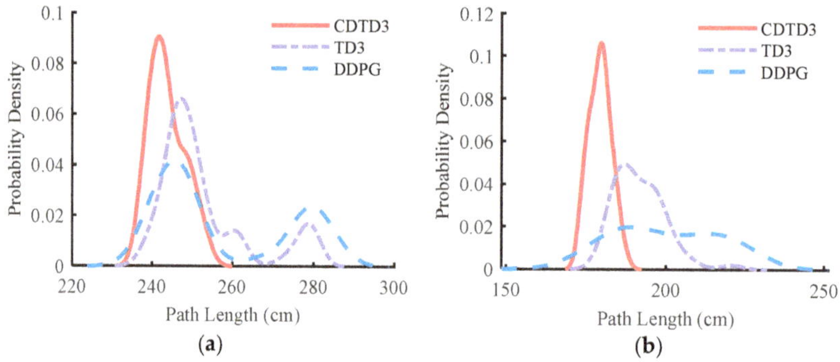

Figure 8. Probability density results of path length. (**a**) The results of the maze space. (**b**) The results of the narrow space.

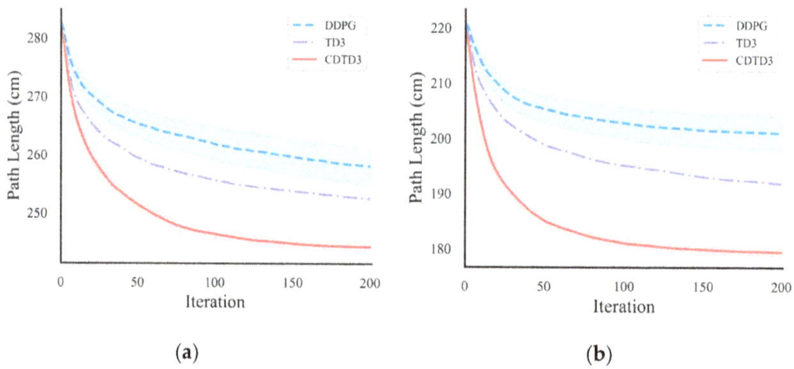

Figure 9. Diagram of the relationship between path length change and the algorithm. (**a**) The results of the maze space. (**b**) The results of the narrow space.

The simulation results in the three environments were analyzed. The mean and variance in Table 1 correspond to the outcomes of 100 experiments. The original path lengths were 154.1 cm, 284.7 cm, and 222.1 cm, respectively, while the path lengths optimized by CDTD3 were 136.6 cm, 244.4 cm, and 179.55 cm, respectively. The success rate of path iteration and the results of path reduction rate are presented in Table 2, where the reduction rate represents the percentage reduction to the RRT* path length. The success rates of CDTD3 performing point-to-point tasks in the three environments were 91.3%, 87.3%, and 95.6%, respectively, which were better than those of TD3 and DDPG, owing to the random exploration mechanism of CDTD3 algorithm.

Table 1. Mean and variance statistics of path length in the 2D simulation environment.

Algorithm	Index	Complex Obstacle	Maze Space	Narrow Space
CDTD3	Mean (cm)	136.60	244.32	179.55
	Variance	2.59	20.51	12.39
TD3	Mean (cm)	142.81	252.56	191.88
	Variance	23.79	127.29	75.65
DDPG	Mean (cm)	145.62	258.12	201.08
	Variance	38.44	278.97	260.86

Table 2. Optimization success rate and path reduction rate of the 2D simulation environment.

Algorithm	Index	Complex Obstacle	Maze Space	Narrow Space
CDTD3	Success rate	91.3%	87.3%	95.6%
	Reduction rate	11.4%	14.2%	19.14%
TD3	Success rate	88.2%	85.5%	98.7%
	Reduction rate	7.4%	10.2%	13.59%
DDPG	Success rate	76.2%	79.5%	86.8%
	Reduction rate	5.5%	73.2%	9.44%

CDTD3 is combined with RRT* and RRT, respectively; and conducted 100 simulations in complex obstacle and maze environments. The settings of RRT remained consistent with those of RRT*, except for the inherent characteristics of their respective. The simulation results are shown in Table 3. Using CDTD3-RRT* could achieve a superior path due to the advantageous features offered by RRT*. This characteristic led to an improved initial path, resulting in a shorter final path. In the maze environment, the reduction rate of CDTD3-RRT was better than that of CDTD3-RRT*, but the final path length was not as good as that of CDTD3-RRT* since the path of the original RRT was longer than that of RRT*. Nevertheless, favorable outcomes can still be achieved using the proposed method.

Table 3. Results of CDTD3-RRT* and CDTD3-RRT in the 2D simulation environment.

Algorithm	Index	Complex Obstacle	Maze Space
CDTD3-RRT*	Mean (cm)	136.6	244.32
	Variance	2.59	20.51
	Reduction rate	11.4%	14.2%
CDTD3-RRT	Mean (cm)	154.9	258.92
	Variance	8.49	51.56
	Reduction rate	10.8%	19.3%

In the complex obstacle space and maze space, CDTD3-RRT* was compared with two more advanced path planning methods, including the artificial potential field with informed RRT* (APF-IRRT*) [36], the adjustable probability and sampling area RRT algorithm (APS-RRT) [37]. Table 4 presents the performance comparison of CDTD3-RRT* under 100 experiments. It can be seen that CDTD3-RRT* has significant advantages under the path length. Owing to the powerful sampling and search ability of RRT* and the optimization ability of CDTD3, APS-RRT and APF-IRRT* use limited exploration range operations in the two complex environments, which is not conducive to obtaining better paths.

Table 4. Comparison of the path planning in the 2D simulation environment.

Algorithm	Index	Complex Obstacle	Maze Space
CDTD3-RRT*	Mean (cm)	136.6	244.32
	Variance	2.59	20.51
APF-IRRT*	Mean (cm)	171.13	312.15
	Variance	200.33	316.21
APS-RRT	Mean (cm)	167.32	315.18
	Variance	191.35	322.69

4.2. Analysis of the 3D Simulation Environment

In order to verify the performance of the CDTD3-RRT* path planning method in 3D complex obstacle environments, a 3D space simulation environment was designed, with size of 100 (cm^3). The environment is shown in Figure 10a, Figure 10b is locally enlarged, and multiple obstacles of different sizes and shapes were set up in the environment.

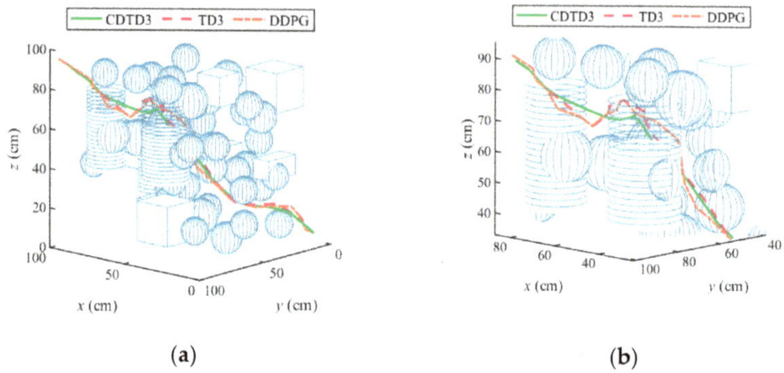

Figure 10. 3D complex obstacle environment: (**a**) complete 3D environment and (**b**) locally enlarged view of the 3D environment.

RRT* was first used for 5000 iterations with the starting position at (6, 6, 6) and the target position at (98, 96, 95). DRL were used to carry out 50 simulation experiments on the DRL-RRT*, each experiment was conducted under different random seeds, and each experiment had 400 iterations. The length of each point-to-point task was 400 steps.

The simulation results are shown in Table 5. The average path lengths for 50 repeated experiments were 186.61 cm, 191.23 cm, and 197.29 cm, respectively. Figure 11 shows the relationship between path length and algorithms. The random policy was added to the CDTD3 to further optimize the performance of the algorithm. In the 3D environment, CDTD3 consistently discovered shorter paths across multiple experiments, leading to reduced path length and higher success rates for D-P2P tasks.

Table 5. Results of the 3D complex obstacle space simulation.

Algorithm	Mean(cm)	Variance	Success Rate	Reduction Rate
CDTD3	186.61	20.09	94.35%	10.6%
TD3	191.23	20.21	91.78%	8.2%
DDPG	197.29	79.93	62.43%	4.4%

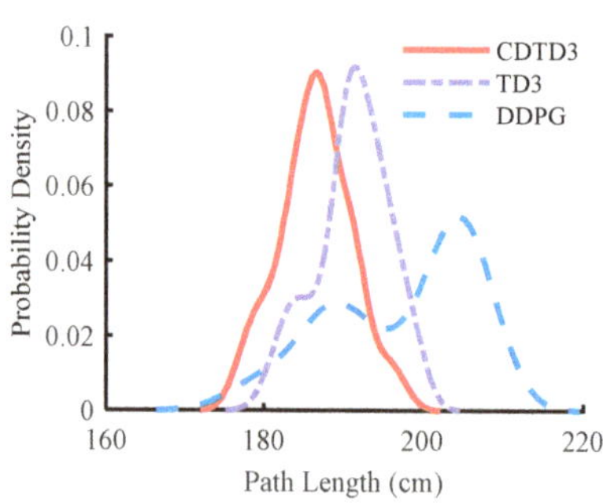

Figure 11. Probability density of the 3D complex obstacle space simulation.

5. Simulation and Experiment of Manipulator Path Planning

5.1. Evaluation Index of Path Planning

The accuracy and stability of path planning algorithms require certain indicators for evaluation. Typically, the root mean square error (RMSE) analysis method is employed to quantify the deviation between the desired value x_i and actual value y_i of the manipulator trajectories. The RMSE is expressed as follows:

$$RMSE = \sqrt{\frac{1}{n}\sum_{i=1}^{n}(x_i - y_i)^2} \tag{13}$$

The success rate of position deviation is used to measure the discrepancy between the actual target position and the set target position. The distance d_i between the actual target (x_i, y_i, z_i) and the set target $(\overline{x}, \overline{y}, \overline{z})$ is calculated as follows:

$$d_i = \sqrt{(x_i - \overline{x})^2 + (y_i - \overline{y})^2 + (z_i - \overline{z})^2} \tag{14}$$

The success rate of the position deviation in the experiment can be expressed by:

$$G = \frac{1}{N}\sum_{i=1}^{n} g_i \tag{15}$$

where N is the number of experiments; $g_i \in \{0,1\}$; if $d_i < \varphi$, then $g_i = 1$, otherwise $g_i = 0$. φ is a threshold that can be used to measure the relationship between d_i and G.

5.2. Experiment and Simulation Research on the Application of Manipulator Path Planning

A manipulator experimental platform was established as shown in Figure 12 to verify the feasibility of the proposed method in the practical manipulator. The hardware of the experimental platform included the 6-DOF manipulator Han's Robot Elfin E05, the supporting tools and the computer were equipped with the Ubuntu20 operating system. ROS Noetic MoveIt 1 is an open-source robotic motion planning framework for robot motion planning and control. It provides collision checking and control functionalities, integrating collision detection libraries such as the Flexible Collision Library (FCL), which is used to detect collisions between the robot and the environment or other objects.

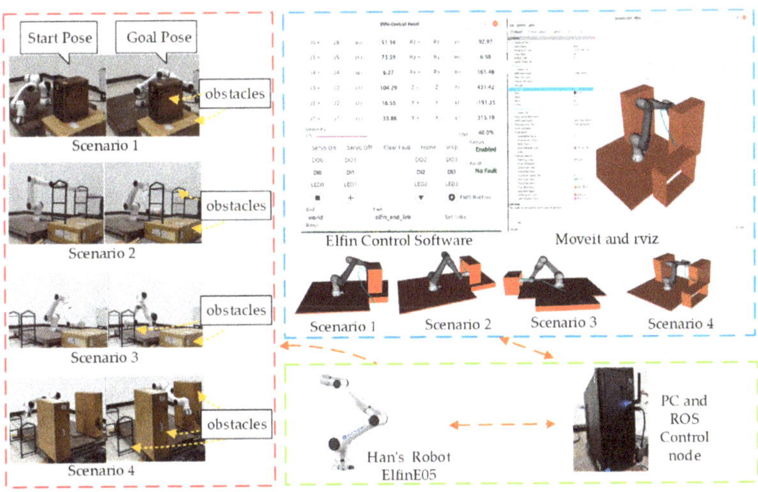

Figure 12. The experimental platform for the manipulator.

Manipulator simulations were conducted to test the path planning. The use of the simulation environment allowed us to perform multiple experiments under different scenarios, conditions, and parameter settings to ensure the rationality and feasibility of the planned paths, reducing the wear and tear on the manipulator in real-world experiments.

The simulation environment shown in Figure 12 was established for the obstacle avoidance path planning task of the manipulator using Rviz and MoveIt1. The simulation environment consisted of four different obstacle environments and various starting positions and target positions, all of which closely resembled real-world scenarios.

In the CDTD3 algorithm, $\lambda = 0.1$, $c = 1$, and $i = 100$. A total of 100 simulation experiments were conducted on the manipulator. The average length of the executed path in the joint space and the success rate of planning execution results are presented in Table 6, the threshold φ in the experiment was 5 (cm). According to the results in Table 6, the paths generated by CDTD3-RRT* had a good performance on different tasks, with shorter paths after planning and execution. In Scenario 3, where the task was relatively simple, the results of CDTD3 and TD3 were similar. Moreover, under four different tasks, three algorithms had a higher success rate in path planning. In the four scenarios, the paths obtained through the CDTD3-RRT* had a better length compared to those obtained through RRT*. The path reduction rates were 11.1%, 31%, 9.0%, and 20.4%, respectively.

Table 6. The results of manipulator simulation.

Algorithm	Index	Scenario 1	Scenario 2	Scenario 3	Scenario 4
CDTD3-RRT*	Mean (cm)	146.3	56.1	108.7	90.7
	Success rate	98%	86%	92%	100%
TD3-RRT*	Mean (cm)	149.7	58.8	108.5	94.1
	Success rate	99%	85%	92%	100%
DDPG-RRT*	Mean (cm)	151.0	58.9	109.4	91.1
	Success rate	98%	87%	93%	100%

The manipulator experimental platform is shown in Figure 12. The process of obtaining the executable path of the manipulator is shown in Figure 13. When conducting path planning experiments on manipulator, given the starting pose, target pose and obstacle information. CDTD3-RRT* is first used to compute a collision-free path based on the environment. Then, the inverse kinematics solution and obstacle avoidance detection were performed using the motion planning method computed cartesian path in MoveIt1, which integrates IKFast and FCL plugin. Thus, the executable path of the joint space was calculated, and information such as joint variables required for the operation of the manipulator was output. After obtaining the executable path information, it was sent to Elfin E05 for execution. CDTD3-RRT* was tested multiple times to verify the performance in different targets. During the operation of the manipulator, a ROS node was utilized to collect real-time joint parameters, which were recorded in the file for further analysis.

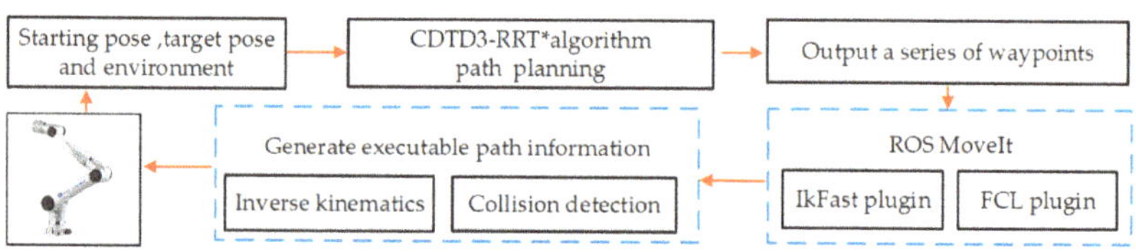

Figure 13. The process of generating the executable path information.

The comparison between the manipulator's end path and the planned trajectory is illustrated in Figure 14. To accurately analyze the error, the trajectory was decomposed into three axes: the x-axis, y-axis, and z-axis. Although there were errors in each axis direction of the manipulator's end path, the error was in a small range, which was consistent with the deviation between the actual experiment and the planned path.

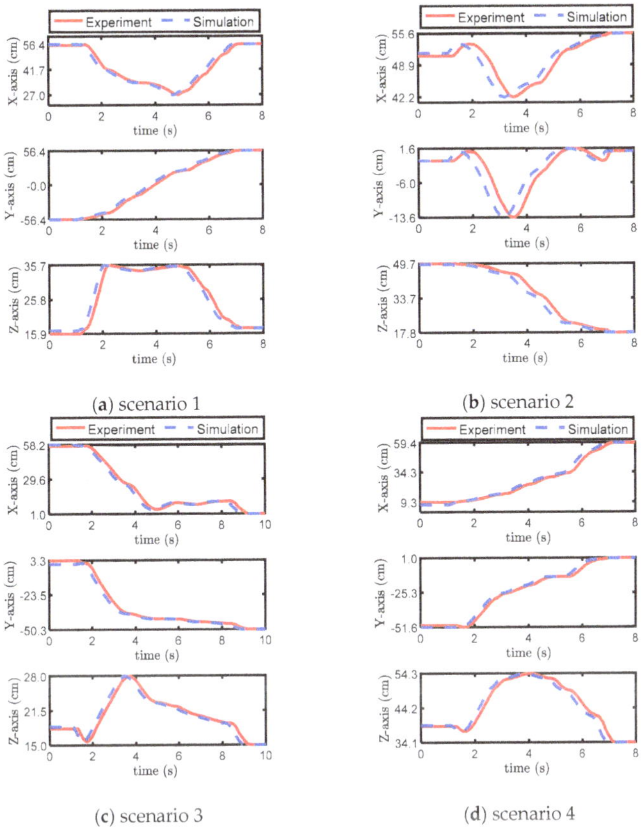

Figure 14. The end trajectory of the manipulator.

According to Equation (13), the RMSE between the end effector trajectories of the four targets and the planned trajectory is calculated, including the results in three axes directions. The results shown in Table 7 are the average of 20 planning experiments. The results show that the proposed algorithm can complete the experimental task of manipulator path planning with high accuracy, and the error between the actual trajectory and the planned trajectory is acceptable. The actual trajectory of the manipulator during operation may have deviations from the planned results due to the inherent instability of the manipulator and the influence of prior information such as environment modeling and motion models in path planning. However, the small magnitude of the errors demonstrates that the algorithm is capable of effectively solving the path planning problem for the manipulator and achieves good effects in various tasks under different scenarios.

Table 7. The RMSE for each direction at the manipulator's end position.

RMSE	Scenario 1	Scenario 2	Scenario 3	Scenario 4
x-axis (cm)	2.1	3.2	2.6	1.4
y-axis (cm)	1.8	2.7	2.8	1.7
z-axis (cm)	3.1	1.3	1.7	3.1

6. Conclusions

Based on the RRT*, this paper introduces the DRL to carry out path planning for the manipulator and seek an optimal path. To enhance the exploration ability of the TD3, an improved method called CDTD3 is proposed. Through simulation verification, this method can effectively improve the insufficient exploration in the early stage of the TD3. Moreover, a path planning method DRL-RRT* was designed that combines the random sampling mechanism of the RRT* and the experience replay mechanism of DRL.

Path planning simulations were designed to validate the optimization ability of the proposed CDTD3-RRT* on the original path. The simulation results demonstrated that in three 2D environments, the original RRT* path achieved a reduction rate of 11.4%, 14.2%, and 19.14%, respectively. The reduction rate in the 3D complex obstacle environment was 10.6%. In addition, the CDTD3 demonstrated a significant improvement in the success rate of iterative optimization and reduction rate compared with the TD3 and DDPG. Finally, an experimental platform for manipulator was established, and the application of path planning methods in obstacle avoidance path planning tasks was analyzed. The results demonstrate that the path length of CDTD3-RRT* was better than that of TD3-RRT*, and DDPG-RRT* in multiple experiments. In the four experimental scenarios, the paths obtained through the CDTD3-RRT* path planning method were more optimal in terms of length compared to the paths obtained through RRT*. The reduction rates of the paths in the four scenarios were 11.1%, 31%, 9.0%, and 20.4%, respectively, and the end path error of the manipulator conformed to the results of planning and actual execution.

Author Contributions: Conceptualization, X.L. and R.C.; methodology, R.C.; software, R.C.; validation R.C.; formal analysis, X.L. and R.C.; investigation, R.C.; resources, X.L.; data curation, R.C.; writing—original draft preparation, R.C.; writing—review and editing, R.C. and X.L.; supervision, X.L. All authors have read and agreed to the published version of the manuscript.

Funding: This work is supported by Innovation Project of Guilin University of Electronic Technology (GUET) Graduate Education (Grant No. 2022YCXS152); Key Laboratory of Automatic Testing Technology and Instruments Foundation of Guangxi (Grant No. YQ19107).

Institutional Review Board Statement: Not applicable.

Informed Consent Statement: Not applicable.

Data Availability Statement: The study data supporting findings are available within this article.

Conflicts of Interest: The authors declare no conflicts of interest.

References

1. Lao, C.; Li, P.; Feng, Y. Path Planning of Greenhouse Robot Based on Fusion of Improved A* Algorithm and Dynamic Window Approach. *Nongye Jixie Xuebao/Trans. Chin. Soc. Agric. Mach.* **2021**, *52*, 14–22.
2. Kavraki, L.E.; Svestka, P.; Latombe, J.C.; Overmars, M.H. Probabilistic roadmaps for path planning in high-dimensional configuration spaces. *IEEE Trans. Robot. Autom.* **1996**, *12*, 566–580. [CrossRef]
3. Qi, J.; Yang, H.; Sun, H. MOD-RRT*: A Sampling-Based Algorithm for Robot Path Planning in Dynamic Environment. *IEEE Trans. Ind. Electron.* **2021**, *68*, 7244–7251. [CrossRef]
4. Viseras, A.; Shutin, D.; Merino, L. Online information gathering using sampling-based planners and GPs: An information theoretic approach. In Proceedings of the 2017 IEEE/RSJ International Conference on Intelligent Robots and Systems (IROS), Vancouver, BC, Canada, 24–28 September 2017; pp. 123–130.
5. Gammell, J.D.; Srinivasa, S.S.; Barfoot, T.D. Informed RRT*: Optimal Sampling-based Path Planning Focused via Direct Sampling of an Admissible Ellipsoidal Heuristic. In Proceedings of the 2014 IEEE/RSJ International Conference on Intelligent Robots and Systems (IROS), Chicago, IL, USA, 14–18 September 2014; pp. 2997–3004.
6. Islam, F.; Nasir, J.; Malik, U.; Ayaz, Y.; Hasan, O. RRT*-Smart: Rapid convergence implementation of RRT* towards optimal solution. In Proceedings of the 2012 IEEE International Conference on Mechatronics and Automation (ICMA), Chengdu, China, 5–8 August 2012; pp. 1651–1656.
7. Lv, H.; Zeng, D.; Li, X. Based on GMM-RRT* Algorithm for Path Planning Picking Kiwifruit Manipulator. In Proceedings of the 2023 42nd Chinese Control Conference (CCC), Tianjin, China, 24–26 July 2023; pp. 4255–4260.
8. Xinyu, W.; Xiaojuan, L.; Yong, G.; Jiadong, S.; Rui, W. Bidirectional Potential Guided RRT* for Motion Planning. *IEEE Access* **2019**, *7*, 95046–95057. [CrossRef]
9. Naderi, K.; Rajamäki, J.; Hämäläinen, P. RT-RRT*: A real-time path planning algorithm based on RRT*. In Proceedings of the Proceedings of the 8th ACM SIGGRAPH Conference on Motion in Games, New York, NY, USA, 16 November 2015; pp. 113–118.
10. Mnih, V.; Kavukcuoglu, K.; Silver, D.; Rusu, A.A.; Veness, J.; Bellemare, M.G.; Graves, A.; Riedmiller, M.; Fidjeland, A.K.; Ostrovski, G.; et al. Human-level control through deep reinforcement learning. *Nature* **2015**, *518*, 529–533. [CrossRef] [PubMed]

11. Dabbaghjamanesh, M.; Moeini, A.; Kavousi-Fard, A. Reinforcement Learning-Based Load Forecasting of Electric Vehicle Charging Station Using Q-Learning Technique. *IEEE Trans. Ind. Inform.* **2021**, *17*, 4229–4237. [CrossRef]
12. Hao, Y.; Chen, M.; Gharavi, H.; Zhang, Y.; Hwang, K. Deep Reinforcement Learning for Edge Service Placement in Softwarized Industrial Cyber-Physical System. *IEEE Trans. Ind. Inform.* **2021**, *17*, 5552–5561. [CrossRef] [PubMed]
13. Shi, H.; Shi, L.; Xu, M.; Hwang, K.S. End-to-End Navigation Strategy with Deep Reinforcement Learning for Mobile Robots. *IEEE Trans. Ind. Inform.* **2020**, *16*, 2393–2402. [CrossRef]
14. Bae, H.; Kim, G.; Kim, J.; Qian, D.; Lee, S. Multi-Robot Path Planning Method Using Reinforcement Learning. *Appl. Sci.* **2019**, *9*, 3057. [CrossRef]
15. Lv, L.; Zhang, S.; Ding, D.; Wang, Y. Path Planning via an Improved DQN-Based Learning Policy. *IEEE Access* **2019**, *7*, 67319–67330. [CrossRef]
16. Lillicrap, T.P.; Hunt, J.J.; Pritzel, A.; Heess, N.; Erez, T.; Tassa, Y.; Silver, D.; Wierstra, D. Continuous control with deep reinforcement learning. *arXiv* **2015**, arXiv:1509.02971.
17. Andrychowicz, M.; Wolski, F.; Ray, A.; Schneider, J.; Fong, R.; Welinder, P.; McGrew, B.; Tobin, J.; Abbeel, P.; Zaremba, W. Hindsight Experience Replay. *arXiv* **2017**, arXiv:1707.01495.
18. Gu, S.; Holly, E.; Lillicrap, T.; Levine, S. Deep reinforcement learning for robotic manipulation with asynchronous off-policy updates. In Proceedings of the 2017 IEEE International Conference on Robotics and Automation (ICRA), Singapore, 29 May–3 June 2017; pp. 3389–3396.
19. Lin, G.; Zhu, L.; Li, J.; Zou, X.; Tang, Y. Collision-free path planning for a guava-harvesting robot based on recurrent deep reinforcement learning. *Comput. Electron. Agric.* **2021**, *188*, 106350. [CrossRef]
20. Yang, Y.; Ni, Z.; Gao, M.; Zhang, J.; Tao, D. Collaborative Pushing and Grasping of Tightly Stacked Objects via Deep Reinforcement Learning. *IEEE/CAA J. Autom. Sin.* **2022**, *9*, 135–145. [CrossRef]
21. Li, X.; Liu, H.; Dong, M. A General Framework of Motion Planning for Redundant Robot Manipulator Based on Deep Reinforcement Learning. *IEEE Trans. Ind. Inform.* **2022**, *18*, 5253–5263. [CrossRef]
22. Kim, M.; Han, D.-K.; Park, J.-H.; Kim, J.-S. Motion Planning of Robot Manipulators for a Smoother Path Using a Twin Delayed Deep Deterministic Policy Gradient with Hindsight Experience Replay. *Appl. Sci.* **2020**, *10*, 575. [CrossRef]
23. Fujimoto, S.; van Hoof, H.; Meger, D. Addressing Function Approximation Error in Actor-Critic Methods. *arXiv* **2018**, arXiv:1802.09477.
24. Pan, L.; Cai, Q.; Huang, L. Softmax Deep Double Deterministic Policy Gradients. *arXiv* **2020**, arXiv:2010.09177.
25. Maoudj, A.; Hentout, A. Optimal path planning approach based on Q-learning algorithm for mobile robots. *Appl. Soft Comput.* **2020**, *97*, 106796. [CrossRef]
26. Chiang, H.T.L.; Faust, A.; Fiser, M.; Francis, A. Learning Navigation Behaviors End-to-End with AutoRL. *IEEE Robot. Autom. Lett.* **2019**, *4*, 2007–2014. [CrossRef]
27. Li, H.; Zhang, Q.; Zhao, D. Deep Reinforcement Learning-Based Automatic Exploration for Navigation in Unknown Environment. *IEEE Trans. Neural Netw. Learn. Syst.* **2020**, *31*, 2064–2076. [CrossRef]
28. Francis, A.; Faust, A.; Chiang, H.T.L.; Hsu, J.; Kew, J.C.; Fiser, M.; Lee, T.W.E. Long-Range Indoor Navigation with PRM-RL. *IEEE Trans. Robot.* **2020**, *36*, 1115–1134. [CrossRef]
29. Kontoudis, G.P.; Vamvoudakis, K.G. Kinodynamic Motion Planning with Continuous-Time Q-Learning: An Online, Model-Free, and Safe Navigation Framework. *IEEE Trans. Neural Netw. Learn. Syst.* **2019**, *30*, 3803–3817. [CrossRef]
30. Gao, J.; Ye, W.; Guo, J.; Li, Z. Deep Reinforcement Learning for Indoor Mobile Robot Path Planning. *Sensors* **2020**, *20*, 5493. [CrossRef]
31. Florensa, C.; Held, D.; Wulfmeier, M.; Zhang, M.; Abbeel, P. Reverse Curriculum Generation for Reinforcement Learning. *arXiv* **2017**, arXiv:1707.05300.
32. Chiang, H.T.L.; Hsu, J.; Fiser, M.; Tapia, L.; Faust, A. RL-RRT: Kinodynamic Motion Planning via Learning Reachability Estimators from RL Policies. *IEEE Robot. Autom. Lett.* **2019**, *4*, 4298–4305. [CrossRef]
33. Uther, W. Markov Decision Processes. In *Encyclopedia of Machine Learning and Data Mining*; Sammut, C., Webb, G.I., Eds.; Springer: Boston, MA, USA, 2017; pp. 793–798.
34. Haarnoja, T.; Zhou, A.; Abbeel, P.; Levine, S. Soft Actor-Critic: Off-Policy Maximum Entropy Deep Reinforcement Learning with a Stochastic Actor. *arXiv* **2018**, arXiv:1801.01290.
35. Silver, D.; Lever, G.; Heess, N.; Degris, T.; Wierstra, D.; Riedmiller, M. Deterministic Policy Gradient Algorithms. In Proceedings of the 31st International Conference on Machine Learning (ICML), Beijing, China, 21–26 June 2014; pp. 387–395.
36. Wu, D.; Wei, L.; Wang, G.; Tian, L.; Dai, G. APF-IRRT*: An Improved Informed Rapidly-Exploring Random Trees-Star Algorithm by Introducing Artificial Potential Field Method for Mobile Robot Path Planning. *Appl. Sci.* **2022**, *12*, 10905. [CrossRef]
37. Li, X.; Tong, Y. Path Planning of a Mobile Robot Based on the Improved RRT Algorithm. *Appl. Sci.* **2024**, *14*, 25. [CrossRef]

applied sciences

MDPI

Article

Characterization of a Rectangular-Cut Kirigami Pattern for Soft Material Tuning

Benigno Muñoz-Barron [1,2], X. Yamile Sandoval-Castro [3], Eduardo Castillo-Castaneda [2] and Med Amine Laribi [4,*]

1 Department of Mechatronics, Tecnológico Nacional de México, Campus Huichapan, Domicilio Conocido Sin número, El Saucillo, Huichapan 42411, Hidalgo, Mexico; bmunoz@iteshu.edu.mx
2 Centro de Investigación en Ciencia Aplicada y Tecnología Avanzada Unidad Querétaro, Mecatrónica, Instituto Politécnico Nacional, Querétaro 76090, Querétaro, Mexico; ecastilloca@ipn.mx
3 School of Engineering and Sciences, Tecnologico de Monterrey, Monterrey 64849, Nuevo León, Mexico; yamile.sandoval@tec.mx
4 Department of GMSC, Pprime Institute CNRS, École Nationale Supérieure de Mécanique et d'aérotechnique, University of Poitiers, 86000 Poitiers, France
* Correspondence: med.amine.laribi@univ-poitiers.fr

Abstract: Kirigami is the art of cutting paper to create three-dimensional figures for primarily aesthetic purposes. However, it can also modify the mechanical behavior of the resulting structure. In the literature, kirigami has been applied to modify the material's structural behavior, such as by changing its elasticity, rigidity, volume, or any other characteristic. This article examines the behavior of a pattern of rectangular kirigami cuts on a thermoplastic polyurethane soft material structure and its influence on the mechanical parameters of the macrostructure. The results demonstrate that rectangular kirigami patterns significantly affect the stiffness of the test specimens, changing from 1635 N/m to 4020 N/m. In elongation, there is a variation from 176.6% to 218% by simply altering the height of the rectangular cut. This enables the adjustment of the soft material structure's stiffness based on the geometry of the propagating kirigami cuts.

Keywords: kirigami; soft; stiffness

check for updates

Citation: Muñoz-Barron, B.; Sandoval-Castro, X.Y.; Castillo-Castaneda, E.; Laribi, M.A. Characterization of a Rectangular-Cut Kirigami Pattern for Soft Material Tuning. *Appl. Sci.* **2024**, *14*, 3223. https://doi.org/10.3390/app14083223

Academic Editors: Qi Song and Qinglei Zhao

Received: 2 March 2024
Revised: 31 March 2024
Accepted: 8 April 2024
Published: 11 April 2024

1. Introduction

Kirigami and origami have been utilized in engineering for various purposes in the design and production of sensors and actuators. The reported cutting patterns of kirigami can be divided into those with an even distribution on the base material and those with a non-uniform distribution. Similarly, the cuts can be classified into simple and complex cuts. Simple cuts consist of linear cuts distributed along the base material, sometimes forming a combination of straight lines, such as trapezoidal cuts and triangular cuts. Complex cuts are defined as cuts forming curves of greater complexity than a simple straight line, such as circular cuts, spiral cuts, and other shapes of greater complexity. The effects of simple cuts on different base materials and geometries have been analyzed in [1–8] with the aim of creating actuators and sensors that exploit material property tuning and deformations. Refs. [9–13] investigate various cutting patterns using more complex geometries arranged in a regular pattern on the base material, with cuts ranging from micrometric dimensions to a few centimeters.

The applications of kirigami are diverse. For instance, in the field of soft robotics, a triangular-cut kirigami pattern was used as a snakeskin on a pneumatic actuator in [14], increasing the actuator's drag capacity through a mechanism inspired by snake scales. In [15] the author proposed a kirigami–like soft elastomeric skin used to cover a snake-arm robot conformed by compliant vertebrae and controlled with cables. Other patterns were also proposed in [16] that, when cut into a flat material and placed around a pneumatic actuator, increase its drag capacity. Flexible robots inspired by kirigami utilizing this type of dragging actuation have been developed for medical applications. Ref. [17] demonstrated the development of a robot in which kirigami patterns facilitate the navigation of robots through cavities of the human body under complex conditions and restrictions using hexagonal kirigami patterns. Meanwhile, ref. [18] showcases the development of a flexible robot for medication dosing within the human body, proposing the combined use of two kirigami patterns, one for navigation and another for medication dosing. Other actuators that use kirigami as a key design element can be found in [19–22].

In the realm of sensing applications, kirigami has been widely applied for the development of sensors. The cuts allow for the deformation of the material and its ability to adapt to complex surfaces, enabling the indirect measurement of stresses or deformations produced through the modification of the material's geometry under external forces. Some of the sensors reported in the literature include a heart rate sensor [23], an ECG signal probe [24], angular deformation sensors for robotic actuators [25], and a biocompatible strain sensor [26].

The pattern of simple linear cuts has been widely studied and applied to various base materials [7–32], even in different cut configurations, with most authors approximating the cut thickness to be equal to zero for simplicity of analysis. The applications of this pattern have been varied, including the development of actuators and sensors [30].

The pattern of rectangular cuts has attracted the attention of researchers in various fields, and its study has been conducted from different perspectives and with different configurations. Ref. [33] presents an experimental study of the rectangular cut configuration manufactured on TPU. In their study, the experiment was conducted on a single column of material upon which a rectangular cut was made followed by bonding material with the next rectangular cut. The experimentation considers five cells of rectangular cuts and obtains the parameters of rupture tension, elongation, and stiffness to characterize the behavior of the cutting pattern. Meanwhile, ref. [34] conducted a study using a pattern inspired by rectangular kirigami cuts distributed on TPU material plates manufactured with thicknesses ranging from 2 to 10 mm. In this study, the walls of the test plate containing the rectangular cuts were fully enclosed, which restricts material deformation, although subsequently, the study was conducted on a cylindrical scheme onto which the rectangular cutting pattern was projected. In this study, material stiffness was estimated against wall thickness, and unit deformation was estimated against the reaction force presented by the plate with cuts. Ref. [35], on the other hand, presented an inflatable actuator, which studies various cutting patterns, including the pattern inspired by rectangular cuts. In their research, they created actuators that, in their uninflated form, are configured with the pattern of rectangular cuts. Subsequently, by joining the edges of two plates with rectangular cuts and the central part forming conduction channels, the actuator was subjected to tensile tests and then inflated to characterize its response, mainly its contraction response, thereby achieving behavior similar to that of muscles. The pattern of rectangular cuts has also been studied for sensor development, primarily in its single-column configuration, as shown in the work developed by [36]. In their study, they used a rectangular cut–bonding material–rectangular cut configuration similar to the configuration used by [33], all distributed on a polyamide base as a substrate with synthesized graphene on a copper sheet via low-pressure CVD. In this study, the aim was to characterize the material's resistance behavior as it deforms since the objective was to use the plate as a stress–strain sensor for use on human skin, and no other mechanical tests were conducted in greater depth.

In this study, we propose an investigation into a pattern inspired by rectangular kirigami cuts to characterize properties such as elasticity and stiffness. This investigation involves varying the width of rectangular cuts applied to a 1 mm thick TPU plate. The configuration of the pattern differs from those previously reported, featuring a greater number of rectangular cutting cells distributed horizontally, with free edges following the rectangular cutting pattern. This approach aims to facilitate the development of mechanisms capable of rapidly adapting to specific needs in applications such as robotics, sensors, and actuators based on thermoplastic polyurethane.

2. Materials and Methods

2.1. Rectangular Pattern Cuts

The proposed pattern of rectangular kirigami cuts is shown in Figure 1. This pattern is defined by the geometry of the rectangular cuts and their distribution on the supporting material. In this work, it is proposed that the cuts have a uniform distribution throughout the entire supporting material, including the edges where the cuts are interrupted by the edge of the base material. The description of the rectangular cuts includes the width (w) and the height (h), with all rectangles having the same dimensions. The distribution of the cuts on the base material is defined by the distance (s) between cuts on the same line and the distance (h1) between lines of cuts. The length (z) of the rectangular cut that coincides with a section of the upper or lower cut is also described.

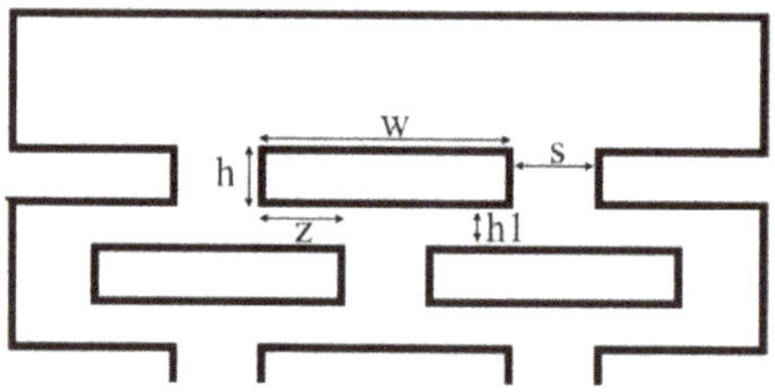

Figure 1. Geometrical description of kirigami rectangular cuts.

The behavior of the kirigami cell can be analyzed by discretizing it and examining the behavior of one of the cuts, which can then be used to propagate the deformation phenomenon throughout the entire base material. To model the deformation that occurs along the kirigami base material, a section surrounded by a dotted line, as shown in Figure 2, is proposed to be considered as a representative section of any section along the cell.

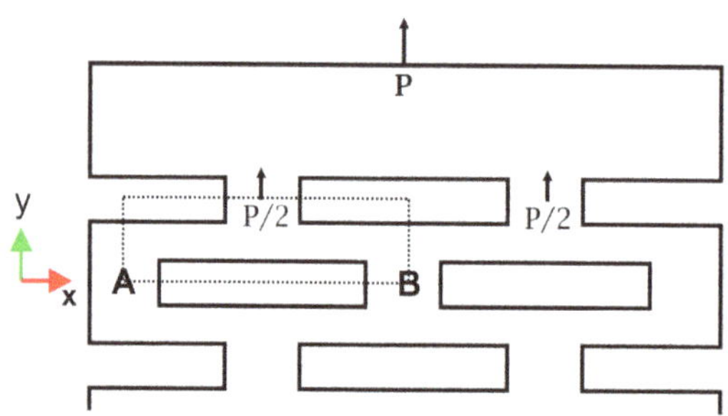

Figure 2. Examined segment of the kirigami pattern delineated by a dashed border.

If it is assumed that when the application of force P begins, supports A and B are fixed, as shown on Figure 3; then, it can be considered as if the element forming the section is a small beam with fixed supports. This type of beam allows the ends to be fixed but can undergo rotation, which is ignored in this analysis.

At the begin of the deformation, this structural element is subjected to a force P applied at the center of the beam. By considering the geometry of the beam, in this case, the section described is analyzed from support point A to the center of the beam with a length of L/2. The material is considered to have a thickness t and a width w. The equation for the vertical displacement at any point between support A and the center of the beam is valid for the range $0 < x < L/2$. The equation for analyzing deformation is derived through an analysis of shear force and bending moment on the selected segment, as explained on [37]. The development of the analysis leads us to Equation (1), considering the moment of inertia as shown in Equation (2).

$$y = \frac{P}{48EI}\left[4x^3 - 3L^2x\right], \quad (1)$$

$$I = \frac{tw^3}{12}, \quad (2)$$

$$y = \frac{P}{Etw^3}\left[x^3 - \frac{3}{4}L^2x\right], \quad (3)$$

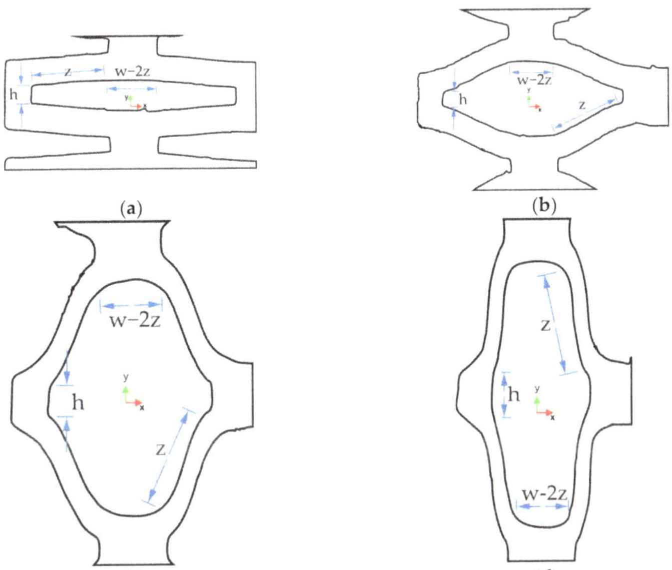

Figure 3. Various stages of deformation of the rectangular kirigami cut: (**a**) rectangular cut at the onset of deformation; (**b**) rectangular cut in a second stage of deformation; (**c**) rectangular cut in an advanced stage of deformation; (**d**) rectangular cut in the maximum deformation stage.

Equation (3) indicates the displacement "y" of the central point of the beam in Figure 2, which is the point where the displacement is at its maximum. This allows us to identify the maximum deformation of the beam under the established considerations. Once the deformation of the bar-like element is determined, we will illustrate the stages through which the rectangular kirigami cut goes, involving other parameters related to the configuration of the rectangular kirigami cut. Figure 3 shows the different states through which the rectangular cut progresses as deformation advances, as obtained through an analysis of images captured during the experimentation with the specimens. The images are presented as illustrations to more effectively emphasize the points of interest, illustrating the location of various geometric parameters from the initial shape to highlight their impact. In Figure 3a, the rectangular kirigami cut is shown in an initial stage of deformation, where it maintains nearly the same shape, illustrating a single cut since, if the distribution of cuts is uniform along the base material, all cuts not on the boundary exhibit similar behavior.

Figure 3b illustrates the same rectangular cut in a subsequent stage of deformation, still revealing the proportions that constitute the cut, albeit with the rectangular shape having deformed into an octagonal form. In Figure 3c, the kirigami cut is depicted in an advanced stage of deformation, where the rectangular shape has been lost, and a more defined octagonal shape has taken its place, with the sides dependent on the original rectangular geometry, including the cut height "h", the cut length "w", and the intersecting length "z" between the lower and upper cells (defined in Figure 1). Finally, Figure 3d presents the kirigami cut at its maximum stage of deformation, where the rectangular form has been completely lost, and an octagonal figure has formed, with its sides delineated by the constraints imposed by the original rectangular cut geometry.

From an analysis of Figure 3, we can observe that the displacement "y" as presented in Equation (3) adds to the distance "h" as we progress toward maximum deformation. Additionally, the maximum deformation is constrained by the distribution of the rectangular cut relative to the upper and lower cuts, as if the distribution of cuts between rows is uniform and symmetrical, and the material between cut lines establishes a limit of "$w-2z$" on the maximum achievable deformation. The maximum vertical deformation attained by a rectangular cut cell transitioning from a height "h" to a maximum height "y_{max}" is

$$y_{\max} \approx h + 2z \tag{4}$$

as the "h" distance increases, it allows for greater freedom of movement for the elements depicted in Figure 2. It is also important to highlight that all rectangular cuts deform uniformly and similarly, except for those at the edges where the uniform distribution over the base material is disrupted. Thus, it becomes possible to sum the contribution of each rectangular cut to estimate the maximum deformation achieved for a plate with rectangular cuts in general.

2.2. FEM Simulation

To compare the proposed model, a finite element simulation was conducted using the SolidWorks 2022 software. The TPU material was described using an Ogden hyperelastic model with the following coefficients: μ1 = −30.921 MPa, α1 = 0.508, μ2 = 10.342 MPa, α2 = 1.375, μ3 = 26.791 MPa, and α3 = −0.482. This model considers the effects of 3D printing, as stated in reference [38]. The applied conditions are depicted in Figure 4. The simulation used a triangular-type meshing, and a normal force was applied at one end while an encastre restriction was applied at the other end to simulate a tensile test. In this study, triangular meshing was employed to optimize simulation execution time. The finite element method (FEM) study utilized a Dell G15 5510 laptop equipped with an Intel Core i5-10500H CPU running at 2.50 GHz and 32 GB of RAM memory, operating on Windows 11 Home Single Language version 22H2.

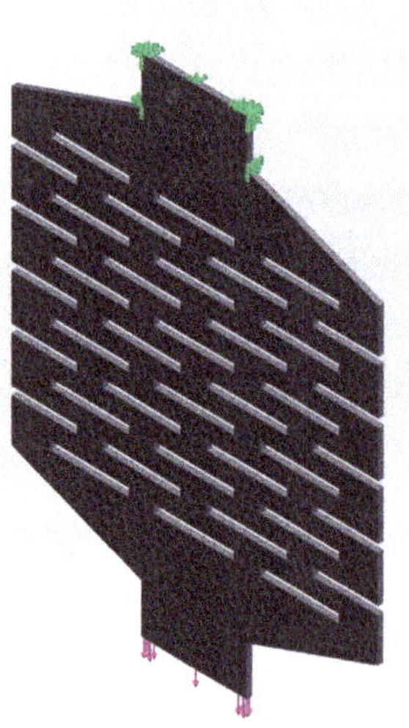

Figure 4. Proposed triangular-type meshing over a test specimen described by FEM. The clamping condition is of the encastre type shown in green in the figure, while the applied force, shown in purple, subjects the specimen to tension.

2.3. Experimental Setup

To compare the behavior of the kirigami pattern with rectangular cuts, an experimental stress test was conducted on a cell with rectangular cuts. During the tensile test, one end of the test specimen was fixed, and a force was applied to the other end, causing deformation similar to that in the FEM simulation as depicted in Figure 5. The test was carried out using the "LLOYD Instruments Texture Analyzer".

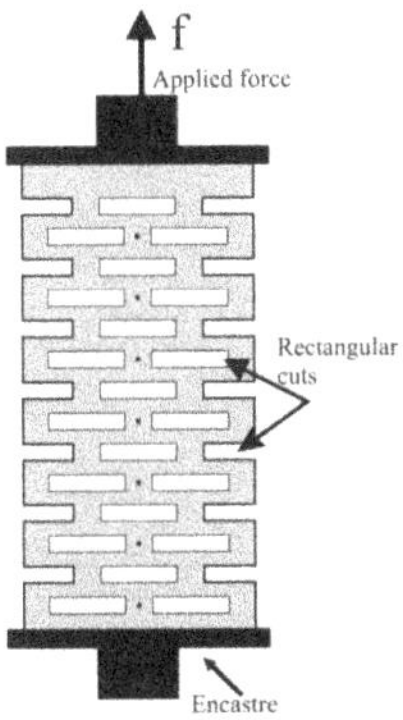

Figure 5. Experimental setup using tensile test device.

The test specimen was made of thermoplastic polyurethane (TPU) using a 3D printing process with a Flash Forge Finder printer and their software, Flash Print 5.3.1. The parameters of the printer were set to the default settings for flexible filaments. For the tests conducted using the texture analyzer equipment, test specimens were fabricated while taking into consideration the workspace constraints of the equipment and the clamping requirements of the equipment's jaws. The usable working area of the test specimens measured 50 mm in height by 70 mm in width. The total height of the test specimen was 95 mm, including the material used for supporting the jaws of the tensile test machine. The specimens were manufactured with a thickness of 1 mm, but with variations in the width of the cuts, resulting in specimens with a minimum cut, 1.0 mm cut, and a 1.5 mm cut. Six specimens test were used for each of the proposed variations in the distribution of rectangular cuts. All tests conducted with the equipment were configured with the parameters shown in Table 1.

Table 1. Configuration parameters of Flash Forge Finder 3D printer for TPU.

Velocity	2.5 mm/s
Specimen Length	95 mm
Specimen Width	70 mm
Thickness	1 mm

The testing equipment allows for the storage of data on applied force, displacement, and time, as well as an estimation of stress.

3. Results

The results are presented below; mathematical modeling illustrates the evolution and maximum estimated deformation. The FEM simulation is used to ascertain whether the behavior aligns with experimental outcomes, demonstrating that the considered parameters are accurate and enabling the simulation-based modeling of structures with rectangular cuts. The experimental results are employed to analyze the behavior of cells with rectangular cuts, as well as to validate the findings of the FEM study and the maximum deformation results obtained through mathematical modeling.

3.1. FEM Results

FEM analysis demonstrates how deformation occurs when force is applied. Figure 6 shows the transitional state of the kirigami-cut test specimen for one configuration. The analysis indicates that all the cuts experience uniform deformation along the test specimen, except for those at the extreme points. Due to varying constraints, these cuts do not reach maximum deformation, in contrast to the central cuts. Deformations outside the plane of the specimen also occur as the force increases, which are estimated by the FEM simulation, as shown in Figure 6a and observed in experimentation. The FEM simulation also provides us with the results of the patterns' behavior concerning the strain–time simulation shown in Figure 6b and the stresses at the nodes distributed along the central points of the specimen. It is important to note that while the FEM study gives us results per node, the conducted experiment considers the entire specimen as a single mechanical entity.

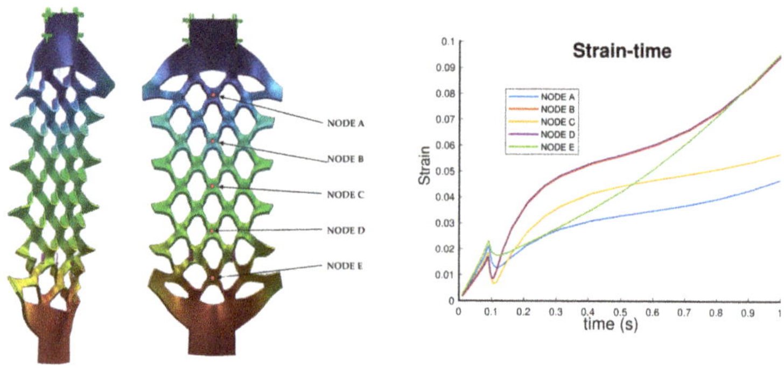

(**a**) (**b**)

Figure 6. FEM results show the evolution of kirigami cut deformation. (**a**) Test specimen deformation shown at transitional state. (**b**) Strain–time evolution on central nodes of the specimen test simulation.

3.2. Experimental Results

The experimental results demonstrate that there is a variation in the behavior of the different specimens in response to changes in the width of the cuts. Similarly, various types of variations can be assessed, such as changes in width, length, and distribution. However, the characterization of the specimens with rectangular cuts allows determining their behavior in response to cut variations. In Figure 7a, the specimen with minimal cuts is shown; these cuts were traced in the 3D printing process and subsequently made directly with a knife. In Figure 7b, the specimen with 1 mm thick cuts is displayed, while in Figure 7c, the specimen with 1.5 mm thick cuts is shown. A total of eighteen test specimens were manufactured for the experimental development, with six specimens for each type of cut and distribution presented. Finally, in Figure 7d, the tension test is shown being conducted on the Lloyd Instruments Texture Analyzer, where it can be observed that one end is secured while a vertical force is applied to the other end, causing displacements and stresses.

Figure 8 displays the final condition before reaching the breaking point of the test specimen. In this state, it is observed that if the distribution of the cuts is uniform, all rectangular cuts deform uniformly along the test specimen, except for those at the extreme points. Therefore, understanding the deformation experienced by a rectangular cut enables the modeling of the behavior of an entire base material with rectangular-cut patterns uniformly distributed on it.

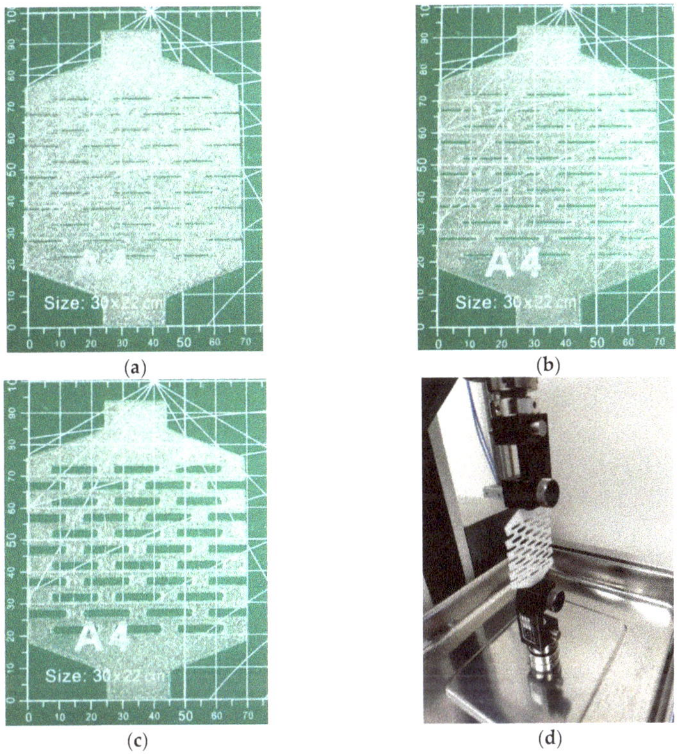

(a) (b) (c) (d)

Figure 7. Different test specimens with a centimeter-scale background: (**a**) test specimen with almost zero width cuts; (**b**) test specimen with 1 mm width cuts; (**c**) test specimen with 1.5 mm cuts; (**d**) test specimen on texture analyzer.

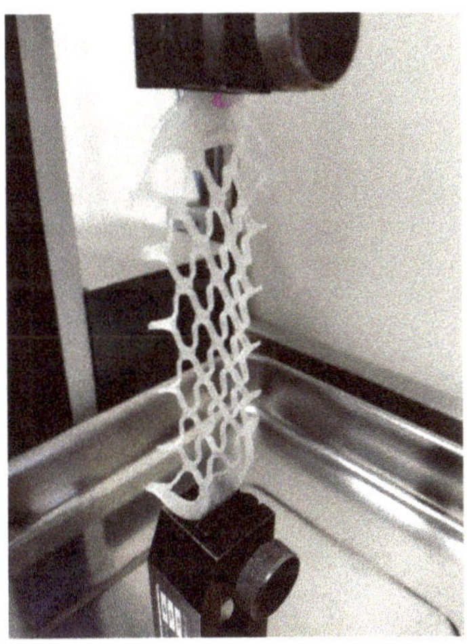

Figure 8. Final state for experimental test on test specimen with rectangular 1 mm cuts.

The variation in maximum elongation obtained with each cutting variation is displayed in Figure 9. For the test specimen in which cuts with the greatest width (1.5 mm) were performed, the highest average elongation of 218 percent was achieved. For the specimen with 1 mm cuts, the average elongation was 188 percent, while for the specimen with the minimum cut, the average obtained elongation was 176 percent. In terms of the variation observed across the tests, it is noted that this variation in final elongation increases with the width of the cut applied to the specimen. Upon reviewing the experimental tests, it was observed that the specimens tended to slip from the grips of the testing device as the displacement increased. Additionally, for tests with a 1 mm wide cut, specimens reached the rupture point at different locations. These variations are believed to be primarily attributed to the manufacturing process.

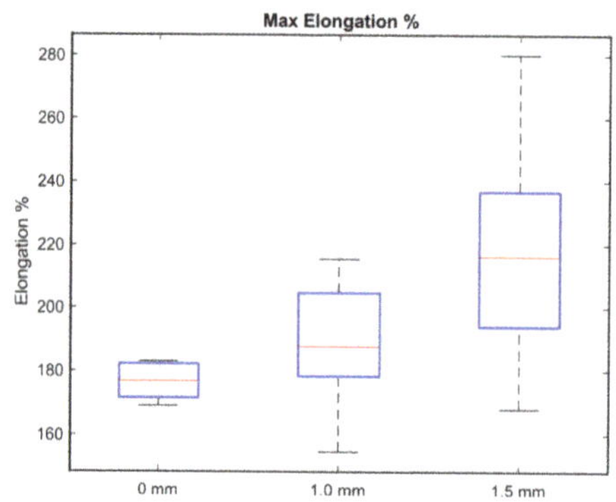

Figure 9. Maximum elongation for each test group.

Furthermore, the stiffness associated with the response exhibited the following variation, as indicated by the results estimated by the testing device, and illustrated in Figure 10. For the test specimen with the widest cuts (1.5 mm), the stiffness was 1635 N/m. In the case of the specimen with 1 mm cuts, the average stiffness measured 3264.84 N/m, while for the specimen with the minimum cut, the average stiffness was 4020.29 N/m.

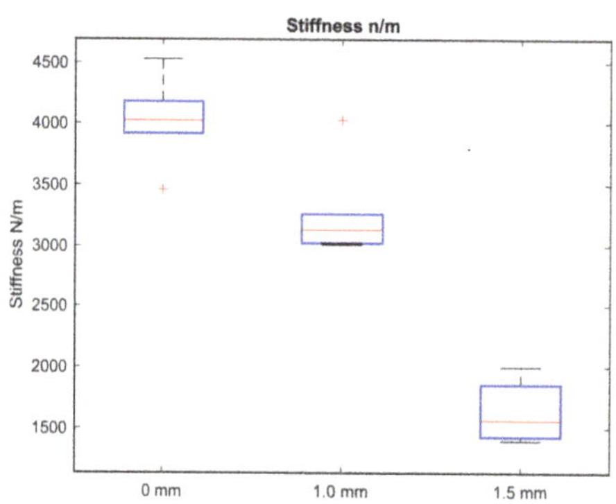

Figure 10. Stiffness distribution of each test group. During the data analysis, outliers are displayed for the group with the minimum cut and for the group with a 1 mm cut. This atypical variation (marked with a '+') in the graphs is associated with the manufacturing process of the specimens.

In Figure 11, the stress–strain relationship is shown for a set of experimental tests. It can be observed that each group of specimens exhibits a behavior trend according to the type of rectangular kirigami cut they had. Specimens with minimal cuts show higher stress against the presented deformation, and similarly, the slope of these curves representing stiffness is also higher. Specimens with a 1 mm wide cut reduced the amount of stress for a similar unitary deformation, and the slope of the trend was also lower. Finally, for specimens with a rectangular cut with a width of 1.5 mm, the stress decreased significantly, and the slope of the trend also decreased compared to that of the previous specimens. It is important to note that in this test, not all specimens reached the breaking point due to the test device's stroke. This can be observed in the vicinity of strain 2, stress 1.5 MPa, where measurements taken by the device are presented once a rupture point occurred in the test specimens with minimal rectangular cuts and with 1 mm wide rectangular cuts. The absence of points related to specimens with 1.5 mm wide cuts is because these specimens did not reach the rupture point during the testing machine's run (trajectories marked with an asterisk on the graph).

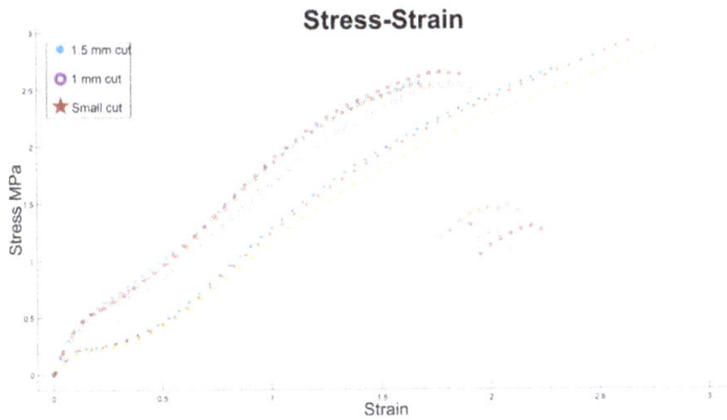

Figure 11. Stress–strain relationship for the conducted experimental tests, where each test group exhibits a trend associated with the kirigami rectangular cutting pattern; when the cut is minimal (with a width close to zero), the stiffness is higher, and as the cut increases, forming the rectangular cell, the stiffness tends to decrease.

Regarding the strain behavior shown in Figure 6b, corresponding to the results of the FEM simulation analyzed at the nodes of the test specimen, Figure 12 provides an enlargement of the experimental tension test results. It illustrates that the same phenomenon occurs experimentally at the onset of the test. Initially, the test exhibits some linear behavior, but beyond a certain point (around a strain of 0.1 in Figure 12), there is a sudden drop in the linear behavior. According to our analysis, this drop occurs because out-of-plane deformations start to appear beyond this point. In the conducted experiment, these out-of-plane deformations would symmetrically manifest along the z-plane. For other cuts in the experimental tests, this change is not as pronounced, which we attribute to the resolution limitations of the testing equipment.

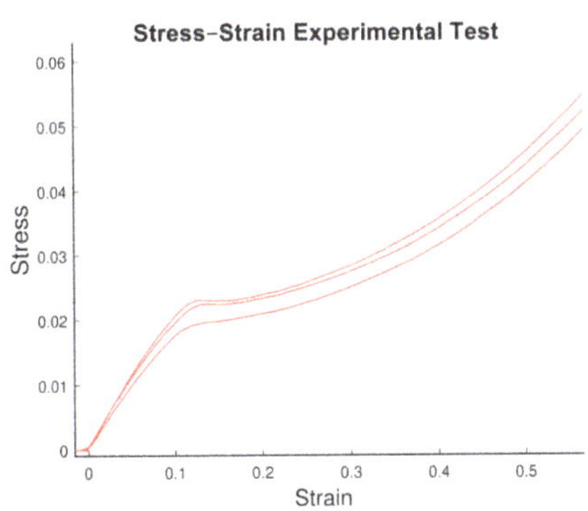

Figure 12. Sudden change observed in the stress response of the specimens during the experimental test. Initially, they exhibit linear behavior, until reaching a point where the stress drops rapidly, followed by a further increase but with a different and less linear slope.

4. Discussion

The effect of rectangular kirigami cuts was primarily evidenced in the tensile tests conducted on each specimen with rectangular cuts. The variations resulting from changes in the size of the cuts were observed in these tests, and the behavior was validated through finite element method (FEM) simulations. This confirms that the results align with the parameters used in the simulation, ensuring consistency between the experimental and simulated behaviors. The slope of the stress–strain curves corresponds to the stiffness of the test specimen being tested. As seen in Figures 9 and 10, for test specimens with equal cuts, the behavior trend is the same, whereas when comparing test specimens with different cuts, there is a variation in behavior. This allows for the adjustment of variables such as total deformation and test specimen stiffness in a macroscopic behavior.

Figures 9 and 10 provide a summary of the behavior in both maximum extension and stiffness for each test group. It can be observed that as the variation in the width of the rectangular cut increases, the total elongation also increases, but stiffness decreases. For tests where the cut width was greater, the maximum elongation averaged 218 percent, whereas for tests with the minimum cut, the maximum elongation decreased to an average of 176.63 percent, representing an approximate 42% variation due to the increased cut width. Regarding stiffness, tests with a rectangular cut width of 1.5 mm had an approximate average stiffness of 1635 N/m, whereas for tests with a 1.0 mm width cut, the average stiffness increased to 3264.84 N/m. Finally, for tests with the minimum cut, stiffness increased to 4020.29 N/m. In other words, as the cut width increases, stiffness decreases while maximum deformation increases.

Finite element simulation validates the way in which strain occurs along the nodes of the specimen, as well as the behavior of stress response as strain increases. Furthermore, the FEM simulation demonstrates that deformation propagates regularly through the test specimen with rectangular cuts, except for those cuts located at the end of the cell just as it occurs in the experimental stage. In other words, the analysis of a specific test specimen can be used to approximate the behavior of an entire test specimen with "n" cuts distributed along its length, and FEM simulation can predict the behavior of a design using a rectangular-cut pattern using TPU. While FEM simulation validates that the variation in the specimen corresponds to the evolution of deformation observed in the experimental tests, the variation observed in the experimental tests is primarily attributed to the manufacturing process of the specimens. Despite being manufactured under the same conditions, the entropy associated with the 3D printing process is reflected in the behavior of the tensile tests.

These structures are proposed for deployment in both deformable robot development and artificial muscle as potential applications. In Figure 13a, a proposed rehabilitation device is depicted, in which a force is being applied to the test specimen with rectangular cuts. This device can be tailored to the rehabilitation needs of the patient. In Figure 13b, the device is shown when no force is being applied (at rest state).

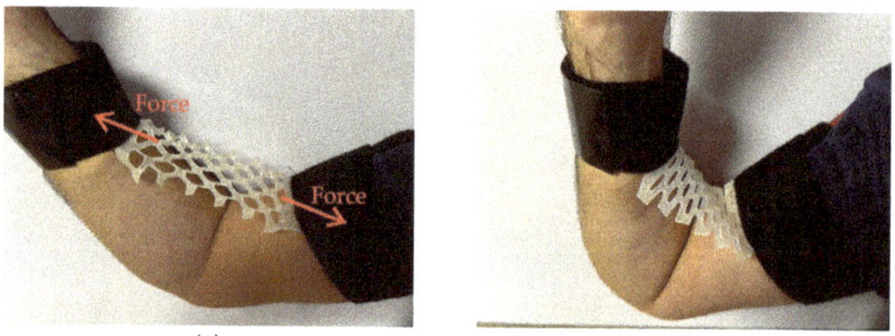

(a) (b)

Figure 13. Potential applications of rectangular kirigami-adjusted actuators. (**a**) Device under an axial load produced by the effort of opening the arm; rectangular cells deform under an axial load similar to the case in the study conducted. (**b**) Device in a resting state; cells return to a resting state close to the rectangular shape, albeit with the effect of residual stresses.

Conversely, the rectangular-cut pattern is viable for incorporation into structural elements constructed with TPU, such as the Fin Ray effect-inspired flexible gripper structures depicted in Figure 14. Investigation will assess the potential enhancement of gripper adaptability and grasping by strategically integrating rectangular cuts. Manufacture of these structures employs flexible materials like TPU through 3D printing. The primary advantage of utilizing flexible materials lies in their ability to absorb impact upon contact with other objects, mitigating damage to the colliding entity.

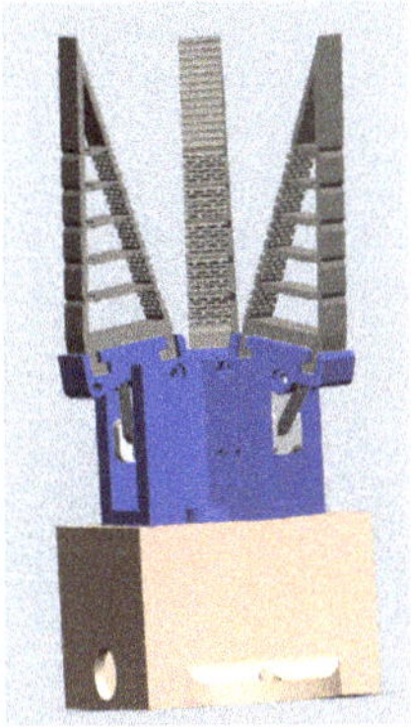

Figure 14. The kirigami rectangular-cut pattern will be projected onto TPU structures to adjust their behavior, as seen in Fin Ray-inspired structures for flexible grippers.

A weakness of the study is the lack of analysis regarding the plastic behavior of the test specimens. This is because, during the conducted experimentation, the test specimens were taken to the point of rupture to understand their behavior in response to variations in the width of the cut. Similarly, although this cutting pattern may be employed away from the rupture zone, this study did not provide a fatigue analysis to determine the material's lifespan under the identified conditions.

5. Conclusions

The kirigami rectangular-cut pattern affects the stiffness and elongation properties of a TPU specimen as the height of the cut is modified. For the studied configuration, stiffness varies by more than 145% from the minimum cut configuration to the condition where the cut height is 1.5 mm, passing through an intermediate state when the cut is 1 mm. The maximum elongation is also affected, varying by 40% for the TPU material with the described configuration when the cut is at a minimum compared to when the cut is 1.5 mm. This feature, enabled using kirigami rectangular patterns, can be employed to fine-tune the properties of a TPU section to adjust the deformation response to an applied load. The consistent behavior of the test groups confirms that the material performance is uniform, even though the specimens were manufactured through 3D printing rather than strictly through material cutting.

The kirigami rectangular-cut pattern on TPU sheets yields different outcomes compared to those reported by [34,35], who proposed studies similar to the one presented in this research. However, they utilized different configurations in terms of the distribution of rectangular cuts, boundary conditions, and the characteristics of the specimens used. For example, in the study conducted in [34], the effect caused by out-of-plane deformations in the stress–strain curves is not observed due to the lack of interaction with other groups of continuous cells. When the pattern of linear cuts (non-rectangular) has been applied to other materials, a greater elongation has been achieved at the expense of significantly reducing the structure's stiffness. In the presented case using rectangular cuts, a reduction in structure stiffness is also observed, but it is not as drastic. On the other hand, the total elongation is not as substantial as that reported in other studies applying the linear cut pattern. This may be due to the TPU specimen not necessarily being a sheet but having a thickness of 1 mm. However, for robotic applications, it is undesirable for the structure's stiffness to decrease to levels where it cannot support itself.

Finally, it is worth mentioning that this study explores the behavior of the kirigami rectangular-cut pattern on a TPU specimen using a different configuration regarding the final distribution and boundary conditions of the test specimens. The interest in studying this cutting pattern applied to TPU material stems from its envisioned integration into actuators and sensors operating within more complex systems related to deformable robotics. It is expected that the development of this study will support the modeling and understanding of applications where this pattern is utilized, as discussed in Section 4. To fully characterize the behavior, fatigue tests and analysis of the plastic zone behavior of the constructed actuators and sensors will be conducted. Additionally, the study of new cut configurations will be carried out, considering that small variations in both the distribution and shape of the cuts could result in significantly different behavior.

Author Contributions: Conceptualization and validation, B.M.-B., E.C.-C. and X.Y.S.-C.; mathematical modelling B.M.-B. and X.Y.S.-C.; methodology, E.C.-C. and M.A.L.; writing—original draft preparation, B.M.-B., E.C.-C., X.Y.S.-C. and M.A.L.; writing—review and editing, X.Y.S.-C. and B.M.-B.; supervision, X.Y.S.-C., M.A.L. and E.C.-C.; project administration, X.Y.S.-C., M.A.L. and E.C.-C. All authors have read and agreed to the published version of the manuscript.

Funding: This research received no external funding.

Institutional Review Board Statement: Not applicable.

Informed Consent Statement: Not applicable.

Data Availability Statement: The data presented in this study are available on request from the corresponding author.

Conflicts of Interest: The authors declare no conflicts of interest.

References

1. Firouzeh, A.; Higashisaka, T.; Nagato, K.; Cho, K.; Paik, J. Stretchable Kirigami Components for Composite Meso-Scale Robots. *IEEE Robot. Autom. Lett.* **2020**, *5*, 1883–1890. [CrossRef]
2. Zheng, M.; Chen, Y.; Liu, Z.; Liu, Y.; Wang, Y.; Liu, P.; Liu, Q.; Bi, K.; Shu, Z.; Zhang, Y.; et al. Kirigami-inspired multiscale patterning of metallic structures via predefined nanotrench templates. *Microsyst. Nanoeng.* **2019**, *5*, 54. [CrossRef]
3. Song, Z.; Wang, X.; Lv, C.; An, Y.; Liang, M.; Ma, T.; He, D.; Zheng, Y.-J.; Huang, S.-Q.; Yu, H.; et al. Kirigami-based stretchable lithium-ion batteries. *Sci. Rep.* **2015**, *5*, 10988. [CrossRef]
4. Shigemune, H.; Maeda, S.; Hara, Y.; Koike, U.; Hashimoto, S. Kirigami robot: Making paper robot using desktop cutting plotter and inkjet printer. In Proceedings of the 2015 IEEE/RSJ International Conference on Intelligent Robots and Systems (IROS), Hamburg, Germany, 28 September–2 October 2015. [CrossRef]
5. Shyu, T.C.; Damasceno, P.F.; Dodd, P.M.; Lamoureux, A.; Xu, L.; Shlian, M.; Shtein, M.; Glotzer, S.C.; Kotov, N.A. A kirigami approach to engineering elasticity in nanocomposites through patterned defects. *Nat. Mater.* **2015**, *14*, 785–789. [CrossRef]
6. Ryu, J.; Tahernia, M.; Mohammadifar, M.; Gao, Y.; Choi, S. Moisture-Responsive Paper Robotics. *J. Microelectromechanical Syst.* **2020**, *29*, 1049–1053. [CrossRef]
7. Wang, X.; Guest, S.D.; Kamien, R.D. Keeping It Together: Interleaved Kirigami Extension Assembly. *Phys. Rev. X* **2020**, *10*, 011013. [CrossRef]
8. Choi, G.P.T.; Dudte, L.H.; Mahadevan, L. Programming shape using kirigami tessellations. *Nat. Mater.* **2019**, *18*, 999–1004. [CrossRef]
9. Chen, S.; Liu, Z.; Du, H.; Tang, C.; Ji, C.-Y.; Quan, B.; Pan, R.; Yang, L.; Li, X.; Gu, C.; et al. Electromechanically reconfigurable optical nano-kirigami. *Nat. Commun.* **2021**, *12*, 1299. [CrossRef]
10. Lin, S.; Xie, Y.M.; Li, Q.; Huang, X.; Zhou, S. A Kirigami Approach to Forming a Synthetic Buckliball. *Sci. Rep.* **2016**, *6*, 33016. [CrossRef] [PubMed]
11. Neville, R.M.; Scarpa, F.; Pirrera, A. Shape morphing Kirigami mechanical metamaterials. *Sci. Rep.* **2016**, *6*, 31067. [CrossRef]
12. Blees, M.K.; Barnard, A.W.; Rose, P.A.; Roberts, S.P.; McGill, K.L.; Huang, P.Y.; Ruyack, A.R.; Kevek, J.W.; Kobrin, B.; Muller, D.A.; et al. Graphene kirigami. *Nature* **2015**, *524*, 204–207. [CrossRef]
13. Bertoldi, K.; Vitelli, V.; Christensen, J.; van Hecke, M. Flexible mechanical metamaterials. *Nat. Rev. Mater.* **2017**, *2*, 17066. [CrossRef]

14. Branyan, C.; Hatton, R.L.; Menguc, Y. Snake-Inspired Kirigami Skin for Lateral Undulation of a Soft Snake Robot. *IEEE Robot. Autom. Lett.* **2020**, *5*, 1728–1733. [CrossRef]
15. Sandoval-Castro, X.Y.; Martinez-Sanchez, D.E.; Cortes-Gonzalez, J.G.; Castillo-Castaneda, E.; Gautreau, E.; Laribi, M.A. A Kirigami–like Soft Elastomeric Skin: Design and Influence Evaluation in the Mobility of a Bio–inspired Snake–Arm Robot. In *Advances in Mechanism and Machine Science*; Okada, M., Ed.; IFToMM WC 2023. Mechanisms and Machine Science; Springer: Cham, Switzerland, 2023; Volume 148. [CrossRef]
16. Liu, B.; Ozkan-Aydin, Y.; Goldman, D.I.; Hammond, F.L. Kirigami Skin Improves Soft Earthworm Robot Anchoring and Locomotion under Cohesive Soil. In Proceedings of the 2019 2nd IEEE International Conference on Soft Robotics (RoboSoft), Seoul, Republic of Korea, 14–18 April 2019. [CrossRef]
17. Babaee, S.; Shi, Y.; Abbasalizadeh, S.; Tamang, S.; Hess, K.; Collins, J.E.; Ishida, K.; Lopes, A.; Williams, M.; Albaghdadi, M.; et al. Kirigami-inspired stents for sustained local delivery of therapeutics. *Nat. Mater.* **2021**, *20*, 1085–1092. [CrossRef]
18. Wang, T.; Ugurlu, H.; Yan, Y.; Li, M.; Li, M.; Wild, A.-M.; Yildiz, E.; Schneider, M.; Sheehan, D.; Hu, W.; et al. Adaptive wireless millirobotic locomotion into distal vasculature. *Nat. Commun.* **2022**, *13*, 4465. [CrossRef]
19. Lamoureux, A.; Lee, K.; Shlian, M.; Forrest, S.R.; Shtein, M. Dynamic kirigami structures for integrated solar tracking. *Nat. Commun.* **2015**, *6*, 8092. [CrossRef]
20. Chen, S.; Chen, J.; Zhang, X.; Li, Z.-Y.; Li, J. Kirigami/origami: Unfolding the new regime of advanced 3D microfabrication/nanofabrication with "folding". *Light Sci. Appl.* **2020**, *9*, 75. [CrossRef]
21. Sedal, A.; Memar, A.H.; Liu, T.; Menguc, Y.; Corson, N. Design of Deployable Soft Robots Through Plastic Deformation of Kirigami Structures. *IEEE Robot. Autom. Lett.* **2020**, *5*, 2272–2279. [CrossRef]
22. Hunt, A.; Freriks, M.; Sasso, L.; Esfahani, P.M.; HosseinNia, S.H. IPMC Kirigami: A Distributed Actuation Concept. In Proceedings of the 2018 International Conference on Manipulation, Automation and Robotics at Small Scales (MARSS), Nagoya, Japan, 4–8 July 2018. [CrossRef]
23. Morikawa, Y.; Yamagiwa, S.; Sawahata, H.; Numano, R.; Koida, K.; Ishida, M.; Kawano, T. Stretchable micro-doughnuts Kirigami bioprobe. In Proceedings of the 2018 IEEE Micro Electro Mechanical Systems (MEMS), Belfast, UK, 21–25 January 2018. [CrossRef]
24. Guo, Z.; Ji, B.; Wang, L.; Yang, B.; Wang, W.; Liu, J. A Flexible and Stretchable Kirigami-Inspired Implantable Neural Probe with Floating Microsites for Electrophysiology Recordings. In Proceedings of the 2020 IEEE 33rd International Conference on Micro Electro Mechanical Systems (MEMS), Vancouver, BC, Canada, 18–22 January 2020. [CrossRef]
25. Miao, L.; Wan, J.; Guo, H.; Wang, H.; Song, Y.; Chen, X.; Zhang, H. Kirigami Cross-Shaped 3D Buckling Active Sensor for Detecting Stretching and Bending. In Proceedings of the 2019 20th International Conference on Solid-State Sensors, Actuators and Microsystems & Eurosensors XXXIII (TRANSDUCERS & EUROSENSORS XXXIII), Berlin, Germany, 23–27 June 2019. [CrossRef]
26. Baldwin, A.; Meng, E. Kirigami Strain Sensors Microfabricated from Thin-Film Parylene C. *J. Microelectromechanical Syst.* **2018**, *27*, 1082–1088. [CrossRef]
27. Hwang, D.-G.; Bartlett, M.D. Tunable Mechanical Metamaterials through Hybrid Kirigami Structures. *Sci. Rep.* **2018**, *8*, 3378. [CrossRef]
28. Tang, Y.; Yin, J. Design of cut unit geometry in hierarchical kirigami-based auxetic metamaterials for high stretchability and compressibility. *Extrem. Mech. Lett.* **2017**, *12*, 77–85. [CrossRef]
29. Han, D.X.; Zhao, L.; Chen, S.H.; Wang, G.; Chan, K.C. Critical transitions in the shape morphing of kirigami metallic glass. *J. Mater. Sci. Technol.* **2021**, *61*, 204–212. [CrossRef]
30. Tao, J.; Khosravi, H.; Deshpande, V.; Li, S. Engineering by Cuts: How Kirigami Principle Enables Unique Mechanical Properties and Functionalities. *Adv. Sci.* **2022**, *10*, 2204733. [CrossRef]
31. Zhang, X.; Ma, J.; Li, M.; You, Z.; Wang, X.; Luo, Y.; Ma, K.; Chen, Y. Kirigami-based metastructures with programmable multistability. *Proc. Natl. Acad. Sci. USA* **2022**, *119*, e2117649119. [CrossRef]
32. Taniyama, H.; Iwase, E. Design of a Kirigami Structure with a Large Uniform Deformation Region. *Micromachines* **2021**, *12*, 76. [CrossRef]
33. Joe, S.; Totaro, M.; Beccai, L. Analysis of Soft Kirigami Unit Cells for Tunable Stiffness Architectures. In Proceedings of the 4th International Conference on Soft Robotics (RoboSoft), New Haven, CT, USA, 12–16 April 2021. [CrossRef]
34. Nakajima, J.; Fayazbakhsh, K.; Teshima, Y. Experimental study on tensile properties of 3D printed flexible kirigami specimens. *Addit. Manuf.* **2020**, *32*, 101100. [CrossRef]
35. Chung, S.; Coutinho, A.; Rodrigue, H. Manufacturing and Design of Inflatable Kirigami Actuators. *IEEE Robot. Autom. Lett.* **2023**, *8*, 25–32. [CrossRef]
36. Yong, K.; De, S.; Hsieh, E.Y.; Leem, J.; Aluru, N.R.; Nam, S. Kirigami-inspired strain-insensitive sensors based on atomically-thin materials. *Mater. Today* **2020**, *34*, 58–65. [CrossRef]
37. Hibbeler, R.C. *Mechanics of Materials*, 10th ed.; Pearson: Boston, MA, USA, 2017.
38. Yap, H.K.; Ng, H.Y.; Yeow, C.-H. High-Force Soft Printable Pneumatics for Soft Robotic Applications. *Soft Robot.* **2016**, *3*, 144–158. [CrossRef]

MDPI
St. Alban-Anlage 66
4052 Basel
Switzerland
www.mdpi.com

Applied Sciences Editorial Office
E-mail: applsci@mdpi.com
www.mdpi.com/journal/applsci

www.ingramcontent.com/pod-product-compliance
Lightning Source LLC
Chambersburg PA
CBHW040834250926
51850CB00007BA/1307